EVOLUTION AND THE FOSSIL RECORD

Readings from
**SCIENTIFIC
AMERICAN**

EVOLUTION AND
THE FOSSIL RECORD

With Introductions by
Léo F. Laporte
University of California, Santa Cruz

W. H. Freeman and Company
San Francisco

For Gosia

Most of the SCIENTIFIC AMERICAN articles in
EVOLUTION AND THE FOSSIL RECORD are available
as separate Offprints. For a complete list of more than
1,000 articles now available as Offprints, write to
W. H. Freeman and Company, 660 Market Street, San
Francisco, California 94104.

Library of Congress Cataloging in Publication Data

Main entry under title:

Evolution and the fossil record.

    Includes bibliographies and index.
    1. Paleontology—Addresses, essays, lectures.
2. Evolution—Addresses, essays, lectures. I. Laporte,
Léo F. II. Scientific American.
QE721.E95   560   77–26073
ISBN 0–7167–0291–6
ISBN 0–7167–0290–8 pbk.

Printed in the United States of America

9 8 7 6 5 4 3 2 1

PREFACE

The earth has not always been as it is now, nor will it remain the same in the future. Rather, our planet had a beginning and will have an end: It formed from cosmic gas and dust almost five billion years ago, and it will be consumed by the expanding sun billions of years hence. As far as we know, life is unique to this planet, and its fate is therefore inextricably linked to it. Life thus had a beginning; it flourishes now; and it will eventually end. Each of us, in our own individual existence, bears for a fleeting instant the fire of life that has come directly down to us through countless generations over billions of years. Think of it. You and I, quite literally, trace our origins directly back through those who came before us: from primate, therapsid reptile, Paleozoic amphibian, lobe-fin fish, primitive echinoderm, late Precambrian wormlike creature, to the simplest of single-celled organisms that floated in the primeval seas.

This collection of readings provides an overview of this evolution and history of life—our life—as recorded by the sequence of fossils preserved in the earth's crust. As with all history, the record becomes dimmer as we reach far backward in time, yet we can see some of the major steps in the origin and early diversification of life. As we come closer to the present, we are able to trace and interpret particular lineages quite clearly. The expansion and contraction of life can now be better integrated with the geological evolution of the planet, thanks to the recent development of the theory of plate tectonics.

The readings have been chosen to give broad representation of the kinds of questions and issues that intrigue paleontologists—people who study ancient life. Some articles are concerned with the big picture, others with specific details; some are interested in fossil organisms for what they tell us about their biology and ecology, others for what they tell us about the inanimate world in which they lived. I have also included articles that explain the phenomenon of evolution itself, which provides the theoretical basis for interpreting the historical documents that the fossils represent.

You will notice that some of these articles go back a generation or so, yet they are as relevant today as the day they were written. Of course, where appropriate, I have provided any updating necessary in the various introductions for each part. Virtually all of the articles are popular classics written by the very people who made many of the original contributions to evolutionary theory and the history of life.

We begin with a discussion of the phenomenon of evolution (Part I) and proceed to earliest evidence for life on earth (Part II). We next consider how fossils occur and the kinds of things they tell us about the ancient world (Part III). Several specific case histories are provided next (Part IV), and we conclude with observations about some of the major patterns seen in the history of life (Part V).

I believe that this reader will not only be useful in elementary paleontology courses but also in evolutionary biology courses, in that students can see some of the *historical* evidence for evolution. As we note in the initial introduction, paleontology is a historical science that has complementary aspects of both theory *and* history. Just as paleontologists need to understand evolutionary theory to interpret fossil history, biologists are well served in their consideration of evolution by examining what in fact happened "way back when."

Finally, I hope that some of the sense of wonder and awe that the evolution and history of life have given me will also grow in you as you use this reader. For, after all, it seems part of the human condition to ask "Who am I? From where have I come?" Or, as a contemporary pop philosopher recently put it, "the ultimate search is still the most fascinating search, what it is all about—why are we here and how big it is and where does it go, what is the system, what is the answer, what is God and all that."*

May you enjoy this reader as much as I did in putting it together!

Léo F. Laporte
Santa Cruz, California

---

*George Lucas, writer and director of the movie *Star Wars,* quoted in *Rolling Stone,* August 25, 1977.

CONTENTS

V   SOME MAJOR PATTERNS IN THE HISTORY OF LIFE

Note on cross-references to SCIENTIFIC AMERICAN *articles:* Articles included in this book are referred to by title and page number; articles not included in this book but available as Offprints are referred to by title and offprint number; articles not included in this book and not available as Offprints are referred to by title and date of publication.

EVOLUTION AND THE FOSSIL RECORD

THE PHENOMENON
OF EVOLUTION

THE PHENOMENON OF EVOLUTION I

INTRODUCTION

Before discussing the fossil record itself, we must first consider the theory that explains the phenomenon of organic evolution. We need such a theoretical framework, because the mere existence of shells, bones, tracks and trails, and silicified wood entombed in crustal rocks can be interpreted in various ways. True, most people nowadays realize, even if only vaguely, that fossils are the remains of ancient animals and plants that once lived on the earth. Some know, too, that these fossils record in some imperfect way the unfolding of life over eons. But these notions, as commonplace as they may be now, are fairly recent in human history. In fact, for most of civilization, fossils were interpreted very differently—as the Creator's mistakes, sports of nature, results of spontaneous generation, tricks of the Devil, or victims of catastrophes that periodically shook the world, including the biblical flood—if they were thought of at all. Like any other historical documents of a bygone age, fossils are objective facts whose interest and significance can vary greatly, depending on the particular theory or conceptual framework in which they are viewed.

Philosophers of science tell us that theory informs us about what is possible and what is not. History, however, goes beyond a consideration of what is possible, to a consideration of "what actually happened." The theory of organic evolution, as developed by Charles Darwin and his successors, gives us as paleontologists a conceptual framework of possibilities within which we try to tell what actually happened in the course of life's history on earth. George Gaylord Simpson, an American paleontologist, has put this another way, by referring to "immanent principles and processes" that are abstract and time-independent (e.g., $E = mc^2$) and "past configurations" that are specific, time-linked examples of those principles and processes (e.g., Hiroshima). Thus, theory explains *how* certain principles and processes operate; history informs about *what, when,* and *where.* Together, theory and history combine in a historical science such as paleontology to tell us *how come.*

In its simplest outline, evolutionary theory states that variations in organisms arise from mutations, which are spontaneous alterations in the genes found in all living cells, and from sexual reproduction, whereby gene contributions from parents are randomly segregated when sex cells form and randomly recombined when sex cells cross-fertilize. These individual variations in genetic structure lead to variations in the way organisms look, grow, behave, reproduce, and a thousand other things, ranging from the trivial to the significant, that permit animals and plants to live out their lives and reproduce themselves. Given the limited resources in nature—especially food and living space—some individuals will be better able to exploit the environment, however slight the advantage, owing to particular genetic variations and, consequently, to rear more offspring than others with less favorable genetic attributes. Over successive generations, the more favorable genes increase proportionally, while the less favorable decrease. It is this shift in gene frequencies that defines evolution. And it is the accumulation of these small changes in gene frequency over the millions, even billions, of years of earth history that has led to the great diversity of animal and plant life. By implication, then, all forms of life today—ants and butterflies, daisies and dandelions, sparrows and horses, jellyfish and whales—share a common ancestry, however far back in time.

Before Charles Darwin, there were those who had come to realize that there was, indeed, a unity to life that was particularly apparent from the way organisms were constructed. In fact, by the early nineteenth century, such concepts as the Scale of Nature and the Chain of Being recognized the graduated links from the simplest to most complex organisms as well as the basic architectural, or morphological, similarities typical of one group of organisms as against another. Such concepts, however, were believed to reflect the overall design in Nature or in the handiwork of God, rather than the result of common ancestry of species over the ages. Some scientists of this time, such as Erasmus Darwin, Charles's grandfather, and Jean Lamarck, a French biologist, did suggest that these similarities among organisms came about from progressive change, in one way or another, over "millions of ages." But the actual mechanism of change eluded them and others who considered the problem.

Charles Darwin's quintessential contribution to evolutionary theory, therefore, is not the idea of evolution, but rather his statement of the mechanism by which animal and plant species change into other distinct species. In fact, the full title to his masterwork succinctly states this: *On the Origin of Species by Means of Natural Selection, or the Preservation of Favoured Races in the Struggle for Life.* How was it, then, that Darwin succeeded where others failed in recognizing the key mechanism in a theory that, on the one hand, explains life's unity and, on the other, life's diversity?

First, Darwin was an excellent naturalist. All during his formal studies, initially in medicine at Edinburgh and later the ministry at Cambridge, Darwin actively pursued his enthusiasm for hunting, beetle collecting, and bird watching, as well as for cultivating friendships with professional scientists such as John Stevens Henslow, a botanist. He also "gloried in the progress of geology," which indeed had been developing rapidly through the efforts of such Scottish geologists as Hall, Hutton, and Playfair, efforts that culminated with his British contemporary, Charles Lyell, whose three volumes on the *Principles of Geology* Darwin read during the long voyage of the *Beagle.* In fact, it was Darwin's accomplishments as a naturalist that led Henslow to recommend him to Captain Robert Fitzroy who was seeking a naturalist for the *Beagle* voyage.

Second, the voyage of H.M.S. *Beagle* certainly came at a most propitious time for Darwin. Having just completed his undergraduate studies at Cambridge and a month's field work with Adam Sedgwick studying Cambrian strata in North Wales, Darwin undertook the voyage at age twenty-two. The voyage included stops on major oceanic islands and the continents of South America, Australia, and southern Africa. All these places had sufficiently exotic floras, faunas, and geology (including fossils) to raise many important questions for Darwin. Among these were: How is it that the giant, extinct armadillos of South America have the same ground plan (general design) as living armadillos? How is it that the ostrichlike species of rhea in Argentina and Uruguay are replaced by a distinctly different species in Patagonia? Why do the birds on the Galápagos Islands more closely resemble the birds of South America than they do of other oceanic islands, such as the Cape Verde Islands, whose birds are African in character? Why are the finches of the various

Galápagos Islands so different, when the physical conditions of the islands are so similar? As Darwin wrote toward the end of the voyage, the facts that pose these questions "would be well worth examining: for such facts would undermine the stability of species." That is, it was beginning to dawn on Darwin that species might not, in fact, be fixed entities, but that they might alter, change, or evolve one into the other.

A third contributing factor to Darwin's eventual success was that he was of independent means; the family fortunes enabled him to spend all his time and energy concentrating on his biological and geological researches. Settling down in Kent, he spent the next several decades reading and experimenting, thinking and writing. In 1838, three years after the end of the *Beagle* voyage, Darwin read Malthus's *Essay on Population,* written 40 years before, and "being well prepared to appreciate the struggle for existence which everywhere goes on . . . [in] animals and plants, it at once struck me that under these circumstances favourable variations would tend to be preserved, and unfavorable ones to be destroyed. The result of this would be the formation of new species" (Darwin's *Autobiography,* 1876).

The papers in this part of the reader develop more fully the theory of organic evolution, starting with Charles Darwin and then discussing how genes through mutation and sexual reproduction generate and transmit hereditable variation over successive generations. For, while Darwin fully appreciated the important role such variation played in the origin of species, it was not until many years later that the source of this variation was recognized and understood. We then turn to a simple, but powerful, present-day example of natural selection in action; one that had been going on even in Darwin's day but was unknown to him. Finally, we will return to a phenomenon observed by Darwin on the *Beagle* voyage, which succinctly and cogently demonstrates evolution by the origin of species.

Loren Eiseley's article "Charles Darwin" recaptures for us the historical and intellectual setting in which Darwin lived and worked and, in particular, how earlier scientific contributions prepared Darwin's way. Lyell's geology, too, coming slightly before and during Darwin's maturity, emphasized the antiquity of the earth, giving the essential element of time so necessary to the Darwinian concept of evolution by small, incremental change. Eiseley points out that time was important to Darwin in another, different sense: Time allowed his ideas to mature slowly and fully. In fact, Darwin would have procrastinated still longer in the publication of his theory had it not been that Alfred Russel Wallace came independently to the same conclusion regarding the crucial role of natural selection in the origin of species.

Eiseley also indicates how unclear Darwin was about human origins, compared to the clarity he brought to other issues, and he attributes this to the poor fossil record of humans then available to Darwin. Except for the Neanderthal skull cap, discovered in Germany in 1856, the human fossil record was virtually unknown during Darwin's life. This illustrates, incidentally, one of the "major objections to the theory" as Darwin himself put it in the *Origin of Species;* namely, where are all the fossilized missing links between species? Darwin, however, was confident that one could argue around this objection by recognizing the "imperfection of the geological record" (Chapter 9 of the *Origin*), invoking all the geologic processes by which organisms are not preserved in rocks. Thus, the "noble science of Geology loses glory from the extreme imperfection of the record. The crust of the earth with its imbedded remains must not be looked at as a well-filled museum, but as a poor collection made at hazard and at rare intervals" (Darwin, 1859, p. 487). But as we shall see further on in this reader, geology has regained its nobility in the century since by unearthing many marvelous examples of Darwinian evolution. Eiseley concludes his article by stressing the importance of oceanic islands in providing living evidence of the theory. This is still true, today, when a number of important evolutionary concepts and principles are derived from studies of isolated island populations.

Dobzhansky, in "The Genetic Basis of Evolution," begins his article by reminding us that the great diversity of living things, as measured by the large numbers of animal and plant species, results from the spatial and temporal variety of environments found on this planet. Thus, each species occupies a very particular place, or "adaptive niche," in the grand economy of nature. He then turns to one of the very few gaps in the original theory of evolution as proposed by Darwin: the source and mechanism of the hereditable variation that Darwin postulated as the raw materials upon which natural selection could operate to produce such diversity. With the birth of the science of genetics in the early twentieth century, we now understand how such variation arises. Describing laboratory experiments with bacteria and fruit flies, Dobzhansky indicates that spontaneous alterations in the structure of genes, found on the chromosomes within a cell's nucleus, can occasionally generate variations in individuals that permit them to survive and reproduce in an otherwise hostile environment. These alterations, called *mutations*, are rare and are random, in the sense that they usually have no correlation with the adaptive needs of the organism; in fact, most mutations are harmful. The reason for this is that adaptive, useful mutations have been already incorporated into the "normal" genetic complex of the species over previous generations. Yet, now and then, genetic novelties do occur that make the organism better able to cope with its environment than before.

At the time that Dobzhansky—himself one of the pioneers in genetics—wrote this article, the chemical basis of the gene's operation was not known. That was to come soon after with the research on DNA and RNA by F. H. C. Crick and J. D. Watson. Note the interesting similarity here. Just as Darwin was able to explain the mechanism of natural selection without knowing of genes, Dobzhansky can explain how genes generate variation without knowing the chemical basis for it.

Besides mutations, genetic variation also arises through sexual reproduction, because the sex cells—male spermatozoa and female eggs—randomly segregate genetic information from the parent when such cells are formed. They randomly recombine this information when the sex cells unite during fertilization of the egg by a spermatozoan. As Dobzhansky points out, sex is a powerful mechanism for ensuring variation within a species. But whatever the source of genetic variation, it provides a species with individuals who can flourish in conditions that change, either over time—seasons, years, or millenia—or over their geographic range. The ability to respond to such changes ensures the survival of the species; conversely, insufficient genetic response to changing conditions results, of course, in eventual extinction.

The importance of genetic variation is made remarkably clear in Kettlewell's article, entitled "Darwin's Missing Evidence." Here we see how industrial atmospheric pollution in Britain resulted in a blackened countryside. The native moth species that spent the daylight hours resting on light-colored backgrounds were no longer camouflaged by their salt-and-pepper coloration from insect-eating birds. The mutant black varieties, however, survived such predation, and their numbers increased over the generations, until today, when they are the predominant variety. Here we see nature selecting for the black varieties under new environmental conditions and selecting against the light-colored forms. Before the Industrial Revolution of the late eighteenth century, the situation was reversed.

The title of Kettlewell's paper alludes to the possibility that this phenomenon was already apparent in Darwin's own day. If so, and had he been aware of it, Darwin would have had marvelous, living, direct proof of natural selection in action. Kettlewell concludes his article by noting that variations in a gene can affect more than one characteristic or attribute of an organism. In the case of his moths, the gene controlling color also influences rate of development of the larval forms and mating preferences. These variations must be integrated with that of color into the total organism, for it is the total organism that nature selects, not just one particular facet that it presents to the selection process.

Finally, we conclude Part I with Lack's "Darwin's Finches," coming back to one of the most striking observations made on the voyage of the *Beagle,* one that illustrates the phenomenon of evolution in the terms we have been considering it. Thus, we will see that the newly formed volcanic islands of the Galápagos were invaded by a species of South American ground finch. Owing to the lack of other native land birds, these finches quickly evolved into the many unoccupied, available adaptive niches, including seed-eating and insect-eating ground finches and tree finches, warblers, and woodpeckers. Presumably we started with a single species, from which more than a dozen have now evolved.

Lack emphasized the initial importance of geographic separation of the founding species, which permitted variant strains, or races, to develop independently on different islands. With time, these geographic races became more and more distinct, to the point that when they later came together they were sufficiently genetically dissimilar that they could no longer interbreed. The capacity to produce fertile offspring through interbreeding defines a biological species, and hence these finches have undergone speciation from a common ancestral species. Biologists today believe that this is how virtually all species arise: by allopatric speciation, with geographic separation of populations of a species, eventually resulting in genetic separation. Sympatric speciation rarely occurs, whereby genetic isolation of species arises without geographic separation. Lack also notes that two species will, over time, increase the differences between them if they overlap in their competition for resources. For example, in two species of seed-eaters and two of insect-eaters found on the same island, differences in beak depth—which is correlated to the size of seed or insect eaten—increase through natural selection to minimize competition; yet these differences strongly overlap when the potential competitors occur alone on other islands. Obviously, when two closely competing species come together, variants in each least like the other are selected *for,* whereas those most like the other are selected *against.* Over the generations, then, natural selection will tend to decrease the overlap in the character highly correlated to interspecific competition.

In summary, the readings in Part I provide a historical and theoretical background for organic evolution, against which we can then see particular episodes in the history of life played out, as revealed by the fossil record.

SUGGESTED FURTHER READING

The following annotated references will allow you to pursue further some of the ideas and information included in the individual readings. The references themselves also have good bibliographies.

Darwin, C. 1859. *On the Origin of Species*. Facsimile of first edition, Cambridge, Mass.: Harvard University Press, 1964. *The* book, of course, that all students of evolution must eventually read to understand Darwin's fundamental contribution to human knowledge.

Darwin, C. 1876. *Autobiography*. Reprint, New York: Crowell-Collier, 1961. Besides reproducing the short and charming autobiography—written for his children, and not the world at large—this volume also contains reminiscences by his son, Sir Francis Darwin; notes and sketches by Charles Darwin during his research that led to the *Origin;* and letters relating to it, including those to Lyell and Hooker asking them what he should do about Wallace's essay, which he had just received and which so neatly duplicated all his own ideas about natural selection.

de Beer, G. 1964. *Charles Darwin: Evolution by Natural Selection*. Garden City, N.Y.: Doubleday. An interesting and insightful book explaining Darwin's geological and biological researches in the context of a scientific biography of the great naturalist.

Dobzhansky, T., Ayala, F. J., Stebbins, G. L., and Valentine, J. W. 1977. *Evolution*. San Francisco: W. H. Freeman and Company. An excellent and up-to-date treatment of evolution written for the advanced student as well as the more general reader, with many examples from the fossil record considered.

Kitts, D. B. 1974. "Continental Drift and Scientific Revolution," *American Association of Petroleum Geologists Bulletin*, vol. 58, pp. 2490–2496. One of a number of articles by a geologist-historian of science that discusses the complementary aspects of theory and history in geology.

Simpson, G. G. 1963. "Historical Science," in *The Fabric of Geology*, C. C. Albritton, Jr., ed. Reading, Mass.: Addison-Wesley, pp. 24–48. A thoughtful essay by a leading paleontologist about the historical perspective that is so much a crucial part of the geological sciences; clarifies the distinction between them and the nonhistorical sciences such as chemistry and physics.

Charles Darwin

1

Loren C. Eiseley
February 1956

*In 1831 this gentle Englishman set forth on his famous
voyage in the Beagle. After 28 years he published
Origin of Species, which revolutionized man's view of
nature and his place in it*

In the autumn of 1831 the past and the future met and dined in London—in the guise of two young men who little realized where the years ahead would take them. One, Robert Fitzroy, was a sea captain who at 26 had already charted the remote, sea-beaten edges of the world and now proposed another long voyage. A religious man with a strong animosity toward the new-fangled geology, Captain Fitzroy wanted a naturalist who would share his experience of wild lands and refute those who used rocks to promote heretical whisperings. The young man who faced him across the table hesitated. Charles Darwin, four years Fitzroy's junior, was a gentleman idler after hounds who had failed at medicine and whose family, in desperation, hoped he might still succeed as a country parson. His mind shifted uncertainly from fox hunting in Shropshire to the thought of shooting llamas in South America. Did he really want to go? While he fumbled for a decision and the future hung irresolute, Captain Fitzroy took command.

"Fitzroy," wrote Darwin later to his sister Susan, "says the stormy sea is exaggerated; that if I do not choose to remain with them, I can at any time get home to England; and that if I like, I shall be left in some healthy, safe and nice country; that I shall always have assistance; that he has many books, all instruments, guns, at my service. . . . There is indeed a tide in the affairs of men, and I have experienced it. Dearest Susan, Goodbye."

They sailed from Devonport December 27, 1831, in H.M.S. *Beagle*, a 10-gun brig. Their plan was to survey the South American coastline and to carry a string of chronometrical measurements around the world. The voyage almost ended before it began, for they at once encountered a violent storm. "The sea ran very high," young Darwin recorded in his diary, "and the vessel pitched bows under and suffered most dreadfully; such a night I never passed, on every side nothing but misery; such a whistling of the wind and roar of the sea, the hoarse screams of the officers and shouts of the men, made a concert that I shall not soon forget." Captain Fitzroy and his officers held the ship on the sea by the grace of God and the cat-o'-nine-tails. With an almost irrational stubbornness Darwin decided, in spite of his uncomfortable discovery of his susceptibility to seasickness, that "I did right to accept the offer." When the *Beagle* was buffeted back into Plymouth Harbor, Darwin did not resign. His mind was made up. "If it is desirable to see the world," he wrote in his journal, "what a rare and excellent opportunity this is. Perhaps I may have the same opportunity of drilling my mind that I threw away at Cambridge."

So began the journey in which a great mind untouched by an old-fashioned classical education was to feed its hunger upon rocks and broken bits of bone at the world's end, and eventually was to shape from such diverse things as bird beaks and the fused wing-cases of island beetles a theory that would shake the foundations of scientific thought in all the countries of the earth.

The Intellectual Setting

The intellectual climate from which Darwin set forth on his historic voyage was predominantly conservative. Insular England had been horrified by the excesses of the French Revolution and was extremely wary of emerging new ideas which it attributed to "French atheists." Religious dogma still held its powerful influence over natural science. True, the 17th-century notion that the world had been created in 4004 B.C. was beginning to weaken in the face of naturalists' studies of the rocks and their succession of life forms. But the conception of a truly ancient and evolving planet was still unformed. No one could dream that the age of the earth was as vast as we now know it to be. And the notion of a continuity of events—of one animal changing by degrees into another—seemed to fly in the face not only of religious beliefs but also of common sense. Many of the greatest biologists of the time—men like Louis Agassiz and Richard Owen—tended to the belief that the successive forms of life in the geological record were all separate creations, some of which had simply been extinguished by historic accidents.

Yet Darwin did not compose the theory of evolution out of thin air. Like so many great scientific generalizations, the theory with which his name is associated had already had premonitory beginnings. All of the elements which were to enter into the theory were in men's minds and were being widely discussed during Darwin's college years. His own grandfather, Erasmus Darwin, who died seven years before Charles was born, had boldly proposed a theory of the "transmutation" of living forms. Jean Baptiste Lamarck had glimpsed a vision of evolutionary continuity. And Sir Charles Lyell—later to be Darwin's lifelong confidant—had opened the way for the evolutionary point of view by demonstrating that the planet must be very old—old enough to allow extremely slow organic change. Lyell dismissed the notion of catastrophic extinction of animal forms on a world-wide scale as impossible, and he made plain that natural forces—the work of wind and frost and water—were sufficient to explain most of the phenomena found in the rocks, provided these forces were seen as operating over enormous periods. Without Lyell's gift of time in immense quantities, Darwin would not have been able to devise the theory of natural selection.

If all the essential elements of the Darwinian scheme of nature were known prior to Darwin, why is he accorded so important a place in biological history? The answer is simple: Almost every great scientific generalization is a supreme act of creative synthesis. There comes a time when an accumulation of smaller discoveries and observations can be combined in some great and comprehensive view of nature. At this point the need is not so much for increased numbers of facts as for a mind of great insight capable of taking the assembled information and rendering it intelligible. Such a synthesis represents the scientific mind at its highest point of achievement. The stature of the discoverer is not diminished by the fact that he has slid into place the last piece of a tremendous puzzle on which many others have worked. To finish the task he must see correctly over a vast and diverse array of data.

Still it must be recognized that Darwin came at a fortunate time. The fact that another man, Alfred Russel Wallace, conceived the Darwinian theory independently before Darwin published it shows clearly that the principle which came to be called natural selection was in the air—was in a sense demanding to be born. Darwin himself pointed out in his autobiography that "innumerable well-observed facts were stored in the minds of naturalists ready to take their proper places as soon as any theory which would receive them was sufficiently explained."

The Voyage

Darwin, then, set out on his voyage with a mind both inquisitive to see and receptive to what he saw. No detail was too small to be fascinating and provocative. Sailing down the South American coast, he notes the octopus changing its color angrily in the waters of a cove. In the dry arroyos of the pampas he observes great bones and shrewdly seeks to relate them to animals of the present. The local inhabitants insist that the fossil bones grew after death, and also that certain rivers have the power of "changing small bones into large." Everywhere men wonder, but they are deceived through their thirst for easy explanations. Darwin, by contrast, is a working dreamer. He rides, climbs, spends long

days on the Indian-haunted pampas in constant peril of his life. Asking at a house whether robbers are numerous, he receives the cryptic reply: "The thistles are not up yet." The huge thistles, high as a horse's back at their full growth, provide ecological cover for bandits. Darwin notes the fact and rides on. The thistles are overrunning the pampas; the whole aspect of the vegetation is altering under the impact of man. Wild dogs howl in the brakes; the common cat, run wild, has grown large and fierce. All is struggle, mutability, change. Staring into the face of an evil relative of the rattlesnake, he observes a fact "which appears to me very curious and instructive, as showing how every character, even though it may be in some degree independent of structure . . . has a tendency to vary by slow degrees."

He pays great attention to strange animals existing in difficult environ-

ments. A queer little toad with a scarlet belly he whimsically nicknames *diabolicus* because it is "a fit toad to preach in the ear of Eve." He notes it lives among sand dunes under the burning sun, and unlike its brethren, cannot swim. From toads to grasshoppers, from pebbles to mountain ranges, nothing escapes his attention. The wearing away of stone, the downstream travel of rock fragments and boulders, the great crevices and upthrusts of the Andes, an earthquake—all confirm the dynamic character of the earth and its great age.

Captain Fitzroy by now is anxious to voyage on. The sails are set. With the towering Andes on their right flank they run north for the Galápagos Islands, lying directly on the Equator 600 miles off the west coast of South America. A one-time refuge of buccaneers, these islands are essentially chimneys of burned-out volcanoes. Darwin remarks that they

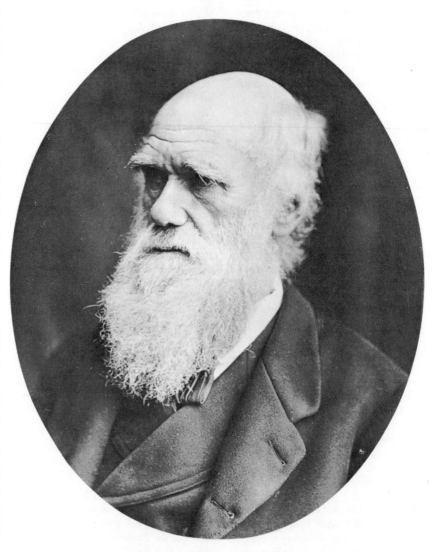

PHOTOGRAPHIC PORTRAIT of Darwin was made some years after the appearance of *Origin of Species.* It is from the collection of George Eastman House in Rochester, N. Y.

THREE IMPORTANT FIGURES in the life of Darwin are shown here and on the following page. They appear in *Portraits of Men of Eminence*, three volumes of which were published be- tween 1863 and 1865. This book is also from George Eastman House. At left is Robert Fitzroy, Captain of the *Beagle*; at right, Charles Lyell, the geologist who was Darwin's lifelong confidant.

remind him of huge iron foundries surrounded by piles of waste. "A little world in itself," he marvels, "with inhabitants such as are found nowhere else." Giant armored tortoises clank through the undergrowth like prehistoric monsters, feeding upon the cacti. Birds in this tiny Eden do not fear men: "One day a mocking bird alighted on the edge of a pitcher which I held in my hand. It began very quietly to sip the water, and allowed me to lift it with the vessel from the ground." Big sea lizards three feet long drowse on the beaches, and feed, fantastically, upon the seaweed. Surveying these "imps of darkness, black as the porous rocks over which they crawl," Darwin is led to comment that "there is no other quarter of the world, where this order replaces the herbivorous mammalia in so extraordinary a manner."

Yet only by degrees did Darwin awake to the fact that he had stumbled by chance into one of the most marvelous evolutionary laboratories on the planet. Here in the Galápagos was a wealth of variations from island to island—among the big tortoises, among plants and es- pecially among the famous finches with remarkably diverse beaks. Dwellers on the islands, notably Vice Governor Lawson, called Darwin's attention to these strange variations, but as he confessed later, with typical Darwinian lack of pretense, "I did not for some time pay sufficient attention to this statement." Whether his visit to the Galápagos was the single event that mainly led Darwin to the central conceptions of his evolutionary mechanism—hereditary change within the organism coupled with external selective factors which might cause plants and animals a few miles apart in the same climate to diverge—is a moot point upon which Darwin himself in later years shed no clear light. Perhaps, like many great men, nagged long after the event for a precise account of the dawn of a great discovery, Darwin no longer clearly remembered the beginning of the intellectual journey which had paralleled so dramatically his passage on the seven seas. Perhaps there had never been a clear beginning at all— only a slowly widening comprehension until what had been seen at first mistily and through a veil grew magnified and clear.

The Invalid and the Book

The paths to greatness are tricky and diverse. Sometimes a man's weaknesses have as much to do with his rise as his virtues. In Darwin's case it proved to be a unique combination of both. He had gathered his material by a courageous and indefatigable pursuit of knowledge that took him through the long vicissitudes of a voyage around the world. But his great work was written in sickness and seclusion. When Darwin reached home after the voyage of the *Beagle*, he was an ailing man, and he remained so to the end of his life. Today we know that this illness was in some degree psychosomatic, that he was anxiety-ridden, subject to mysterious headaches and nausea. Shortly after his voyage Darwin married his cousin Emma Wedgwood, granddaughter of the founder of the great pottery works, and isolated himself and his family in a little village in Kent. He avoided travel like the plague,

THE THIRD IMPORTANT FIGURE, shown above, is Thomas Huxley, who defended Darwin in debate.

save for brief trips to watering places for his health. His seclusion became his strength and protected him; his very fears and doubts of himself resulted in the organization of that enormous battery of facts which documented the theory of evolution as it had never been documented before.

Let us examine the way in which Darwin developed his great theory. The nature of his observations has already been indicated—the bird beaks, the recognition of variation and so on. But it is an easier thing to perceive that evolution has come about than to identify the mechanism involved in it. For a long time this problem frustrated Darwin. He was not satisfied with vague references to climatic influence or the inheritance of acquired characters. Finally he reached the conclusion that since variation in individual characteristics existed among the members of any species, selection of some individuals and elimination of others must be the key to organic change.

This idea he got from the common recognition of the importance of selective breeding in the improvement of domestic plants and livestock. He still did not understand, however, what selective force could be at work in wild nature. Then in 1838 he chanced to read Thomas Malthus, and the solution came to him.

Malthus had written in 1798 a wide[ly] read population study in which h[e] pointed out that the human populatio[n] tended to increase faster than its foo[d] supply, precipitating in consequence [a] struggle for existence.

Darwin applied this principle to th[e] whole world of organic life and argue[d] that the struggle for existence und[er] changing environmental conditions wa[s] what induced alterations in the physic[al] structure of organisms. To put it in oth[er] words, fortuitous and random variatio[n] occurred in living things. The strugg[le] for life perpetuated advantageous vari[a]tions by means of heredity. The wea[k] and unfit were eliminated and those wit[h] the best heredity for any given enviro[n]ment were "selected" to be the paren[ts] of the next generation. Since neither li[fe] nor climate nor geology ever cease[d] changing, evolution was perpetual. N[o] organ and no animal was ever in com[plete equilibrium with its surrounding[s].

This, briefly stated, is the crux of th[e] Darwinian argument. Facts which ha[d] been known before Darwin but had n[ot] been recognized as parts of a sing[le] scheme—variation, inheritance of vari[a]tion, selective breeding of domest[ic] plants and animals, the struggle for e[x]istence—all suddenly fell into place a[s] "natural selection," as "Darwinism."

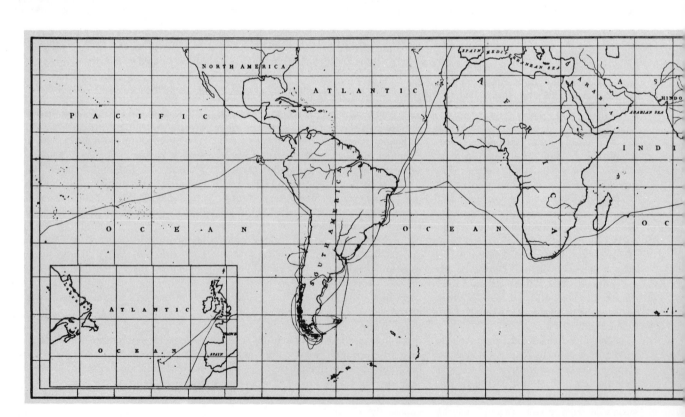

VOYAGE OF THE BEAGLE is traced in this map from Fitzroy's *Narrative of the Surveying Voyage of His Majesty's Ships Adven-* *ture and Beagle*. The *Beagle*'s course on her departure from and return to England is at lower left. The ship made frequent stops at

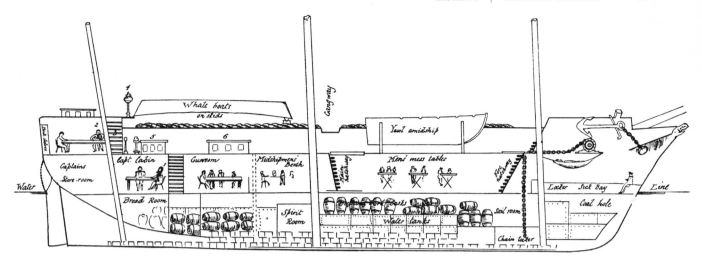

H. M. S. BEAGLE was drawn in cross section many years after the voyage by Philip Gidley King, who accompanied Darwin when he was ashore during the voyage. Darwin is shown in two places: the captain's cabin (*small figure 1 at upper left*) and poop cabin (*2*).

Procrastination

While he developed his theory and marshaled his data, Darwin remained in seclusion and retreat, hoarding the secret of his discovery. For 22 years after the *Beagle's* return he published not one word beyond the bare journal of his trip (later titled *A Naturalist's Voyage around the World*) and technical monographs on his observations.

Let us not be misled, however, by Darwin's seclusiveness and illness. No more lovable or sweet-tempered invalid ever lived. Visitors, however beloved, always aggravated his illness, but instead of the surly misanthropy which afflicts most people under similar circumstances, the result in Darwin's case was merely nights of sleeplessness. Throughout the long night hours his restless mind went on working with deep concentration; more than once, walking alone in the dark hours of winter, he met the foxes trotting home at dawn.

Darwin's gardener is said to have responded once to a visitor who inquired about his master's health: "Poor man, he just stands and stares at a yellow flower for minutes at a time. He would be better off with something to do." Darwin's work was of an intangible nature which eluded people around him. Much of it consisted in just such standing and staring as his gardener reported. It was a kind of magic at which he excelled. On a visit to the Isle of Wight he watched thistle seed wafted about on offshore winds and formulated theories of plant dispersal. Sometimes he engaged in activities which his good wife must surely have struggled to keep from reaching the neighbors. When a friend sent him a half ounce of locust dung from Africa, Darwin triumphantly grew seven plants from the specimen. "There is no error," he assured Lyell, "for I dissected the seeds out of the middle of the pellets." To discover how plant seeds traveled, Darwin would go all the way down a grasshopper's gullet, or worse, without embarrassment. His eldest son Francis spoke amusedly of his father's botanical experiments: "I think he personified each seed as a small demon trying to elude him by getting into the wrong heap, or jumping away all together; and this gave to the work the excitement of a game."

The point of his game Darwin kept largely to himself, waiting until it should be completely finished. He piled up vast stores of data and dreamed of presenting his evolution theory in a definitive, monumental book, so large that it would certainly have fallen dead and unreadable from the press. In the meantime, Robert Chambers, a bookseller and journalist, wrote and brought out anonymously a modified version of Lamarckian evolution, under the title *Vestiges of the Natural History of Creation*. Amateurish in some degree, the book drew savage onslaughts from the critics, including Thomas Huxley, but it caught the public fancy and was widely read. It passed through numerous editions both in England and America—evidence that *sub rosa* there was a good deal more interest on the part of the public in the "development hypothesis," as evolution was then called, than the fulminations of critics would have suggested.

Throughout this period Darwin remained stonily silent. Many explanations of his silence have been ventured by his biographers: that he was busy accumulating materials; that he did not wish to affront Fitzroy; that the attack on the *Vestiges* had intimidated him; that he thought it wise not to write upon so controversial a subject until he had first acquired a reputation as a professional naturalist of the first rank. Primarily, however, the basic reason lay in his personality—a nature reluctant to face the storm that publication would bring about his ears. It was pleasanter to procrastinate, to talk of the secret to a few chosen companions such as Lyell

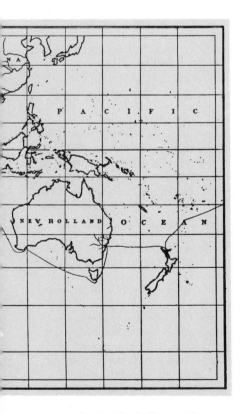

oceanic islands. The Galápagos Islands are on the Equator to the west of South America.

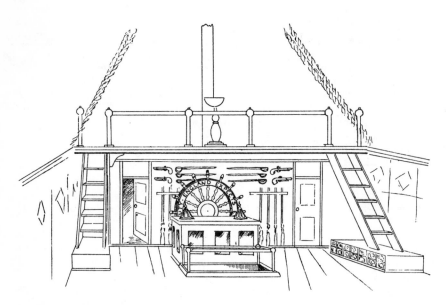

QUARTER-DECK of the *Beagle* is depicted in this drawing by King. In the center is the wheel, the circumference of which is inscribed: "England expects every man to do his duty."

and the great botanist Joseph Hooker.

The Darwin family had been well-to-do since the time of grandfather Erasmus. Charles was independent, in a position to devote all his energies to research and under no academic pressure to publish in haste.

"You will be anticipated," Lyell warned him. "You had better publish." That was in the spring of 1856. Darwin promised, but again delayed. We know that he left instructions for his wife to see to the publication of his notes in the event of his death. It was almost as if present fame or notoriety were more than he could bear. At all events he continued to delay, and this situation might very well have continued to the end of his life, had not Lyell's warning suddenly come true and broken his pleasant dream.

Alfred Russel Wallace, a comparatively unknown, youthful naturalist, had divined Darwin's great secret in a moment of fever-ridden insight while on a collecting trip in Indonesia. He, too, had put together the pieces and gained a clear conception of the scheme of evolution. Ironically enough, it was to Darwin, in all innocence, that he sent his manuscript for criticism in June of 1858. He sensed in Darwin a sympathetic and traveled listener.

Darwin was understandably shaken.

The work which had been so close to his heart, the dream to which he had devoted 20 years, was a private secret no longer. A newcomer threatened his priority. Yet Darwin, wanting to do what was decent and ethical, had been placed in an awkward position by the communication. His first impulse was to withdraw totally in favor of Wallace. "I would far rather burn my whole book," he insisted, "than that he or any other man should think that I had behaved in a paltry spirit." It is fortunate for science that before pursuing this quixotic course Darwin turned to his friends Lyell and Hooker, who knew the many years he had been laboring upon his *magnum opus*. The two distinguished scientists arranged for the delivery of a short summary by Darwin to accompany Wallace's paper before the Linnaean Society. Thus the theory of the two men was announced simultaneously.

Publication

The papers drew little comment at the meeting but set in motion a mild undercurrent of excitement. Darwin, though upset by the death of his son Charles, went to work to explain his views more fully in a book. Ironically he called it *An Abstract of an Essay on the*

Origin of Species and insisted it would be only a kind of preview of a much larger work. Anxiety and devotion to his great hoard of data still possessed him. He did not like to put all his hopes in this volume, which must now be written at top speed. He bolstered himself by references to the "real" book—that Utopian volume in which all that could not be made clear in his abstract would be clarified.

His timidity and his fears were totally groundless. When the *Origin of Species* (the title distilled by his astute publisher from Darwin's cumbersome and halfhearted one) was published in the fall of 1859, the first edition was sold in a single day. The book which Darwin had so apologetically bowed into existence was, of course, soon to be recognized as one of the great books of all time. It would not be long before its author would sigh happily and think no more of that huge, ideal volume which he had imagined would be necessary to convince the public. The public and his brother scientists would find the *Origin* quite heavy going enough. His book to end all books would never be written. It did not need to be. The world of science in the end could only agree with the sharp-minded Huxley, whose immediate reaction upon reading the *Origin* was: "How extremely stupid not to have thought of that!" And so it frequently seems in science, once the great synthesizer has done his work. The ideas were not new, but the synthesis was. Men would never again look upon the world in the same manner as before.

No great philosophic conception ever entered the world more fortunately. Though it is customary to emphasize the religious and scientific storm the book aroused—epitomized by the famous debate at Oxford between Bishop Wilberforce and Thomas Huxley—the truth is that Darwinism found relatively easy acceptance among scientists and most of the public. The way had been prepared by the long labors of Lyell and the wide popularity of Chambers' book, the *Vestiges*. Moreover, Darwin had won the support of the great Hooker and of Huxley, the most formidable scientific debater of all time. Lyell, though more cautious, helped to publicize Darwin and at no time attacked him. Asa Gray, one of America's leading botanists, came to his defense. His codiscoverer, Wallace, as generous-hearted as Darwin, himself advanced the word "Darwinism" for Darwin's theory, and minimized his own part in the elaboration of the theory as "one week to 20 years."

This sturdy band of converts assumed

the defense of Darwin before the public, while Charles remained aloof. Sequestered in his estate at Down, he calmly answered letters and listened, but not too much, to the tumult over the horizon. "It is something unintelligible to me how anyone can argue in public like orators do," he confessed to Hooker, though he was deeply grateful for the verbal swordplay of his cohorts. Hewett Watson, another botanist of note, wrote to him shortly after the publication of the *Origin:* "Your leading idea will assuredly become recognized as an established truth in science, *i.e.*, 'Natural Selection.' It has the characteristics of all great natural truths, clarifying what was obscure, simplifying what was intricate, adding greatly to previous knowledge. You are the greatest revolutionist in natural history of this century, if not of all centuries."

Watson's statement was clairvoyant. Not a line of his appraisal would need to be altered today. Within 10 years the *Origin* and its author were known all over the globe, and evolution had become the guiding motif in all biological studies.

Summing up the achievement of this book, we may say today, first, that Darwin had proved the reality of evolutionary change beyond any reasonable doubt, and secondly, that he had demonstrated, in natural selection, a principle capable of wide, if not universal, application. Natural selection dispelled the confusions that had been introduced into biology by the notion of individual creation of species. The lad who in 1832 had noted with excited interest "that there are three sorts of birds which use their wings for more purposes than flying; the Steamer [duck] as paddles, the Penguin as fins, and the Ostrich (*Rhea*) spreads its plumes like sails" now had his answer—"descent with modification." "If you go any considerable lengths in the admission of modification," warned Darwin, "I can see no possible means of drawing the line, and saying here you must stop." Rung by rung, was his plain implication, one was forced to descend down the full length of life's mysterious ladder until one stood in the brewing vats where the thing was made. And similarly, rung by rung, from mudfish to reptile and mammal, the process ascended to man.

A Small Place for Man

Darwin had cautiously avoided direct references to man in the *Origin of Species*. But 12 years later, after its triumph was assured, he published a study of human evolution entitled *The Descent of Man*. He had been preceded in this field by Huxley's *Evidences as to Man's Place in Nature* (1863). Huxley's brief work was written with wonderful clarity and directness. By contrast, the *Descent of Man* has some of the labored and inchoate quality of Darwin's overfull folios of data. It is contradictory in spots, as

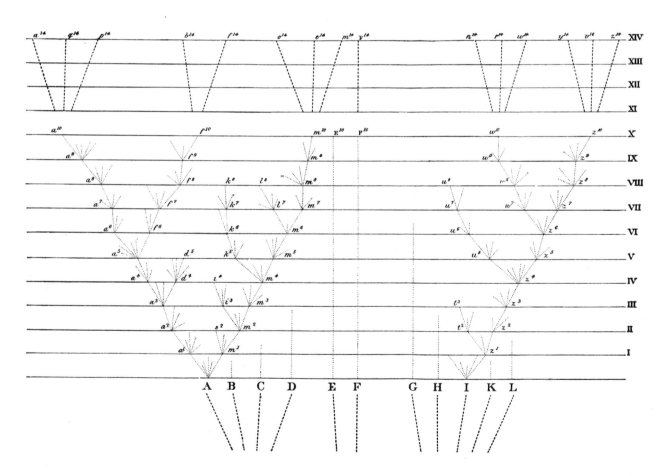

NATURAL SELECTION through the divergence of characters is illustrated in *Origin of Species*. The capital letters at the bottom of the illustration represent different species of the same genus. Each horizontal line, labeled with a Roman numeral at the right, represents 1,000 or more generations. Darwin believed that some of the original species, such as A, would diverge more than others. After many generations they would give rise to new varieties, such as a^1 and m^1. These new varieties would diverge in turn. After thousands of generations the new varieties would give rise to entirely new species, such as a^{14}, q^{14}, p^{14} and so on. The original species would meantime have died out. Darwin thought that only some species of the original genus would diverge sufficiently to give rise to new species. Some of the species, such as F, would remain much the same. Others, such as B, C and D, would die out.

though the author simply poured his notes together and never fully read the completed manuscript to make sure it was an organic whole.

One of its defects is Darwin's failure to distinguish consistently between biological inheritance and cultural influences upon the behavior and evolution of human beings. In this, of course, Darwin was making a mistake common to biologists of the time. Anthropology was then in its infancy. In the biological realm, the *Descent of Man* did make plain in a general way that man was related to the rest of the primate order, though the precise relationship was left ambiguous. After all, we must remember that no one had yet unearthed any clear fossils of early man. A student of evolution had to content himself largely with tracing morphological similarities between living man and the great apes. This left considerable room for speculation as to the precise nature of the human ancestors. It is not surprising that they were occasionally visualized as gorilloid beasts with huge canine teeth, nor that Darwin wavered between this and gentler interpretations.

An honest biographer must record the fact that man was not Darwin's best subject. In the words of a 19th-century critic, his "was a world of insects and pigeons, apes and curious plants, but man as he exists, had no place in it." Allowing for the hyperbole of this religious opponent, it is nonetheless probable that Darwin did derive more sheer delight from writing his book on earthworms than from any amount of contemplation of a creature who could talk back and who was apt stubbornly to hold ill-founded opinions. In any case, no man afflicted with a weak stomach and insomnia has any business investigating his own kind. At least it is best to wait until they have undergone the petrification incident to becoming part of a geological stratum.

Darwin knew this. He had fled London to work in peace. When he dealt with the timid gropings of climbing plants, the intricacies of orchids or the calculated malice of the carnivorous sundew, he was not bedeviled by metaphysicians, by talk of ethics, morals or the nature of religion. Darwin did not wish to leave man an exception to his system, but he was content to consider man simply as a part of that vast, sprawling, endlessly ramifying ferment called "life." The rest of him could be left to the philosophers. "I have often," he once complained to a friend, "been made wroth (even by Lyell) at the confidence with which people speak of the intro-

duction of man, as if they had seen him walk on the stage and as if in a geological sense it was more important than the entry of any other mammifer."

Darwin's fame as the author of the theory of evolution has tended to obscure the fact that he was, without doubt, one of the great field naturalists of all time. His capacity to see deep problems in simple objects is nowhere better illustrated than in his study of movement in plants, published some two years before his death. He subjected twining plants, previously little studied, to a series of ingenious investigations of pioneer importance in experimental botany. Perhaps Darwin's intuitive comparison of plants to animals accounted for much of his success in this field. There is an entertaining story that illustrates how much more perceptive than his contemporaries he was here. To Huxley and another visitor, Darwin was trying to explain the remarkable behavior of *Drosera*, the sundew plant, which catches insects with grasping, sticky hairs. The two visitors listened to Darwin as one might listen politely to a friend who is slightly "touched." But as they watched the plant, their tolerant poise suddenly vanished and Huxley cried out in amazement: "Look, it *is* moving!"

The Islands

As one surveys the long and tangled course that led to Darwin's great discovery, one cannot but be struck by the part played in it by oceanic islands. It is a part little considered by the general public today. The word "evolution" is commonly supposed to stand for something that occurred in the past, something involving fossil apes and dinosaurs, something pecked out of the rocks of eroding mountains—a history of the world largely demonstrated and proved by the bone hunter. Yet, paradoxically, in Darwin's time it was this very history that most cogently challenged the evolutionary point of view. Paleontology was not nearly so extensively developed as today, and the record was notable mainly for its gaps. "Where are the links?" the critics used to rail at Darwin. "Where are the links between man and ape— between your lost land animal and the whale? Show us the fossils; prove your case." Darwin could only repeat: "This is the most obvious and gravest objection which can be urged against my theory. The explanation lies, as I believe, in the extreme imperfection of the geological record." The evidence for the continuity of life must be found else-

where. And it was the oceanic islands that finally supplied the clue.

Until Darwin turned his attention to them, it appears to have been generally assumed that island plants and animals were simply marooned evidences of a past connection with the nearest continent. Darwin, however, noted that whole classes of continental life were absent from the islands; that certain plants which were herbaceous (nonwoody) on the mainland had developed into trees on the islands; that island animals often differed from their counterparts on the mainland.

Above all, the fantastically varied finches of the Galápagos particularly amazed and puzzled him. The finches diverged mainly in their beaks. There were parrot-beaks, curved beaks for probing flowers, straight beaks, small beaks—beaks for every conceivable purpose. These beak variations existed nowhere but on the islands; they must have evolved there. Darwin had early observed: "One might really fancy that, from an original paucity of birds in this archipelago, one species had been taken and modified for different ends." The birds had become transformed, through the struggle for existence on their little islets, into a series of types suited to particular environmental niches where, properly adapted, they could obtain food and survive. As the ornithologist David Lack has remarked: "Darwin's finches form a little world of their own, but one which intimately reflects the world as a whole" [see the article "Darwin's Finches," by David Lack, beginning on page 35].

Darwin's recognition of the significance of this miniature world, where the forces operating to create new beings could be plainly seen, was indispensable to his discovery of the origin of species. The island worlds reduced the confusion of continental life to more simple proportions; one could separate the factors involved with greater success. Over and over Darwin emphasized the importance of islands in his thinking. Nothing would aid natural history more, he contended to Lyell, "than careful collecting and investigating of *all the productions* of the most isolated islands. . . . Every sea shell and insect and plant is of value from such spots."

Darwin was born in precisely the right age even in terms of the great scientific voyages. A little earlier the story the islands had to tell could not have been read; a little later much of it began to be erased. Today all over the globe the populations of these little worlds are vanishing, many without ever having

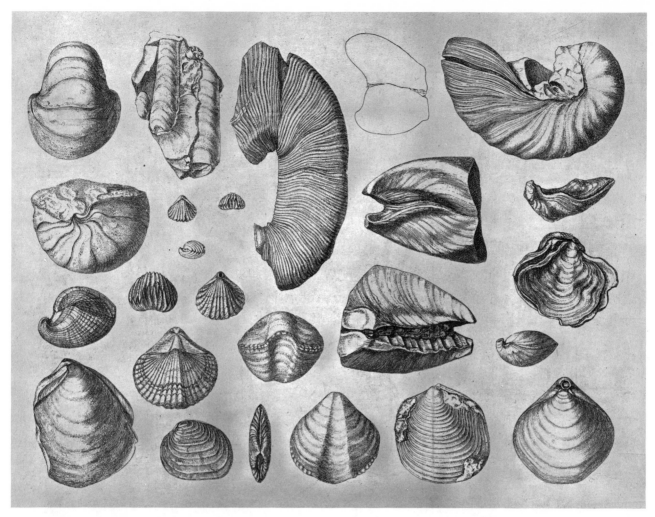

FOSSIL SHELLS were depicted in this engraving from Darwin's *Geological Observations on the Volcanic Islands and Parts of South* *America Visited during the Voyage of H. M. S. Beagle.* This was a technical work published by Darwin before *Origin of Species.*

been seriously investigated. Man, breaking into their isolation, has brought with him cats, rats, pigs, goats, weeds and insects from the continents. In the face of these hardier, tougher, more aggressive competitors, the island faunas—the rare, the antique, the strange, the beautiful— are vanishing without a trace. The giant Galápagos tortoises are almost extinct, as is the land lizard with which Darwin played. Some of the odd little finches and rare plants have gone or will go. On the island of Madagascar our own remote relatives, the lemurs, which have radiated into many curious forms, are now being exterminated through the destruction of the forests. Even that continental island Australia is suffering from the decimation wrought by man. The Robinson Crusoe worlds where

small castaways could create existences idyllically remote from the ravening slaughter of man and his associates are about to pass away forever. Every such spot is now a potential air base where the cries of birds are drowned in the roar of jets, and the crevices once frequented by bird life are flattened into the long runways of the bombers. All this would not have surprised Darwin, one would guess.

Of Darwin's final thoughts in the last hours of his life in 1882, when he struggled with a weakening heart, no record remains. One cannot but wonder whether this man who had no faith in Paradise may not have seen rising on his dying sight the pounding surf and black slag heaps of the Galápagos, those islands called by Fitzroy "a fit shore for Pande-

monium." None would ever see them again as Darwin had seen them—smoldering sullenly under the equatorial sun and crawling with uncertain black reptiles lost from some earlier creation. Once he had cried out suddenly in anguish: "What a book a devil's chaplain might write on the clumsy, wasteful, blundering, low and horribly cruel works of nature!" He never spoke or wrote in quite that way again. It was more characteristic of his mind to dwell on such memories as that Eden-like bird drinking softly from the pitcher held in his hand. When the end came, he remarked with simple dignity, "I am not in the least afraid of death."

It was in that spirit he had ventured upon a great voyage in his youth. It would suffice him for one more journey.

CONTROLLED ENVIRONMENT for the study of fruit-fly populations is a glass-covered box. In bottom of the box are cups of food that are filled in rotation to keep food a constant factor in environment.

The Genetic Basis of Evolution

2

by Theodosius Dobzhansky
January 1950

The rich variety of living and dead species of plants and animals is the result of subtle interplay between the hereditary mechanism and a diversified environment

THE living beings on our planet come in an incredibly rich diversity of forms. Biologists have identified about a million species of animals and some 267,000 species of plants, and the number of species actually in existence may be more than twice as large as the number known. In addition the earth has been inhabited in the past by huge numbers of other species that are now extinct, though some are preserved as fossils. The organisms of the earth range in size from viruses so minute that they are barely visible in electron microscopes to giants like elephants and sequoia trees. In appearance, body structure and ways of life they exhibit an endlessly fascinating variety.

What is the meaning of this bewildering diversity? Superficially considered, it may seem to reflect nothing more than the whims of some playful deity, but one soon finds that it is not fortuitous. The more one studies living beings the more one is impressed by the wonderfully effective adjustment of their multifarious body structures and functions to their varying ways of life. From the simplest to the most complex, all organisms are constructed to function efficiently in the environments in which they live. The body of a green plant can build itself from food consisting merely of water, certain gases in the air and some mineral salts taken from the soil. A fish is a highly efficient machine for exploiting the organic food resources of water, and a bird is built to get the most from its air en-

vironment. The human body is a complex, finely coordinated machine of marvelously precise engineering, and through the inventive abilities of his brain man is able to control his environment. Every species, even the most humble, occupies a certain place in the economy of nature, a certain adaptive niche which it exploits to stay alive.

The diversity and adaptedness of living beings were so difficult to explain that during most of his history man took the easy way out of assuming that every species was created by God, who contrived the body structures and functions of each kind of organism to fit it to a predestined place in nature. This idea has now been generally replaced by the less easy but intellectually more satisfying explanation that the living things we see around us were not always what they are now, but evolved gradually from very different-looking ancestors; that these ancestors were in general less complex, less perfect and less diversified than the organisms now living; that the evolutionary process is still under way, and that its causes can therefore be studied by observation and experiment in the field and in the laboratory.

The origins and development of this theory, and the facts that finally convinced most people of its truth beyond reasonable doubt, are too long a story to be presented here. After Charles Darwin published his convincing exposition and proof of the theory of evolution in 1859, two main currents developed in evolu-

tionary thought. Like any historical process, organic evolution may be studied in two ways. One may attempt to infer the general features and the direction of the process from comparative studies of the sequence of events in the past; this is the method of paleontologists, comparative anatomists and others. Or one may attempt to reconstruct the causes of evolution, primarily through a study of the causes and mechanisms that operate in the world at present; this approach, which uses experimental rather than observational methods, is that of the geneticist and the ecologist. This article will consider what has been learned about the causes of organic evolution through the second approach.

Darwin attempted to describe the causes of evolution in his theory of natural selection. The work of later biologists has borne out most of his basic contentions. Nevertheless, the modern theory of evolution, developed by a century of new discoveries in biology, differs greatly from Darwin's. His theory has not been overthrown; it has evolved. The authorship of the modern theory can be credited to no single person. Next to Darwin, Gregor Mendel of Austria, who first stated the laws of heredity, made the greatest contribution. Within the past two decades the study of evolutionary genetics has developed very rapidly on the basis of the work of Thomas Hunt Morgan and Hermann J. Muller of the U. S. In these developments the principal contributors have been C. D.

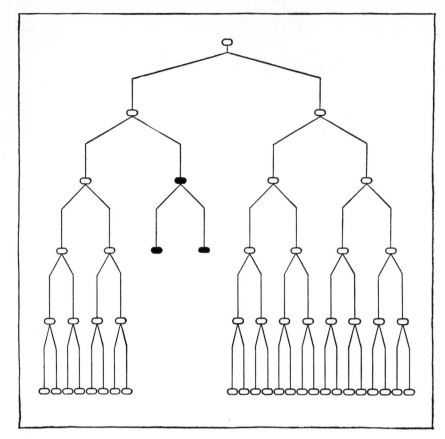

IN NORMAL ENVIRONMENT the common strain of the bacterium *Escherichia coli* (*white bacteria*) multiplies. A mutant strain resistant to streptomycin (*black bacteria*) remains rare because the mutation is not useful.

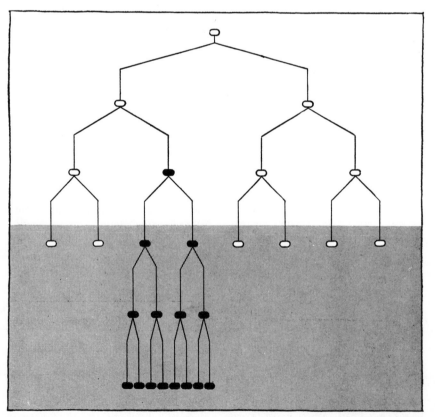

IN CHANGED ENVIRONMENT produced by the addition of streptomycin (*gray area*) the streptomycin-resistant strain is better adapted than the common strain. The mutant strain then multiplies and the common one dies.

Darlington, R. A. Fisher, J. B. S. Haldane, J. S. Huxley and R. Mather in England; B. Rensch and N. W. Timofeeff-Ressovsky in Germany; S. S. Chetverikov, N. P. Dubinin and I. I. Schmalhausen in the U.S.S.R.; E. Mayr, J. T. Patterson, C. G. Simpson, G. L. Stebbins and Sewall Wright in the U. S., and some others.

Evolution in the Laboratory

Evolution is generally so slow a process that during the few centuries of recorded observations man has been able to detect very few evolutionary changes among animals and plants in their natural habitats. Darwin had to deduce the theory of evolution mostly from indirect evidence, for he had no means of observing the process in action. Today, however, we can study and even produce evolutionary changes at will in the laboratory. The experimental subjects of these studies are bacteria and other low forms of life which come to birth, mature and yield a new generation within a matter of minutes or hours, instead of months or years as in most higher beings. Like a greatly speeded-up motion picture, these observations compress into a few days evolutionary events that would take thousands of years in the higher animals.

One of the most useful bacteria for this study is an organism that grows, usually harmlessly, in the intestines of practically every human being: *Escherichia coli*, or colon bacteria. These organisms can easily be cultured on a nutritive broth or nutritive agar. At about 98 degrees Fahrenheit, bacterial cells placed in a fresh culture medium divide about every 20 minutes. Their numbers increase rapidly until the nutrients in the culture medium are exhausted; a single cell may yield billions of progeny in a day. If a few cells are placed on a plate covered with nutritive agar, each cell by the end of the day produces a whitish speck representing a colony of its offspring.

Now most colon bacteria are easily killed by the antibiotic drug streptomycin. It takes only a tiny amount of streptomycin, 25 milligrams in a liter of a nutrient medium, to stop the growth of the bacteria. Recently, however, the geneticist Milislav Demerec and his collaborators at the Carnegie Institution in Cold Spring Harbor, N. Y., have shown that if several billion colon bacteria are placed on the streptomycin-containing medium, a few cells will survive and form colonies on the plate. The offspring of these hardy survivors are able to multiply freely on a medium containing streptomycin. A mutation has evidently taken place in the bacteria; they have now become resistant to the streptomycin that was poisonous to their sensitive ancestors.

How do the bacteria acquire their

resistance? Is the mutation caused by their exposure to streptomycin? Demerec has shown by experimental tests that this is not so; in any large culture a few resistant mutants appear even when the culture has not been exposed to streptomycin. Some cells in the culture undergo mutations from sensitivity to resistance regardless of the presence or absence of streptomycin in the medium. Demerec found that the frequency of mutation was about one per billion; *i.e.,* one cell in a billion becomes resistant in every generation. Streptomycin does not induce the mutations; its role in the production of resistant strains is merely that of a selecting agent. When streptomycin is added to the culture, all the normal sensitive cells are killed, and only the few resistant mutants that happened to be present before the streptomycin was added survive and reproduce. Evolutionary changes are controlled by the environment, but the control is indirect, through the agency of natural or artificial selection.

What governs the selection? If resistant bacteria arise in the absence of streptomycin, why do sensitive forms predominate in all normal cultures; why has not the whole species of colon bacteria become resistant? The answer is that the mutants resistant to streptomycin are at a disadvantage on media free from this drug. Indeed, Demerec has discovered the remarkable fact that about 60 per cent of the bacterial strains derived from streptomycin-resistant mutants become dependent on streptomycin; they are unable to grow on media free from it!

On the other hand one can reverse the process and obtain strains of bacteria that can live without streptomycin from cultures predominantly dependent on the drug. If some billions of dependent bacteria are plated on nutrient media free of the drug, all dependent cells cease to multiply and only the few mutants independent of the drug reproduce and form colonies. Demerec estimates the frequency of this "reverse" mutation at about 37 per billion cells in each generation.

Evolutionary changes of the type described in colon bacteria have been found in recent years in many other bacterial species. The increasing use of antibiotic drugs in medical practice has made such changes a matter of considerable concern in public health. As penicillin, for example, is used on a large scale against bacterial infections, the strains of bacteria that are resistant to penicillin survive and multiply, and the probability that they will infect new victims is increased. The mass application of antibiotic drugs may lead in the long run to increased incidence of cases refractory to treatment. Indications exist that this has already happened in some instances: in certain cities penicillin-resistant gonorrhea has become more frequent than it was.

The same type of evolutionary change has also been noted in some larger organisms. A good example is the case of DDT and the common housefly, *Musca domestica.* DDT was a remarkably effective poison for houseflies when first introduced less than 10 years ago. But already reports have come from places as widely separated as New Hampshire, New York, Florida, Texas, Italy and Sweden that DDT sprays in certain localities have lost their effectiveness. What has happened, of course, is that strains of houseflies relatively resistant to DDT have become established in these localities. Man has unwittingly become an agent of a selection process which has led to evolutionary changes in housefly populations. Similar changes are known to have occurred in other insects; *e.g.,* in some orchards of California where hydrocyanic gas has long been used as a fumigant to control scale insects that prey on citrus fruits, strains of these insects that are resistant to hydrocyanic gas have developed.

Obviously evolutionary selection can take place only if nature provides a supply of mutants to choose from. Thus no bacteria will survive to start a new strain resistant to streptomycin in a culture in which no resistant mutant cells were present in the first place, and housefly races resistant to DDT have not appeared everywhere that DDT is used. Adaptive changes are not mechanically forced upon the organism by the environment. Many species of past geological epochs died out because they did not have a supply of mutants which fitted changing environments. The process of mutation furnishes the raw materials from which evolutionary changes are built.

Mutations

Mutations arise from time to time in all organisms, from viruses to man. Perhaps the best organism for the study of mutations is the now-famous fruit fly, Drosophila. It can be bred easily and rapidly in laboratories, and it has a large number of bodily traits and functions that are easy to observe. Mutations affect the color of its eyes and body, the size and shape of the body and of its parts, its internal anatomical structures, its fecundity, its rate of growth, its behavior, and so on. Some mutations produce differences so minute that they can be detected only by careful measurements; others are easily seen even by beginners; still others produce changes so drastic that death occurs before the development is completed. The latter are called lethal mutations.

The frequency of any specific mutation is usually low. We have seen that in colon bacteria a mutation to resistance to streptomycin occurs in only about one cell per billion in every generation, and the reverse mutation to independence of streptomycin is about 37 times more frequent. In Drosophila and in the corn plant mutations have been found to range in frequency from one in 100,000 to one in a million per generation. In man, according to estimates by Haldane in England and James Neel in the U. S., mutations that produce certain hereditary diseases, such as hemophilia and Cooley's anemia, arise in one in 2,500 to one in 100,000 sex cells in each generation. From this it may appear that man is more mutable than flies and bacteria, but it should be remembered that a generation in man takes some 25 years, in flies two weeks, and in bacteria 25 minutes. The frequency of mutations per unit of time is actually greater in bacteria than in man.

A single organism may of course produce several mutations, affecting different features of the body. How frequent are all mutations combined? For technical reasons, this is difficult to determine; for example, most mutants produce small changes that are not detected unless especially looked for. In Drosophila it is estimated that new mutants affecting one part of the body or another are present in between one and 10 per cent of the sex cells in every generation.

In all organisms the majority of mutations are more or less harmful. This may seem a very serious objection against the theory which regards them as the mainspring of evolution. If mutations produce incapacitating changes, how can adaptive evolution be compounded of them? The answer is: a mutation that is harmful in the environment in which the species or race lives may become useful, even essential, if the environment changes. Actually it would be strange if we found mutations that improve the adaptation of the organism in the environment in which it normally lives. Every kind of mutation that we observe has occurred numerous times under natural conditions, and the useful ones have become incorporated into what we call the "normal" constitution of the species. But when the environment changes, some of the previously rejected mutations become advantageous and produce an evolutionary change in the species. The writer and B. A. Spassky have carried out certain experiments in which we intentionally disturbed the harmony between an artificial environment and the fruit flies living in it. At first the change in environment killed most of the flies, but during 50 consecutive generations most strains showed a gradual improvement of viability, evidently owing to the environment's selection of the better-adapted variants.

This is not to say that every mutation will be found useful in some environment somewhere. It would be difficult to imagine environments in which such human mutants as hemophilia or the absence of hands and feet might be useful. Most mutants that arise in any species are, in effect, degenerative changes; but some, perhaps a small minority, may be beneficial in some environments. If the environment were forever constant, a species might conceivably reach a summit of adaptedness and ultimately suppress the mutation process. But the environment is never constant; it varies not only from place to place but from time to time. If no mutations occur in a species, it can no longer become adapted to changes and is headed for eventual extinction. Mutation is the price that organisms pay for survival. They do not possess a miraculous ability to produce only useful mutations where and when needed. Mutations arise at random, regardless of whether they will be useful at the moment, or ever; nevertheless, they make the species rich in adaptive possibilities.

The Genes

To understand the nature of the mutation process we must inquire into the nature of heredity. A man begins his individual existence when an egg cell is fertilized by a spermatozoon. From an egg cell weighing only about a 20-millionth of an ounce, he grows to an average weight at maturity of some 150 pounds—a 48-billionfold increase. The material for this stupendous increase in mass evidently comes from the food consumed, first by the mother and then by the individual himself. But the food becomes a constituent part of the body only after it is digested and assimilated, *i.e.*, transformed into a likeness of the assimilating body. This body, in turn, is a likeness of the bodies of the individual's ancestors. Heredity is, then, a process whereby the organism reproduces itself in its progeny from food materials taken in from the environment. In short, heredity is self-reproduction.

The units of self-reproduction are called genes. The genes are borne chiefly in chromosomes of the cell nucleus, but certain types of genes called plasmagenes are present in the cytoplasm, the part of the cell outside the nucleus. The chemical details of the process of self-reproduction are unknown. Apparently a gene enters into some set of chemical reactions with materials in its surroundings; the outcome of these reactions is the appearance of two genes in the place of one. In other words, a gene synthesizes a copy of itself from nongenic materials. The genes are considered to be stable because the copy is a true likeness of the original in the overwhelming majority of cases; but occasionally the copying process is faulty, and the new gene that emerges differs from its model. This is a mutation. We can increase the frequency of mutations in experimental animals by treating the genes with X-rays, ultraviolet rays, high temperature or certain chemical substances.

Can a gene be changed by the environment? Assuredly it can. But the important point is the kind of change produced. The change that is easiest to make is to treat the gene with poisons or heat in such a way that it no longer reproduces itself. But a gene that cannot produce a copy of itself from other materials is no longer a gene; it is dead. A mutation is a change of a very special kind: the altered gene can reproduce itself, and the copy produced is like the changed structure, not like the original. Changes of this kind are relatively rare. Their rarity is not due to any imperviousness of the genes to influences of the environment, for genic materials are probably the most active chemical constituents of the body; it is due to the fact that genes are by nature self-reproducing, and only the rare changes that preserve the genes' ability to reproduce can effect a lasting alteration of the organism.

Changes in heredity should not be confused, as they often are, with changes in the manifestations of heredity. Such expressions as "gene for eye color" or "inheritance of musical ability" are figures of speech. The sex cells that transmit heredity have no eyes and no musical ability. What genes determine are patterns of development which result in the emergence of eyes of a certain color and of individuals with some musical abilities. When genes reproduce themselves from different food materials and in different environments, they engender the development of different "characters" or "traits" in the body. The outcome of the development is influenced both by heredity and by environment.

In the popular imagination, heredity is transmitted from parents to offspring through "blood." The heredity of a child is supposed to be a kind of alloy or solution, resulting from the mixture of the paternal and maternal "bloods." This blood theory became scientifically untenable as long ago as Mendel's discovery of the laws of heredity in 1865. Heredity is transmitted not through miscible bloods but through genes. When different variants of a gene are brought together in a single organism, a hybrid, they do not fuse or contaminate one another; the genes keep their integrity and separate when the hybrid forms sex cells.

Genetics and Mathematics

Although the number of genes in a single organism is not known with precision, it is certainly in the thousands, at least in the higher organisms. For Drosophila, 5,000 to 12,000 seems a reasonable estimate, and for man the figure is, if anything, higher. Since most or all genes suffer mutational changes from time to time, populations of virtually every species must contain mutant variants of many genes. For example, in the human species there are variations in the skin, hair and eye colors, in the shape and distribution of hair, in the form of the head, nose and lips, in stature, in body proportions, in the chemical composition of the blood, in psychological traits, and so on. Each of these traits is influenced by several or by many genes. To be conservative, let us assume that the human species has only 1,000 genes and that each gene has only two variants. Even on this conservative basis, Mendelian segregation and recombination would be capable of producing 2^{1000} different gene combinations in human beings.

The number 2^{1000} is easy to write but is utterly beyond comprehension. Compared with it, the total number of electrons and protons estimated by physicists to exist in the universe is negligibly small! It means that except in the case of identical twins no two persons now living, dead, or to live in the future are at all likely to carry the same complement of genes. Dogs, mice and flies are as individual and unrepeatable as men are. The mechanism of sexual reproduction, of which the recombination of genes is a part, creates ever new genetic constitutions on a prodigious scale.

One might object that the number of possible combinations does not greatly matter; after all, they will still be combinations of the same thousand gene variants, and the way they are combined is not significant. Actually it is: the same gene may have different effects in combinations with different genes. Thus Timofeeff-Ressovsky showed that two mutants in Drosophila, each of which reduced the viability of the fly when it was present alone, were harmless when combined in the same individual by hybridization. Natural selection tests the fitness in certain environments not of single genes but of constellations of genes present in living individuals.

Sexual reproduction generates, therefore, an immense diversity of genetic constitutions, some of which, perhaps a small minority, may prove well attuned to the demands of certain environments. The biological function of sexual reproduction consists in providing a highly efficient trial-and-error mechanism for the operation of natural selection. It is a reasonable conjecture that sex became established as the prevalent method of reproduction because it gave organisms the greatest potentialities for adaptive and progressive evolution.

Let us try to imagine a world providing a completely uniform environment. Suppose that the surface of our planet were absolutely flat, covered everywhere with the same soil; that instead of summer and winter seasons we had eternally constant temperature and humidity; that

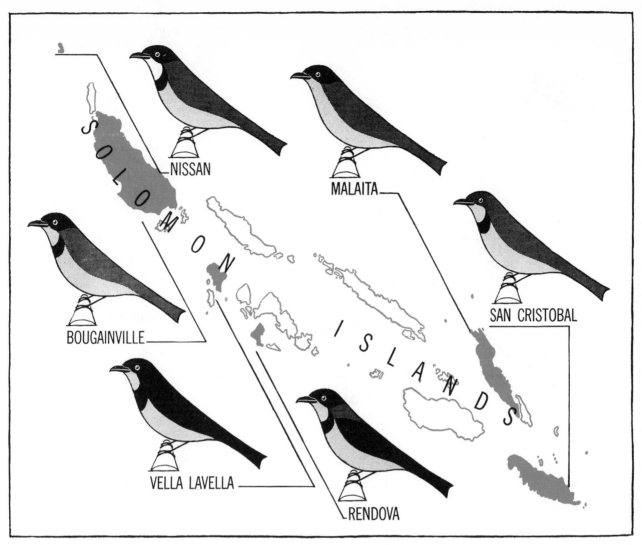

CONCEPT OF RACE is illustrated by the varieties of the golden whistler (*Pachycephala pectoralis*) of the Solomon Islands. The races are kept distinct principally by geographical isolation. They differ in their black and white and colored markings. Dark gray areas are symbol for green markings; light gray for yellow.

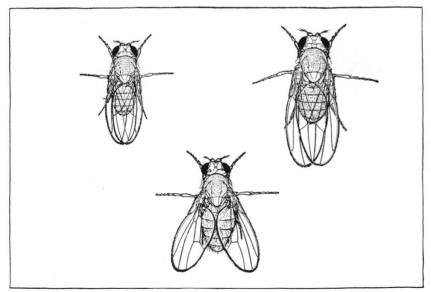

SPECIES OF DROSOPHILA and some other organisms tend to remain separate because their hybrid offspring are often weak and sterile. At left is *D. pseudoobscura*; at right *D. miranda*. Their hybrid descendant is at bottom.

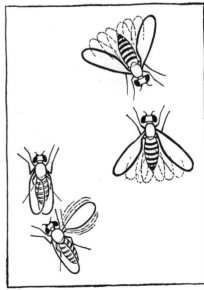

RITUALS OF MATING in *D. nebulosa* (*top*) and *D. willistoni* are example of factor that separates species.

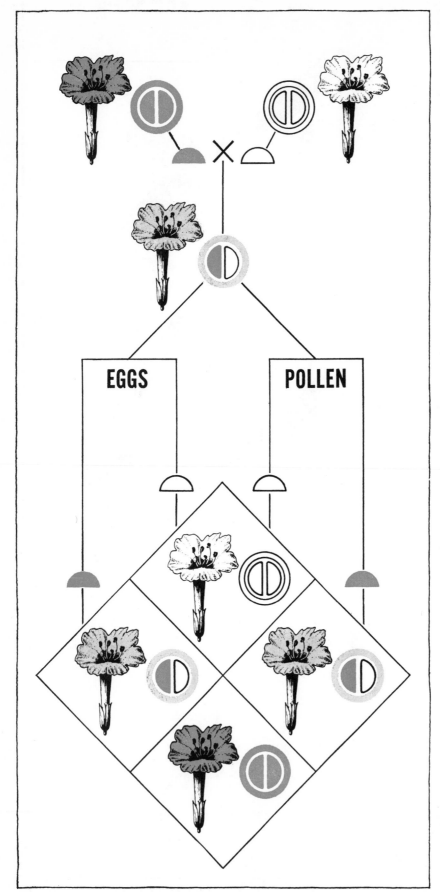

MENDELIAN SEGREGATION is illustrated by the four o'clock (*Mirabilis jalapa*). The genes of red and white flowers combine in a pink hybrid. Genes are segregated in the cross-fertilized descendants of pink flowers.

instead of the existing diversity of foods there was only one kind of energy-yielding substance to serve as nourishment. The Russian biologist Gause has pointed out that only a single kind of organism could inhabit such a tedious world. If two or more kinds appeared in it, the most efficient form would gradually crowd out and finally eliminate the less efficient ones, remaining the sole inhabitant. In the world of reality, however, the environment changes at every step. Oceans, plains, hills, mountain ranges, regions where summer heat alternates with winter cold, lands that are permanently warm, dry deserts, humid jungles —these diverse environments have engendered a multitude of responses by protoplasm and a vast proliferation of distinct species of life through the evolutionary process.

Some Adaptations

Many animal and plant species are polymorphic, *i.e.*, represented in nature by two or more clearly distinguishable kinds of individuals. For example, some individuals of the ladybird beetle *Adalia bipunctata* are red with black spots while others are black with red spots. The color difference is hereditary, the black color behaving as a Mendelian dominant and red as a recessive. The red and black forms live side by side and interbreed freely. Timofeeff-Ressovsky observed that near Berlin, Germany, the black form predominates from spring to autumn, and the red form is more numerous during the winter. What is the origin of these changes? It is quite improbable that the genes for color are transformed by the seasonal variations in temperature; that would mean epidemics of directed mutations on a scale never observed. A much more plausible view is that the changes are produced by natural selection. The black form is, for some reason, more successful than the red in survival and reproduction during summer, but the red is superior to the black under winter conditions. Since the beetles produce several generations during a single season, the species undergoes cyclic changes in its genetic composition in response to the seasonal alterations in the environment. This hypothesis was confirmed by the discovery that black individuals are more frequent among the beetles that die during the rigors of winter than among those that survive.

The writer has observed seasonal changes in some localities in California in the fly *Drosophila pseudoobscura*. Flies of this species in nature are rather uniform in coloration and other external traits, but they are very variable in the structure of their chromosomes, which can be seen in microscopic preparations. In the locality called Piñon Flats, on Mount San Jacinto in southern Califor-

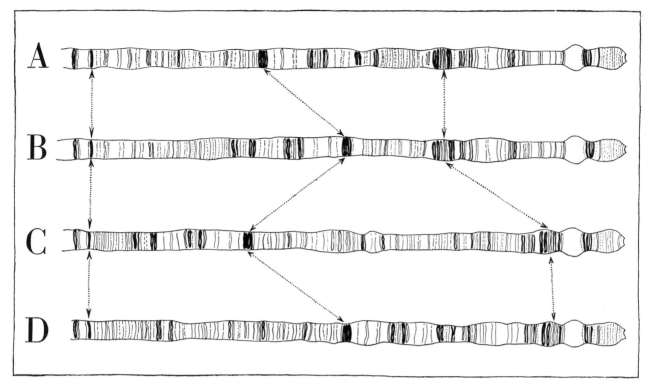

FOUR VARIETIES of the species *Drosophila pseudo-obscura* are revealed by differences in the structure of their chromosomes. Under the microscope similar markings may be observed at different locations (*arrows*).

nia, the fruit-fly population has four common types of chromosome structure, which we may, for simplicity, designate as types A, B, C and D. From 1939 to 1946, samples of flies were taken from this population in various months of the year, and the chromosomes of these flies were examined. The relative frequencies of the chromosomal types, expressed in percentages of the total, varied with the seasons as follows:

Month	A	B	C	D
March	52	18	23	7
April	40	28	28	4
May	34	29	31	6
June	28	28	39	5
July	42	22	31	5
Aug.	42	28	26	4
Sept.	48	23	26	3
Oct.-Dec.	50	26	20	4

Thus type A was common in winter but declined in the spring, while type C waxed in the spring and waned in summer. Evidently flies carrying chromosomes of type C are somehow better adapted than type A to the spring climate; hence from March to June, type A decreases and type C increases in frequency. Contrariwise, in the summer type A is superior to type C. Types B and D show little consistent seasonal variation.

Similar changes can be observed under controlled laboratory conditions. Populations of Drosophila flies were kept in a very simple apparatus consisting of a wood and glass box, with openings in the bottom for replenishing the nutrient medium on which the flies lived—a kind of pudding made of Cream of Wheat, molasses and yeast. A mixture of flies of which 33 per cent were type A and 67 per cent type C was introduced into the apparatus and left to multiply freely, up to the limit imposed by the quantity of food given. If one of the types was better adapted to the environment than the other, it was to be expected that the better-adapted type would increase and the other decrease in relative numbers. This is exactly what happened. During the first six months the type A flies rose from 33 to 77 per cent of the population, and type C fell from 67 to 23 per cent. But then came an unexpected leveling off: during the next seven months there was no further change in the relative proportions of the flies, the frequencies of types A and C oscillating around 75 and 25 per cent respectively.

If type A was better than type C under the conditions of the experiment, why were not the flies with C chromosomes crowded out completely by the carriers of A? Sewall Wright of the University of Chicago solved the puzzle by mathematical analysis. The flies of these types interbreed freely, in natural as well as in experimental populations. The populations therefore consist of three kinds of individuals: 1) those that obtained chromosome A from father as well as from mother, and thus carry two A chromosomes (AA); 2) those with two C chromosomes (CC); 3) those that re-ceived chromosomes of different types from their parents (AC). The mixed type, AC, possesses the highest adaptive value; it has what is called "hybrid vigor." As for the pure types, under the conditions that obtain in nature AA is superior to CC in the summer. Natural selection then increases the frequency of A chromosomes in the population and diminishes the C chromosomes. In the spring, when CC is better than AA, the reverse is true. But note now that in a population of mixed types neither the A nor the C chromosomes can ever be entirely eliminated from the population, even if the flies are kept in a constant environment where type AA is definitely superior to type CC. This, of course, is highly favorable to the flies as a species, for the loss of one of the chromosome types, though it might be temporarily advantageous, would be prejudicial in the long run, when conditions favoring the lost type would return. Thus a polymorphic population is better able than a uniform one to adjust itself to environmental changes and to exploit a variety of habitats.

Races

Populations of the same species which inhabit different environments become genetically different. This is what a geneticist means when he speaks of races. Races are populations within a species that differ in the frequencies of some genes. According to the old concept of race, which is based on the notion that

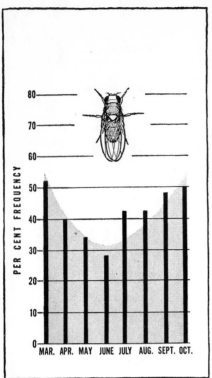
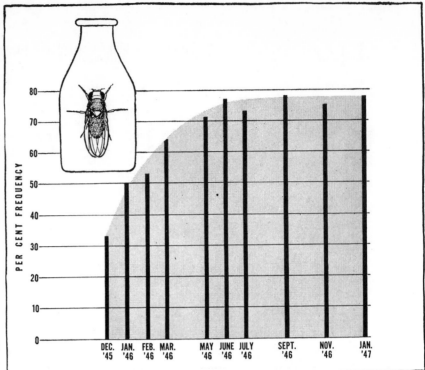

NUMBER OF FLIES of one chromosomal type varies in nature (*left*) and in the laboratory. In seasonal environ- ment of nature the type increases and decreases regular- ly; in constant environment of laboratory it levels off.

heredity is transmitted through "blood" and which still prevails among those ignorant of modern biology, the hereditary endowment of an isolated population would become more and more uniform with each generation, provided there was no interbreeding with other tribes or populations. The tribe would eventually become a "pure" race, all members of which would be genetically uniform. Scientists misled by this notion used to think that at some time in the past the human species consisted of an unspecified number of "pure" races, and that intermarriage between them gave rise to the present "mixed" populations.

In reality, "pure" races never existed, nor can they possibly exist in any species, such as man, that reproduces by sexual combination. We have seen that all human beings except identical twins differ in heredity. In widely differing climatic environments the genetic differences may be substantial. Thus populations native in central Africa have much higher frequencies of genes that produce dark skin than do European populations. The frequency of the gene for blue eye color progressively diminishes southward from Scandinavia through central Europe to the Mediterranean and Africa. Nonetheless some blue-eyed individuals occur in the Mediterranean region and even in Africa, and some brown-eyed ones in Norway and Sweden.

It is important to keep in mind that races are populations, not individuals. Race differences are relative and not absolute, since only in very remote races

do all members of one population possess a set of genes that is lacking in all members of another population. It is often difficult to tell how many races exist in a species. For example, some anthropologists recognize only two human races while others list more than 100. The difficulty is to know where to draw the line. If, for example, the Norwegians are a "Nordic race" and the southern Italians a "Mediterranean race," to what race do the inhabitants of Denmark, northern Germany, southern Germany, Switzerland and northern Italy belong? The frequencies of most differentiating traits change rather gradually from Norway to southern Italy. Calling the intermediate populations separate races may be technically correct, but this confuses the race classification even more, because nowhere can sharp lines of demarcation between these "races" be drawn. It is quite arbitrary whether we recognize 2, 4, 10, or more than 100 races—or finally refuse to make any rigid racial labels at all.

The differences between human races are, after all, rather small, since the geographic separation between them is nowhere very marked. When a species is distributed over diversified territories, the process of adaptation to the different environments leads to the gradual accumulation of more numerous and biologically more and more important differences between races. The races gradually diverge. There is, of course, nothing fatal about this divergence, and under some circumstances the divergence may

stop or even be turned into convergence. This is particularly true of the human species. The human races were somewhat more sharply separated in the past than they are today. Although the species inhabits almost every variety of environment on earth, the development of communications and the increase of mobility, especially in modern times, has led to much intermarriage and to some genetic convergence of the human races.

The diverging races become more and more distinct with time, and the process of divergence may finally culminate in transformation of races into species. Although the splitting of species is a gradual process, and it is often impossible to tell exactly when races cease to be races and become species, there exist some important differences between race and species which make the process of species formation one of the most important biological processes. Indeed, Darwin called his chief work *The Origin of Species*.

Races of sexually reproducing organisms are fully capable of intercrossing; they maintain their distinction as races only by geographical isolation. As a rule in most organisms no more than a single race of any one species inhabits the same territory. If representatives of two or more races come to live in the same territory, they interbreed, exchange genes, and eventually become fused into a single population. The human species, however, is an exception. Marriages are influenced by linguistic, religious, social, economic and other cultural factors.

Hence cultural isolation may keep populations apart for a time and slow down the exchange of genes even though the populations live in the same country. Nevertheless, the biological relationship proves stronger than cultural isolation, and interbreeding is everywhere in the process of breaking down such barriers. Unrestricted interbreeding would not mean, as often supposed, that all people would become alike. Mankind would continue to include at least as great a diversity of hereditary endowments as it contains today. However, the same types could be found anywhere in the world, and races as more or less discrete populations would cease to exist.

The Isolationism of Species

Species, on the contrary, can live in the same territory without losing their identity. F. Lutz of the American Museum of Natural History found 1,402 species of insects in the 75-by-200-foot yard of his home in a New Jersey suburb. This does not mean that representatives of distinct species never cross. Closely related species occasionally do interbreed in nature, especially among plants, but these cases are so rare that the discovery of one usually merits a note in a scientific journal.

The reason distinct species do not interbreed is that they are more or less completely kept apart by isolating mechanisms connected with reproduction, which exist in great variety. For example, the botanist Carl C. Epling of the University of California found that two species of sage which are common in southern California are generally separated by ecological factors, one preferring a dry site, the other a more humid one. When the two sages do grow side by side, they occasionally produce hybrids. The hybrids are quite vigorous, but their seed set amounts to less than two per cent of normal; i.e., they are partially sterile. Hybrid sterility is a very common and effective isolating mechanism. A classic example is the mule, hybrid of the horse and donkey. Male mules are always sterile, females usually so. There are, however, some species, notably certain ducks, that produce quite fertile hybrids, not in nature but in captivity.

Two species of Drosophila, *pseudoobscura* and *persimilis*, are so close together biologically that they cannot be distinguished by inspection of their external characteristics. They differ, however, in the structure of their chromosomes and in many physiological traits. If a mixed group of females of the two species is exposed to a group of males of one species, copulations occur much more frequently between members of the same species than between those of different species, though some of the latter do take place. Among plants, the flowers of related species may differ so much in structure that they cannot be pollinated by the same insects, or they may have such differences in smell, color and shape that they attract different insects. Finally, even when cross-copulation or cross-pollination can occur, the union may fail to result in fertilization or may produce offspring that cannot live. Often several isolating mechanisms, no one of which is effective separately, combine to prevent interbreeding. In the case of the two fruit-fly species, at least three such mechanisms are at work: 1) the above-mentioned disposition to mate only with their own kind, even when they are together; 2) different preferences in climate, one preferring warmer and drier places than the other; 3) the fact that when they do interbreed the hybrid males that result are completely sterile and the hybrid females, though fertile, produce offspring that are poorly viable. There is good evidence that no gene exchange occurs between these species in nature.

The fact that distinct species can coexist in the same territory, while races generally cannot, is highly significant. It permits the formation of communities of diversified living beings which exploit the variety of habitats present in a territory more fully than any single species, no matter how polymorphic, could. It is responsible for the richness and colorfulness of life that is so impressive to biologists and non-biologists alike.

Evolution v. Predestination

Our discussion of the essentials of the modern theory of evolution may be concluded with a consideration of the objections raised against it. The most serious objection is that since mutations occur by "chance" and are undirected, and since selection is a "blind" force, it is difficult to see how mutation and selection can add up to the formation of such complex and beautifully balanced organs as, for example, the human eye. This, say critics of the theory, is like believing that a monkey pounding a typewriter might accidentally type out Dante's *Divine Comedy*. Some biologists have preferred to suppose that evolution is directed by an "inner urge toward perfection," or by a "combining power which amounts to intentionality," or by "telefinalism" or the like. The fatal weakness of these alternative "explanations" of evolution is that they do not explain anything. To say that evolution is directed by an urge, a combining power, or a telefinalism is like saying that a railroad engine is moved by a "locomotive power."

The objection that the modern theory of evolution makes undue demands on chance is based on a failure to appreciate the historical character of the evolutionary process. It would indeed strain credulity to suppose that a lucky sudden combination of chance mutations produced the eye in all its perfection. But the eye did not appear suddenly in the offspring of an eyeless creature; it is the result of an evolutionary development that took many millions of years. Along the way the evolving rudiments of the eye passed through innumerable stages, all of which were useful to their possessors, and therefore all adjusted to the demands of the environment by natural selection. Amphioxus, the primitive fish-like darling of comparative anatomists, has no eyes, but it has certain pigment cells in its brain by means of which it perceives light. Such pigment cells may have been the starting point of the development of eyes in our ancestors.

We have seen that the "combining power" of the sexual process is staggering, that on the most conservative estimate the number of possible gene combinations in the human species alone is far greater than that of the electrons and protons in the universe. When life developed sex, it acquired a trial-and-error mechanism of prodigious efficiency. This mechanism is not called upon to produce a completely new creature in one spectacular burst of creation; it is sufficient that it produces slight changes that improve the organism's chances of survival or reproduction in some habitat. In terms of the monkey-and-typewriter analogy, the theory does not require that the monkey sit down and compose the *Divine Comedy* from beginning to end by a lucky series of hits. All we need is that the monkey occasionally form a single word, or a single line; over the course of eons of time the environment shapes this growing text into the eventual masterpiece. Mutations occur by "chance" only in the sense that they appear regardless of their usefulness at the time and place of their origin. It should be kept in mind that the structure of a gene, like that of the whole organism, is the outcome of a long evolutionary development; the ways in which the genes can mutate are, consequently, by no means indeterminate.

Theories that ascribe evolution to "urges" and "telefinalisms" imply that there is some kind of predestination about the whole business, that evolution has produced nothing more than was potentially present at the beginning of life. The modern evolutionists believe that, on the contrary, evolution is a creative response of the living matter to the challenges of the environment. The role of the environment is to provide opportunities for biological inventions. Evolution is due neither to chance nor to design; it is due to a natural creative process.

3   Darwin's Missing Evidence

by H. B. D. Kettlewell
March 1959

In his time certain species of moths were light in color. Today in many areas these species are largely dark. If he had noticed the change occurring, he would have observed evolution in action

Charles Darwin's *Origin of Species,* the centenary of which we celebrate in 1959, was the fruit of 26 years of laborious accumulation of facts from nature. Others before Darwin had believed in evolution, but he alone produced a cataclysm of data in support of it. Yet there were two fundamental gaps in his chain of evidence. First, Darwin had no knowledge of the mechanism of heredity. Second, he had no visible example of evolution at work in nature.

It is a curious fact that both of these gaps could have been filled during Darwin's lifetime. Although Gregor Mendel's laws of inheritance were not discovered by the community of biologists until 1900, they had first been published in 1866. And before Darwin died in 1882, the most striking evolutionary change ever witnessed by man was taking place around him in his own country.

The change was simply this. Less than a century ago moths of certain species were characterized by their light coloration, which matched such backgrounds as light tree trunks and lichen-covered rocks, on which the moths passed the daylight hours sitting motionless. Today in many areas the same species are predominantly dark! We now call this reversal "industrial melanism."

It happens that Darwin's lifetime coincided with the first great man-made change of environment on earth. Ever since the Industrial Revolution commenced in the latter half of the 18th century, large areas of the earth's surface have been contaminated by an insidious and largely unrecognized fallout of smoke particles. In and around industrial areas the fallout is measured in tons per square mile per month; in places like Sheffield in England it may reach 50 tons or more. It is only recently that we have begun to realize how widely the

lighter smoke particles are dispersed, and to what extent they affect the flora and fauna of the countryside.

In the case of the flora the smoke particles not only pollute foliage but also kill vegetative lichens on the trunks and boughs of trees. Rain washes the pollutants down the boughs and trunks until they are bare and black. In heavily polluted districts rocks and the very ground itself are darkened.

Now in England there are some 760 species of larger moths. Of these more than 70 have exchanged their light color and pattern for dark or even all-black coloration. Similar changes have occurred in the moths of industrial areas of other countries: France, Germany, Poland, Czechoslovakia, Canada and the U. S. So far, however, such changes have not been observed anywhere in the tropics. It is important to note here that industrial melanism has occurred only among those moths that fly at night and spend the day resting against a background such as a tree trunk.

These, then, are the facts. A profound change of color has occurred among hundreds of species of moths in industrial areas in different parts of the world. How has the change come about? What underlying laws of nature have produced it? Has it any connection with one of the normal mechanisms by which one species evolves into another?

In 1926 the British biologist Heslop Harrison reported that the industrial melanism of moths was caused by a special substance which he alleged was present in polluted air. He called this substance a "melanogen," and suggested that it was manganous sulfate or lead nitrate. Harrison claimed that when he fed foliage impregnated with these salts to the larvae of certain species of light-

colored moths, a proportion of their offspring were black. He also stated that this "induced melanism" was inherited according to the laws of Mendel.

Darwin, always searching for missing evidence, might well have accepted Harrison's Lamarckian interpretation, but in 1926 biologists were skeptical. Although the rate of mutation of a hereditary characteristic can be increased in the laboratory by many methods, Harrison's figures inferred a mutation rate of 8 per cent. One of the most frequent mutations in nature is that which causes the disease hemophilia in man; its rate is in the region of .0005 per cent, that is, the mutation occurs about once in 50,000 births. It is, in fact, unlikely that an increased mutation rate has played any part in industrial melanism.

At the University of Oxford during the past seven years we have been attempting to analyze the phenomenon of industrial melanism. We have used many different approaches. We are in the process of making a survey of the present frequency of light and dark forms of each species of moth in Britain that exhibits industrial melanism. We are critically examining each of the two forms to see if between them there are any differences in behavior. We have fed large numbers of larvae of both forms on foliage impregnated with substances in polluted air. We have observed under various conditions the mating preferences and relative mortality of the two forms. Finally we have accumulated much information about the melanism of moths in parts of the world that are far removed from industrial centers, and we have sought to link industrial melanism with the melanics of the past.

Our main guinea pig, both in the field and in the laboratory, has been the peppered moth *Biston betularia* and its melanic form *carbonaria*. This species occurs throughout Europe, and is probably identical with the North American *Amphidasis cognataria*. It has a one-year life cycle; the moth appears from May to August. The moth flies at night and passes the day resting on the trunks or on the underside of the boughs of rough-barked deciduous trees such as the oak. Its larvae feed on the foliage of such trees from June to late October; its pupae pass the winter in the soil.

The dark form of the peppered moth was first recorded in 1848 at Manchester in England. Both the light and dark forms appear in each of the photographs at right and on the next page. The background of each photograph is noteworthy. In the photograph on the next page the background is a lichen-encrusted oak trunk of the sort that today is found only in unpolluted rural districts. Against this background the light form is almost invisible and the dark form is conspicuous. In the photograph at right the background is a bare and blackened oak trunk in the heavily polluted area of Birmingham. Here it is the dark form which is almost invisible, and the light form which is conspicuous. Of 621 wild moths caught in these Birmingham woods in 1953, 90 per cent were the dark form and only 10 per cent the light. Today this same ratio applies in nearly all British industrial areas and far outside them.

We decided to test the rate of survival of the two forms in the contrasting types of woodland. We did this by releasing known numbers of moths of both forms. Each moth was marked on its underside with a spot of quick-drying cellulose paint; a different color was used for each day. Thus when we subsequently trapped large numbers of moths we could identify those we had released and established the length of time they had been exposed to predators in nature.

In an unpolluted forest we released 984 moths: 488 dark and 496 light. We recaptured 34 dark and 62 light, indicating that in these woods the light form had a clear advantage over the dark. We then repeated the experiment in the polluted Birmingham woods, releasing 630 moths: 493 dark and 137 light. The result of the first experiment was completely reversed; we recaptured proportionately twice as many of the dark form as of the light.

For the first time, moreover, we had witnessed birds in the act of taking moths from the trunks. Although Britain has more ornithologists and bird watch-

DARK AND LIGHT FORMS of the peppered moth were photographed on the trunk of an oak blackened by the polluted air of the English industrial city of Birmingham. The light form (*Biston betularia*) is clearly visible; the dark form (*carbonaria*) is well camouflaged.

ers than any other country, there had been absolutely no record of birds actually capturing resting moths. Indeed, many ornithologists doubted that this happened on any large scale.

The reason for the oversight soon became obvious. The bird usually seizes the insect and carries it away so rapidly that the observer sees nothing unless he is keeping a constant watch on the insect. This is just what we were doing in the course of some of our experiments. When I first published our findings, the editor of a certain journal was sufficiently rash as to question whether birds took resting moths at all. There was only one thing to do, and in 1955 Niko Tinbergen of the University of Oxford filmed a repeat of my experiments. The film not only shows that birds capture and eat resting moths, but also that they do so selectively.

These experiments lead to the following conclusions. First, when the environment of a moth such as *Biston betularia* changes so that the moth cannot hide by day, the moth is ruthlessly eliminated by predators unless it mutates to a form that is better suited to its new environment. Second, we now have visible proof that, once a mutation has occurred, natural selection alone can be responsible for its rapid spread. Third, the very fact that one form of moth has replaced another in a comparatively short span of years indicates that this evolutionary mechanism is remarkably flexible.

The present status of the peppered moth is shown in the map on the opposite page. This map was built up from more than 20,000 observations made by 170 voluntary observers living in various parts of Britain. The map makes the fol-

lowing points. First, there is a strong correlation between industrial centers and a high percentage of the dark form of the moth. Second, populations consisting entirely of the light form are found today only in western England and northern Scotland. Third, though the counties of eastern England are far removed from industrial centers, a surprisingly high percentage of the dark form is found in them. This, in my opinion, is due to the long-standing fallout of smoke particles carried from central England by the prevailing southwesterly winds.

Now in order for the dark form of a moth to spread, a mutation from the light form must first occur. It appears that the frequency with which this happens—that is, the mutation rate—varies according to the species. The rate at which the light form of the peppered

SAME TWO FORMS of the peppered moth were photographed against the lichen-encrusted trunk of an oak in an unpolluted area.   Here it is the dark form which may be clearly seen. The light form, almost invisible, is just below and to the right of the dark form.

moth mutates to the dark form seems to be fairly high; the rate at which the mutation occurs in other species may be very low. For example, the light form of the moth *Procus literosa* disappeared from the Sheffield area many years ago, but it has now reappeared in its dark form. It would seem that a belated mutation has permitted the species to regain lost territory. Another significant example is provided by the moth *Tethea ocularis*. Prior to 1947 the dark form of this species was unknown in England. In that year, however, many specimens of the dark form were for the first time collected in various parts of Britain; in some districts today the dark form now comprises more than 50 per cent of the species. There is little doubt that this melanic arrived in Britain not by mutation but by migration. It had been known for a considerable time in the industrial areas of northern Europe, where presumably the original mutation occurred.

The mutation that is responsible for industrial melanism in moths is in the majority of cases controlled by a single gene. A moth, like any other organism that reproduces sexually, has two genes for each of its hereditary characteristics: one gene from each parent. The mutant gene of a melanic moth is inherited as a Mendelian dominant; that is, the effect of the mutant gene is expressed and the effect of the other gene in the pair is not. Thus a moth that inherits the mutant gene from only one of its parents is melanic.

The mutant gene, however, does more than simply control the coloration of the moth. The same gene (or others closely linked with it in the hereditary material) also gives rise to physiological and even behavioral traits. For example, it appears that in some species of moths the caterpillars of the dark form are hardier than the caterpillars of the light form. Genetic differences are also reflected in mating preference. On cold nights more males of the light form of the peppered moth appear to be attracted to light females than to dark. On warm nights, on the other hand, significantly more light males are attracted to dark females.

There is evidence that, in a population of peppered moths that inhabits an industrial area, caterpillars of the light form attain full growth earlier than caterpillars of the dark form. This may be due to the fact that the precipitation of pollutants on leaves greatly increases late in the autumn. Caterpillars of the dark form may be hardier in the presence

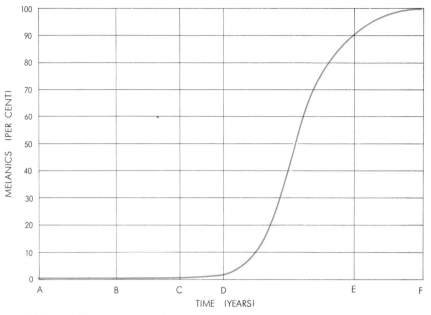

SPREAD OF MUTATION from the light form to the dark (melanic) is expressed by this curve, discussed in detail in the text. The mutation occurs in the period AB, spreads slowly during BD and spreads rapidly during DE. During EF the light form is either gradually eliminated, as indicated by the curve, or remains at a level of about 5 per cent of the population.

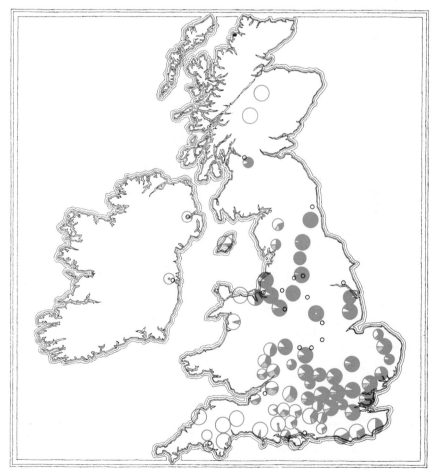

PROPORTION OF FORMS of the peppered moth at various locations in the British Isles is indicated on this map. The open area within a colored circle represents the proportion of the light form *Biston betularia* recorded; the solid colored area, the proportion of the dark form *carbonaria*; the hatched colored area, the proportion of another dark form, *insularia*. Small black circles on the map indicate the location of major industrial centers.

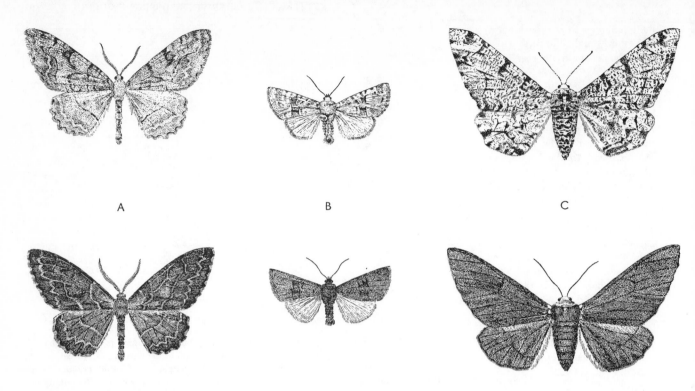

FIVE SPECIES OF MOTH that have both light and dark forms are depicted in their actual size. In each case the light form is at the top and the dark form is at the bottom. The species are: *Cleora repandata* (A), *Procus literosa* (B), *Biston betularia* (C), *Ectropis*

of such pollution than caterpillars of the light form. In that case natural selection would favor light-form caterpillars which mature early over light-form caterpillars which mature late. For the hardier caterpillars of the dark form, on the other hand, the advantages of later feeding and longer larval life might outweigh the disadvantages of feeding on increasingly polluted leaves. Then natural selection would favor those caterpillars which mature late.

Another difference between the behavior of *B. betularia* and that of its dark form *carbonaria* is suggested by our experiments on the question of whether each form can choose the "correct" background on which to rest during the day. We offered light and dark backgrounds of equal area to moths of both forms, and discovered that a significantly large proportion of each form rested on the correct background. Before these results can be accepted as proven, the experiments must be repeated on a larger scale. If they are proven, the behavior of both forms could be explained by the single mechanism of "contrast appreciation." This mechanism assumes that one segment of the eye of a moth senses the color of the background and that another segment senses the moth's own color; thus the two colors could be compared. Presumably if they were the same, the moth would remain on its background; but if they were different,

"contrast conflict" would result and the moth would move off again. That moths tend to be restless when the colors conflict is certainly borne out by recent field observations.

It is evident, then, that industrial melanism is much more than a simple change from light to dark. Such a change must profoundly upset the balance of hereditary traits in a species, and the species must be a long time in restoring that balance. Taking into account all the favorable and unfavorable factors at work in this process, let us examine the spread of a mutation similar to the dark form of the peppered moth. To do so we must consult the diagram at the top of the preceding page.

According to the mutation rate and the size of the population, the new mutation may not appear in a population for a period varying from one to 50 years. This is represented by AB on the diagram. Let us now assume the following: that the original successful mutation took place in 1900, that subsequent new mutations failed to survive, that the total local population was one million, and that the mutant had a 30-per-cent advantage over the light form. (By a 30-per-cent advantage for the dark form we mean that, if in one generation there were 100 light moths and 100 dark, in the next generation there would be 85 light moths and 115 dark.)

On the basis of these assumptions there would be one melanic moth in 1,000 only in 1929 (BC). Not until 1938 would there be one in 100 (BD). Once the melanics attain this level, their rate of increase greatly accelerates.

In the period between 1900 and 1938 (BD) natural selection is complicated by other forces. Though the color of the dark form gives it an advantage over the light, the new trait is introduced into a system of other traits balanced for the light form; thus the dark form is at first at a considerable physiological disadvantage. In fact, when moths of the dark form were crossed with moths of the light form 50 years ago, the resulting broods were significantly deficient in the dark form. When the same cross is made today, the broods contain more of the dark form than one would expect. The system of hereditary traits has become adjusted to the new trait.

There is evidence that other changes take place during the period BD. Specimens of the peppered moth from old collections indicate that the earliest melanics were not so dark as the modern dark form: they retained some of the white spots of the light form. Today a large proportion of the moths around a city such as Manchester are jet black. Evidently when the early melanics inherited one gene for melanism, the gene was not entirely dominant with respect to the gene for light coloration. As the

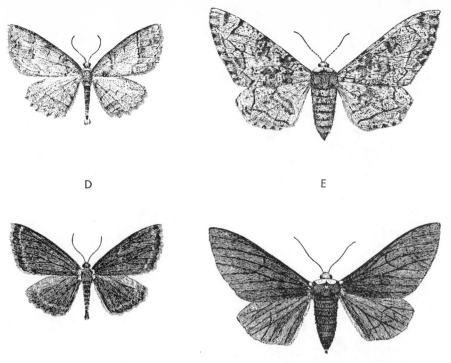

D

E

consonaria (D) and *Amphidasis cognataria* (E). All of the species are European except the last, which occurs in North America and may be identical with *Biston betularia*.

gene complex adjusted to the mutation, however, the new gene became almost entirely dominant.

When the dark form comprises about 10 per cent of the population, it may jump to 90 per cent in as little as 15 or 20 years. This is represented by period DE on the graph. Thereafter the proportion of the dark form increases at a greatly reduced rate.

Eventually one of two things must happen: either the light form will slowly be eliminated altogether, or a balance will be struck so that the light form continues to appear as a small but definite proportion of the population. This is due to the fact that the moths which inherit one gene for dark coloration and one for light (heterozygotes) have an advantage over the moths which inherit two genes for dark coloration (homozygotes). And when two heterozygotes mate, a quarter of their offspring will have two genes for light coloration, *i.e.*, they will be light. Only after a very long period of time, therefore, could the light forms (and with them the gene for light coloration) be entirely eliminated. This period of removal, represented by EF on the diagram, might be more than 1,000 years. Indications so far suggest, however, that complete removal is unlikely, and that a balance of the two forms would probably occur. In this balance the light form would represent about 5 per cent of the population.

The mechanisms I have described are without doubt the explanation of industrial melanism: normal mutation followed by natural selection resulting in an insect of different color, physiology and behavior. Industrial melanism involves no new laws of nature; it is governed by the same mechanisms which have brought about the evolution of new species in the past.

There remains, however, one major unsolved problem. Why is it that, in almost all industrial melanics, the gene for melanism is dominant? Many geneticists would agree that dominance is achieved by natural selection, that it is somehow related to a successful mutation in the distant past. With these thoughts in mind I recently turned my attention away from industrial centers and collected moths in one of the few remaining pieces of ancient Caledonian pine forest in Britain: the Black Wood of Rannoch. Located in central Scotland far from industrial centers, the Black Wood is probably very similar to the forests that covered Britain some 4,000 years ago. The huge pines of this forest are only partly covered with lichens. Here I found no fewer than seven species of moths with melanic forms.

I decided to concentrate on the species *Cleora repandata*, the dark form of which is similar to the dark form of the same species that has swept through central England. This dark form, like the industrial melanics, is inherited as a Mendelian dominant. Of just under 500 specimens of *C. repandata* observed, 10 per cent were dark.

C. repandata spends the day on pine trunks, where the light form is almost invisible. The dark form is somewhat more easily seen. By noting at dawn the spot where an insect had come to rest, and then revisiting the tree later in the day, we were able to show that on some days more than 50 per cent of the insects had moved. Subsequently we found that because of disturbances such as ants or hot sunshine they had had to fly to another tree trunk, usually about 50 yards away. I saw large numbers of these moths on the wing, and three other observers and I agreed that the dark form was practically invisible at a distance of more than 20 yards, and that the light form could be followed with ease at a distance of up to 100 yards. In fact, we saw birds catch three moths of the light form in flight. It is my belief that when it is on the wing in these woods the dark form has an advantage over the light, and that when it is at rest the reverse is true.

This may be one of many ways in which melanism was useful in the past. It may also explain the balance between the light and dark forms of *Cleora repandata* in the Black Wood of Rannoch. In this case a melanic may have been preserved for one evolutionary reason but then have spread widely for another.

The melanism of moths occurs in many parts of the world that are not industrialized, and in environments that are quite different. It is found in the mountain rain forest of New Zealand's South Island, which is wet and dark. It has been observed in arctic and subarctic regions where in summer moths must fly in daylight. It is known in very high mountains, where dark coloration may permit the absorption of heat and make possible increased activity. In each case recurrent mutation has provided the source of the change, and natural selection, as postulated by Darwin, has decided its destiny.

Melanism is not a recent phenomenon but a very old one. It enables us to appreciate the vast reserves of genetic variability which are contained within each species, and which can be summoned when the occasion arises. Had Darwin observed industrial melanism he would have seen evolution occurring not in thousands of years but in thousands of days—well within his lifetime. He would have witnessed the consummation and confirmation of his life's work.

Darwin's Finches

by David Lack
April 1953

*These drab but famous little birds of the Galapagos
Islands are a living case study in evolution. Isolated in
the South Pacific, they have developed 14 species from
a common ancestor*

ON THE Galapagos Islands in the Pacific Charles Darwin in 1835 saw a group of small, drab, finchlike birds which were to change the course of human history, for they provided a powerful stimulus to his speculations on the origin of species—speculations that led to the theory of evolution by natural selection. In the study of evolution the animals of remote islands have played a role out of all proportion to their small numbers. Life on such an island approaches the conditions of an experiment in which we can see the results of thousands of years of evolutionary development without outside intervention. The Galapagos finches are an admirable case study.

These volcanic islands lie on the Equator in the Pacific Ocean some 600 miles west of South America and 3,000 miles east of Polynesia. It is now generally agreed that they were pushed up out of the sea by volcanoes more than one million years ago and have never been connected with the mainland. Whatever land animals they harbor must have come over the sea, and very few species have established themselves there: just two kinds of mammals, five reptiles, six songbirds and five other land birds.

Some of these animals are indistinguishable from the same species on the mainland; some are slightly different; a few, such as the giant land-tortoises and the mockingbirds, are very different. The latter presumably reached the Galapagos a long time ago. In addition, there are variations from island to island among the local species themselves, in-

dicating that the colonists diverged into variant forms after their arrival. Darwin's finches go further than this: not only do they vary from island to island but up to 10 different species of them can be found on a single island.

The birds themselves are less dramatic than their story. They are dull in color, unmusical in song and, with one exception, undistinguished in habits. This dullness is in no way mitigated by their dreary surroundings. Darwin in his diary succinctly described the islands: "The country was compared to what we might imagine the cultivated parts of the Infernal regions to be." This diary, it is interesting to note, makes no mention of the finches, and the birds received only a brief mention in the first edition of his book on the voyage of the *Beagle*. Specimens which Darwin brought home, however, were recognized by the English systematist and bird artist, John Gould, as an entirely new group of birds. By the time the book reached its second edition, the ferment had begun to work, and Darwin added that "one might really fancy that from an original paucity of birds in this archipelago, one species had been taken and modified for different ends." Thus obscurely, as an afterthought in a travel book, man received a first intimation that he might once have been an ape.

THERE ARE 13 species of Darwin's finches in the Galapagos, plus one on Cocos Island to the northwest. A self-contained group with no obvious relations elsewhere, these finches are usually

placed in a subfamily of birds named the *Geospizinae*. How did this remarkable group evolve? I am convinced, from my observations in the islands in 1938-39 and from subsequent studies of museum specimens, that the group evolved in much the same way as other birds. Consequently the relatively simple story of their evolution can throw valuable light on the way in which birds, and other animals, have evolved in general. Darwin's finches form a little world of their own, but a world which differs from the one we know only in being younger, so that here, as Darwin wrote, we are brought nearer than usual "to that great fact—that mystery of mysteries—the first appearance of new beings on this earth."

The 14 species of Darwin's finches fall into four main genera. First, there are the ground-finches, embracing six species, nearly all of which feed on seeds on the ground and live in the arid coastal regions. Secondly, there are the tree-finches, likewise including six species, nearly all of which feed on insects in trees and live in the moist forests. Thirdly, there is the warbler-like finch (only one species) which feeds on small insects in bushes in both arid and humid regions. Finally, there is the isolated Cocos Island species which lives on insects in a tropical forest.

Among the ground-finches, four species live together on most of the islands: three of them eat seeds and differ from each other mainly in the size of their beaks, adapted to different sizes of seeds; the fourth species feeds largely on prickly pear and has a much longer

THE 14 SPECIES of Darwin's finches are arranged at the left to suggest the evolutionary tree of their development. Grayish brown to black, all belong to the subfamily *Geospizinae*, divided broadly into ground finches (*Geospiza*), closest to the primitive form, and tree finches (mainly *Camarhynchus*), which evolved later. Of the tree species, 1 is a woodpecker-like finch (*C. pallidus*), 2 inhabits mangrove swamps (*C. heliobates*), 3, 4 and 5 are large, medium and small insect-

eating birds (*C. psittacula, pauper* and *parvulus*), 6 is a vegetarian (*C. crassirostris*), 7 is a single species of warbler-finch (*Certhidea*) and 8 an isolated species of Cocos Island finch (*Pinaroloxias*). The ground-finches, mainly seed-eaters, run thus: 9, 10 and 11 are large, medium and small in size (*G. magnirostris, fortis* and *fuliginosa*), 12 is sharp-beaked (*G. difficilis*), 13 and 14 are cactus eaters (*G. conirostris* and *scandens*). All of the species in the drawing are shown about half-size.

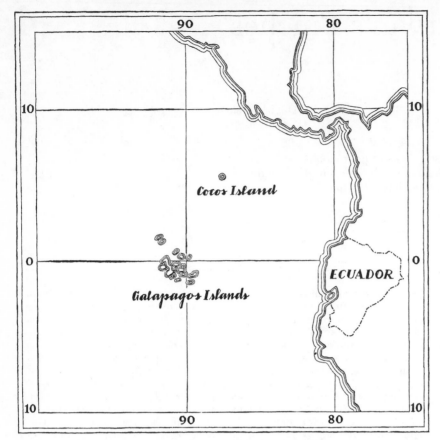

THE GALAPAGOS are shown some 600 miles west of Ecuador, above, and close up below. Cocos Island is not in the group, but it has developed one species of finch, presumed to have come originally from the mainland.

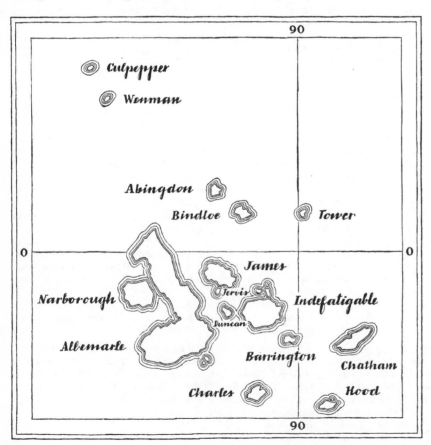

and more pointed beak. The two remaining species of ground-finches, one large and one small, live chiefly on the outlying islands, where some supplement their seed diet with cactus, their beaks being appropriately modified.

Of the tree-finches, one species is vegetarian, with a parrot-like beak seemingly fitted to its diet of buds and fruits. The next three species are closely alike, differing primarily in body size and in the size of their beaks, presumably scaled to the size of the insects they take. A fifth species eats insects in mangrove swamps. The sixth species of tree-finch is one of the most remarkable birds in the world. Like a woodpecker, it climbs tree trunks in search of insects, which it excavates from the bark with its chisel-shaped beak. While its beak approaches a woodpecker's in shape, it has not evolved the long tongue with which a woodpecker probes insects from crannies. Instead, this tree-finch solves the problem in another way: it carries about a cactus spine or small twig which it pokes into cracks, dropping the stick to seize any insect that emerges. This astonishing practice is one of the few recorded cases of the use of tools by any animal other than man or the apes.

The warbler-like finch is in its own way as remarkable as the Galapagos attempt at a woodpecker. It has no such wonderful habit, but in its appearance and character it has evolved much closer to a warbler than the other finch has to a woodpecker. Thus its beak is thin and pointed like that of a warbler; its feeding methods and actions are similar, and it even has the warbler-like habit of flicking its wings partly open as it hunts for food. For nearly a century it was classified as a warbler, but its internal anatomy, the color of its eggs, the shape of its nest and other characteristics clearly place it among the finches.

The close resemblance among Darwin's finches in plumage, calls, nests, eggs and display suggests that they have not yet had time to diverge far from one another. The only big difference is in their beaks, adapted to their different diets. It is reasonably certain that all the Galapagos finches evolved from one original colonizing form. What is unusual about them is the existence of several distinct species on the same island. In this we may have an indirect clue to how separate species establish themselves.

LET US consider first how new forms of an animal may originate from a common ancestor. When a member of the original species moves into a new environment, it is likely to evolve new features adapted to the new local conditions. Such geographical variations among animals are commonly found; in the Galapagos, for instance, the land birds other than finches vary from island to island, with only one form on each island. These forms are not distinct species but subspecies, or geographical races. Their differences, however, are hereditary and not trivial or accidental. There are several examples of such geographical variation among Darwin's finches. Three common species of the ground-finch, for instance, are found on most of the islands; they are large, medium and small, feeding on large, medium and small seeds respectively. Now on two southern islands the large species is missing, and here the medium species has a rather larger beak than elsewhere, presumably an adaptation to the large seeds available to it in the absence of the large species. Again, on another islet the small ground-finch is absent, and the medium species fills the gap by being rather smaller than elsewhere. On still other islets the medium species is missing and the small species is rather larger than elsewhere.

It seems clear that the beak differences among the subspecies of Darwin's finches are adaptive. Further, some of these differences are as great as those distinguishing true species.

What is likely to happen if a subspecies evolved in isolation on one island later spreads to an island occupied by another race of the same species? If the two populations have not been isolated for long and differ in only minor ways, they may interbreed freely and so merge with each other. But evidence from the study of insects suggests that if two populations have been isolated for a long time, so many hereditary differences will have accumulated that their genes will not combine well. Any hybrid offspring will not survive as well as the parent types. Hence natural selection will tend to intensify the gap between the two forms, and they will continue to evolve into two distinct species.

DARWIN'S finches provide circumstantial evidence for the origin of a new species by means of geographical isolation. Consider three different forms of the large insectivorous tree-finch. On the most southerly Galapagos island is a small dark form with a comparatively small beak. On another island to the northwest is a rather larger and less barred form. On the central islands is a yet larger and paler type with a larger, more parrot-like beak. Evidently these three forms had a common ancestor and evolved their differences in geographical isolation. The differences among them do not seem great enough to set them apart as separate species, and they would be classed as subspecies but for one curious circumstance: on the southernmost island the two extremes—the small dark form and the largest pale form—live side by side without merging.

Clearly these must be truly separate species. It seems likely that the large pale form spread from the central islands to the southern island in comparatively recent times, after both it and the small dark form had evolved into distinct species.

If differentiated forms are to persist alongside each other as separate species, two conditions must be met. First, they must avoid interbreeding. In birds this is usually taken care of by differences in appearance (generally in the color pattern) and in the song. It is no accident that bird-watchers find male birds so easy to recognize: correct identification is even more important for the female bird! Darwin's finches recognize each other chiefly by the beak. We have often seen a bird start to chase another from behind and quickly lose interest when a front view shows that the beak is that of a species other than its own.

The second requirement for the existence of two species together is that they must not compete for the same food. If they tend to eat similar food, the one that is better adapted to obtain that food will usually eliminate the other. In those cases where two closely related species live side by side, investigation shows that they have in fact evolved differences in diet. Thus the beak differences among the various Galapagos finches are not just an insular curiosity but are adapted to differences in diet and are an essential factor in their persistence together. It used to be supposed that related species of birds overlapped considerably in their feeding habits. A walk through a wood in summer may suggest that many of the birds have similar habits. But having established the principle of food differentiation in Darwin's finches, I studied many other examples of closely related species and found that most, if not all, differ from one another in the places where they feed, in their feeding methods or in the size of the food items they can take. The appearance of overlap was due simply to inadequate knowledge.

NOW the key to differentiation is geographical isolation. Probably one form can establish itself alongside another only after the two have already evolved some differences in separate places. Evolutionists used to believe that new species evolved by becoming adapted to different habitats in the same area. But there is no positive evidence for that once popular theory, and it is now thought that geographical isolation is the only method by which new species originate, at least among birds. One of

THE WOODPECKER-FINCH is the most remarkable of Darwin's finches. It has evolved the beak but not the long tongue of a woodpecker, hence carries a twig or cactus spine to dislodge insects from bark crevices.

Darwin's species of finches provides an interesting illustration of this. The species on Cocos Island is so different from the rest that it must have been isolated there for a long time. Yet despite this long isolation, along with a great variety of foods and habitats and a scarcity of other bird competitors, the Cocos finch has remained a single species. This is because Cocos is an isolated island, and so does not provide the proper opportunities for differentiation. In the Galapagos, differentiation was possible because the original species could scatter and establish separate homes on the various islands of the archipelago. It is significant that the only other group of birds which has evolved in a similar way, the sicklebills of Hawaii, are likewise found in an archipelago.

Why is it that this type of evolution has been found only in the Galapagos and Hawaii? There are other archipelagoes in the world, and geographical isolation is also possible on the continents. The ancestor of Darwin's finches, for instance, must formerly have lived on the American mainland, but it has not there given rise to a group of species similar to those in the Galapagos. The answer is, probably, that on the mainland the available niches in the environment were already occupied by efficient species of other birds. Consider the woodpecker-like finch on the Galapagos, for example. It would be almost impossible for this type to evolve in a land which already possessed true woodpeckers, as the latter would compete with it and eliminate it. In a similar way the warbler-like finch would, at least in its intermediate stages, have been less efficient than a true warbler.

Darwin's finches may well have been the first land birds to arrive on the Galapagos. The islands would have provided an unusual number of diverse, and vacant, environmental niches in which the birds could settle and differentiate. The same may have been true of Hawaii. In my opinion, however, the type of evolution that has occurred in those two groups of islands is not unique. Similar developments could have taken place very long ago on the continents; thus our own finches, warblers and woodpeckers may have evolved from a common ancestor on the mainland. What is unique about the Galapagos and Hawaii is that the birds' evolution there occurred so recently that we can still see the evidence of the differentiations.

MUCH MORE is still to be learned from the finches. Unfortunately the wonderful opportunities they offer may not long remain available. Already one of the finches Darwin found in the Galapagos is extinct, and so are several other animals peculiar to the islands. With man have come hunters, rats, dogs and other predators. On some islands men and goats are destroying the native vegetation. This last is the most serious threat of all to Darwin's finches. Unless we take care, our descendants will lose a treasure which is irreplaceable.

II

EARLIEST TRACES
OF LIFE

INTRODUCTION

Given the phenomenon of evolution, we naturally next ask, When and how did it all begin? Paleontologists and biologists postulate that the process of speciation and diversification has gone on for a very long time, and can be projected backward through the Cenozoic, Mesozoic, and into the dawn of the Paleozoic. But until fairly recently, scientists have been puzzled by this abrupt and relatively late boundary between the rocks of the lower Paleozoic, with its abundant and diverse, although primitive, fossils, and the rocks of the Precambrian, which seemed virtually to lack fossils. The apparent puzzle was that the Precambrian spanned about seven-eighths of earth history, whereas the Phanerozoic (Paleozoic, Mesozoic, and Cenozoic eras) included but one-eighth. Moreover, the shelly invertebrates found in basal Cambrian strata included trilobites, who were members of the highly complex and advanced phylum of the arthropods.

It was argued that surely the Precambrain must have been a time of gradual evolution and diversification of life, leading up to the Cambrian fossil faunas. But where were these Precambrian fossils? Except for fossilized algal mounds and scattered tracks and trails, Precambrian rocks failed to reveal suitable presumptive ancestors of early Paleozoic life. Various ideas were offered to explain away this problem: Perhaps Precambrian rocks are too metamorphosed to allow preservation of fossils; maybe the environments in which these primitive organisms lived were not recorded in the rocks of the Precambrian; or, if the fossils and environments were initially preserved, perhaps there was a subsequent period of widespread erosion that removed the crucial evidence; or maybe Precambrian organisms just lacked hard parts that could be readily fossilized.

Increasingly, however, attempts to rationalize the absence of a good Precambrian fossil record seemed weaker and weaker. Among other things, examples of relatively unmetamorphosed rocks of many different sedimentary environments became known. In addition, a number of places in the world showed that there was no significant erosion separating upper Precambrian and lower Cambrian rocks. As for the lack of hard parts, one would have expected that somewhere within the Precambrian impressions or traces ought to be found, just as they are, even if rarely, in younger rocks. Moreover, to some paleontologists it seemed biologically unlikely that soft-bodied trilobites or brachiopods existed, because their very anatomy would require mineralized supporting structures.

Gradually, it dawned on people that maybe we should accept the Precambrian fossil record, such as it is, as a true reflection of what in fact occurred during this period of evolution. One immediate implication of this new way of viewing the fossil record was the need to explain the sharp biological discontinuity between late Precambrian (without Cambrian precursors) and the abundant and diverse shelly invertebrates of the Cambrian. A related consequence of this new approach was the need to explain why this discontinuity happened so relatively late in the history of the planet.

Just about this time, during the 1950s, geologists began discovering a diverse, but *microscopic*, assemblage of primitive algae and bacterialike organisms in various parts of the Precambrian on several different continents, as well as evidence of soft-bodied, primitive invertebrates near the end of the Precambrian. These discoveries illustrate an interesting sidelight about the nature of scientific inquiry. As long as paleontologists expected to find macroscopic examples of prototrilobites, protobrachiopods, or protomolluscs in the Precambrian—as suggested by earlier ideas about how Precambrian evolution must have taken place—they would not only not find these fossils (because they didn't exist), but they also would fail to observe what was really there. However, as opinions changed about what the record might look like, important discoveries were quickly made and, more significantly, were appreciated as such. This episode demonstrates that what we see or do not see in our scientific investigations is often influenced by what we think we should be seeing. So much for scientific "objectivity."

"The Oldest Fossils," by Barghoorn, describes microscopic bacterial and algal remains found about one, two, and three *billion* years ago in Precambrian rocks of Australia, Canada, and Africa. These fossils support and date the commonly held view today regarding the early evolution of life on earth. During the first billion and half years of earth history, inorganic synthesis of complex organic molecules from the early reducing atmosphere—mostly carbon monoxide and dioxide with varying amounts of methane, ammonia, and perhaps hydrogen cyanide—led to the formation of the first self-replicating cells. Although geological data and laboratory experiments support various parts of the theory of chemical evolution during this time, we have no fossil evidence for it whatsoever. As Barghoorn points out, once this threshold was passed, we then did begin to pick up evidence of the subsequent life forms. These consist of bacterialike and algalike structures in three-billion-year-old rocks of South Africa. As we would expect, these organisms are simply constructed, and thus it is difficult to be sure that they are genuine fossils, rather than artifacts of some geologic process, let alone really "bacteria" or "algae." (See the article by Cloud, 1976, cited at the end of this introduction, for a discussion of this problem.)

In Canada, rocks almost two billion years old yield a still more diverse and convincing assemblage of primitive, photosynthesizing organisms preserved in cherts. Apparently, then, early algae were present several billion years ago; the result of their photosynthesis was the slow and inexorable generation of an oxidizing atmosphere. The presence of thick accumulations of banded iron ores, ranging in age from 1.8 to 3.2 billion years, in all the Precambrian continental shields suggests that the early formed photosynthetic oxygen swept the seas of their soluble ferrous iron into the virtually insoluble ferric oxides. (Most of the world's steel comes from this early biological activity.) Free oxygen in the atmosphere also resulted in the ozone layer, which screens out much of the biologically damaging, solar ultraviolet radiation. Shallow sea life was restricted and terrestrial life prevented until this protective layer became established.

Another important milestone was passed about one billion years ago, when the eukaryote cell appeared, differentiated in part from the earlier prokaryote cells by being able to undergo sexual reproduction. Fossils documenting this have been found in rocks from Australia that strongly suggest various steps in meiosis, the process whereby a nucleated cell undergoes nuclear duplication before subsequent cell division. As we noted in the introduction to Part I, sexual reproduction creates much genetic variability and so may have provided the trigger for the rapid evolution of life, particularly animals, soon after.

Glaessner's "Precambrian Animals" picks up the story where Barghoorn leaves off. Australian rocks formed some 650 million years ago contain excellent fossil impressions along clay partings in shallow marine sandstones. These impressions are of soft-bodied animals reminiscent of modern jellyfish, sea pens (cousins to corals), flat worms, and segmented worms. Since this first discovery, these distinctive fossils have been found in many other parts of the world in rocks of about the same age and are now judged to be representatives of an "Ediacaran fauna" (named after the Australian locality where first described) that was ancestral to the diverse early Cambrian faunas about one hundred million years later.

The Precambrian fossil record thus supports present-day ideas about evolution during this interval of earth history, which are outlined as follows (refer to the chart near the end of the Barghoorn reading and to updated versions in Table 1 and Figures 1 and 2 of Cloud, 1976).

1. Formation of the earth, about 4.5 billion years ago.
2. Outgassing creates a reducing atmosphere; chemical evolution leading to earliest life forms; 3.5 to 4.5 billion years ago.
3. Heterotrophs appear and "feed" on inorganically synthesized compounds. As this reservoir becomes depleted, photosynthesis—the process by which organisms make their own food, using carbon dioxide and water—evolves the so-called autotrophs; about 3.5 million years ago.
4. Diverse autotrophic prokaryote cells; increasing free oxygen in the atmosphere and consequent development of ozone layer; about 2 billion years ago.
5. Rise of eukaryote cell and sexual reproduction with greatly increased genetic variation; almost 1 billion years ago, perhaps going back to 1.5 billion.
6. Appearance of soft-bodied multicellular animals into basic ground plans followed by the higher invertebrates; about 700 million years ago.
7. Diversification of invertebrates and beginning of skeletonization for protection and support; about 600 million years ago, the beginning of the Cambrian period.

SUGGESTED FURTHER READING

Cloud, P. 1976. "Beginnings of Biosphere Evolution and their Biogeochemical Consequences," *Paleobiology*, vol. 2, pp. 351–387. An excellent review of the state of knowledge about the major events leading to the evolution of higher forms of life, with special emphasis on the interactions between the biosphere and the atmosphere, hydrosphere, and the rock record. Cloud carefully points out the different degrees of confidence to be placed on various facts and their interpretations regarding early life on earth. The bibliography of 172 articles provides a comprehensive entry into the subject.

5   The Oldest Fossils

by Elso S. Barghoorn
May 1971

*The remains of ancient bacteria and algae, some of
them more than three billion years old, have been
found in Africa, Australia and Canada. They provide
evidence on the earliest stages of evolution*

How did life originate on the earth? It was not so long ago that attempts to answer this question defined more areas of uncertainty than of agreement. Today a fossil record that was virtually unknown before the 1950's has been found to bear witness to three of the key events in the earliest stages of organic evolution. The fossils come from widely separated parts of the earth. All are preserved in unusual rocks of the Precambrian, the first and by far the longest interval in geologic history. The earliest of the fossils are more than three billion years old.

All that is known or conjectured about the terrestrial origin of life suggests that the first appearance of living organisms was preceded by the gradual development of a complex chemical environment. This environment is usually pictured as a kind of primordial broth, filled with such "organic" molecules as amino acids, sugars and other biologically important substances that came into existence through nonorganic processes. Millions of years must have been required for the accumulation, elaboration and differentiation of the broth. That period may be called the time of chemical evolution, a concept that owes much to the work of a leading student of abiotic synthesis, Cyril A. Ponnamperuma of the Ames Research Center of the National Aeronautics and Space Administration. As a preliminary stage in the total process of organic evolution, chemical evolution of course reaches its climax when lifeless organic molecules are assembled by chance into a living organism. This first form of life is what the Russian biochemist A. I. Oparin calls a "protobiont."

Judging from the various forms of life we know today, the first protobionts were probably microscopic in size and single-celled in structure, perhaps resembling the modern coccoid, or spherical, bacteria. Rather than imagining some specific organism, however, let us consider this first form of life more abstractly and give it a name that describes how it lives rather than how it looks. Call it a heterotroph, which is to say an organism that cannot manufacture all its own nutrients but must feed on organic molecules in the broth that surrounds it. (This assumes that the organism is immersed in an aqueous medium or at least rests on a wet surface; a supply of water is essential to protoplasmic life.) It seems reasonable to suppose that the organism was a heterotroph; to have at this stage a full-fledged autotroph, an organism that can manufacture its organic nutrients out of inorganic substances, would be asking too much of chance.

We can, however, demand that heterotrophs rather promptly give rise to autotrophs. Otherwise, as Preston Cloud of the University of California at Santa Barbara puts it, once the heterotrophs had "gobbled up all the goodies" within reach they would die off, making another accident of biosynthesis necessary in order to get things going again.

One way or another, then, autotrophs must evolve from an original heterotrophic population. This event in the development of life marks the time when the organic nutrients of the primordial broth are depleted and photosynthesis—the most plausible form of organic self-nutrition—must have been invented. Such an assumption is not based solely on logical considerations of thermodynamics and physiology. It is also supported by geological evidence that early organisms depended on photosynthesis to sustain themselves. Among the most ancient formations in the geological record are some showing signs that small amounts of free oxygen—the gaseous by-product of photosynthesis—were present early in the history of the earth. This chemical evidence coincides with indications in the fossil record that some of the most primitive forms of terrestrial life, organisms that resembled modern bacteria and blue-green algae, were then increasing in numbers and diversity.

Many modern bacteria are photosynthetic, although they do not produce free oxygen; all modern blue-green algae are photosynthetic and all produce free oxygen. The evidence suggesting that the first autotrophs were organisms of the same primitive kind is significant for an additional reason. Bacteria and blue-green algae are alone among living things in the simplicity of their cells. They have neither a membrane-enclosed nucleus nor such specialized cellular organelles as mitochondria. Their genetic material is diffused throughout the cell, and they are incapable of either mitosis (body-cell division) or meiosis (germ-cell division). Both kinds of cell division require that the genetic material be organized into chromosomes.

Bacteria and blue-green algae are thus fundamentally different from other organisms: all other plants, all animals and the many forms that are neither entirely plant nor entirely animal. These other organisms have cells with nuclei and specialized organelles or specialized intracellular structures; they are called eukaryotic (truly nucleated), whereas bacteria and blue-green algae are prokaryotic (prenuclear). It would be surprising if the autotrophs on the lowest rungs of the evolutionary ladder were anything but prokaryotic.

The prokaryotic cell is notable for still another reason. Any organism whose genetic material is diffused throughout the

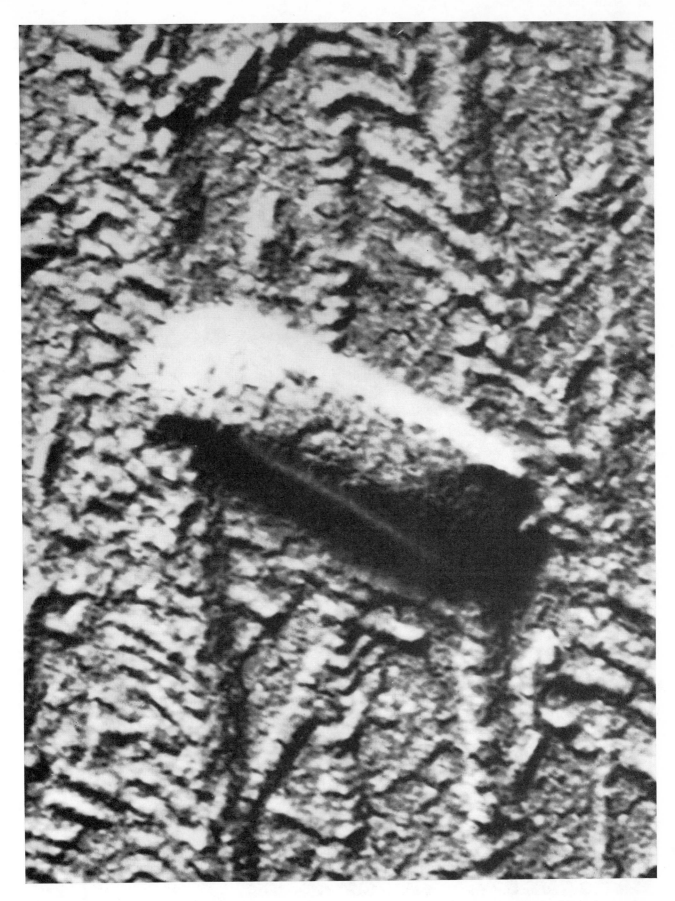

OLDEST KNOWN BACTERIUM, one of two primitive forms of life preserved in a Precambrian rock formation in South Africa, appears as a raised rectangular shape in this electron micrograph. What is seen is a carbon replica of the polished surface of a rock sample, shadowed with heavy metal. The fossil bacteria come from cherts of the Fig Tree formation; they are from .5 to .75 micron long and about .25 micron wide. The organisms are some 3.2 billion years old and have been given the name *Eobacterium isolatum.*

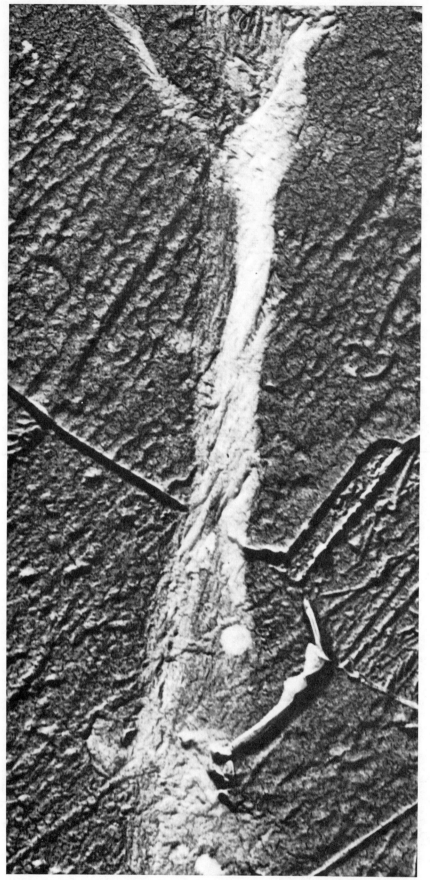

THREADLIKE FILAMENT of organic matter resembling decomposed plant tissue is another kind of fossil that appears in electron micrographs of the Fig Tree cherts. Some specimens are nine microns long. Not identifiable with any known organism, the filaments might conceivably be polymerized abiotic molecules from the "primordial broth."

cell and whose reproduction does not involve a recombination of parental genes is genetically conservative. In such an organism random mutations, instead of being preserved when they benefit the organism, tend to be damped out in a few generations. The blue-green algae are an outstanding example of such genetic conservatism. Many living species are almost indistinguishable in structure from species that flourished a billion or more years ago.

Against this background of fact and conjecture, how many of the events in the early stages of organic evolution can be labeled as being outstanding? There appear to be three such events, each of them a kind of threshold-crossing. The first, obviously the *sine qua non*, is successful biosynthesis: a crossing of the threshold separating the initial period of abiotic chemical evolution from the subsequent organic period. Perhaps someday fossil evidence of the earth's earliest organisms, the heterotrophs that crossed this first threshold, will be discovered. In the meantime the remains of their successors, the early photosynthetic autotrophs, provide the necessary proof that the first threshold had been passed.

The next threshold can be characterized as the threshold of diversification. As evolution progresses the first photosynthetic organism should not be limited to a few similar forms. Instead they should develop differences in shape and structure indicative of roles in a variety of ecological niches.

The third threshold divides prokaryotic organisms from all others. A world populated solely by bacteria and blue-green algae is conceivable; indeed, that is apparently the way it was. Viewed from our present vantage such a world seems poor in possibilities for further evolution. The extraordinary variety of plant and animal life that has arisen on the earth over the past 600 million years is due entirely to the invention of the eukaryotic cell, with its potential for genetic diversity.

Thanks to a lucky accident of fossil preservation, we now have evidence that each of these thresholds was successfully crossed during the vast span of Precambrian times. The Precambrian represents nearly four billion of the earth's first 4.5 billion years. It began with the formation of the earth and ended some 600 million years ago with the dawn of the Paleozoic era [*see illustration on page 55*]. Nothing is known of the first billion years or so; the world's oldest-known rocks, found in Africa, are not much more than three billion years old.

Precambrian rocks are found not only in Africa but also on every other continent. The best-known areas are the Canadian Shield in North America and the Fenno-Scandian Shield in Europe, but more than a third of Australia is also a Precambrian shield and there are sizable Precambrian areas in South America and Asia. Some Precambrian rocks are of igneous origin and some are of sedimentary origin. Most of the sediments have been heavily metamorphosed: changed in form and chemical composition by heat and pressure. In these metamorphic rocks not only fossils but also the faintest traces of organic matter have been obliterated.

A few Precambrian sediments have escaped substantial alteration. Extensive deposits of black shale, black chert and other stratified sediments are scattered through the major shield areas in virtually unmetamorphosed condition. These carbon-rich rocks—for example the formations in the Lake Superior region of North America, in the Transvaal of South Africa and in parts of western Australia—look even to the experienced eye very much like certain sediments of the Carboniferous epoch that are a mere 300 million years old.

The oldest-known group of Precambrian sediments is located in the border region between the Republic of South Africa and Swaziland. The formation is called the Swaziland Sequence; its stratified rocks are thousands of feet thick. One series of strata in the middle of the sequence, known as the Fig Tree formation, is well exposed in the Barberton Mountain Land, a gold-mining district near the town of Barberton in the eastern Transvaal. The Fig Tree rocks consist of black, gray and greenish cherts, interbedded with jaspers, ironstones, slates, shales and graywackes. In places the chert beds are 400 feet thick. The chert is usually fractured, but it is cemented together and the fractures are filled with quartz; parts of the formation show little evidence of metamorphism. The Fig Tree cherts contain traces of organic matter and a few microfossils.

The Barberton Mountain Land has long been an active mining area, and its geology has been studied in considerable detail. The age of the graywackes and shales has been determined independently by several laboratories using a method based on the decay of radioactive strontium and rubidium. This particular radioactive clock started to run 3.1 billion years ago, but there is good reason to believe the sediments were deposited

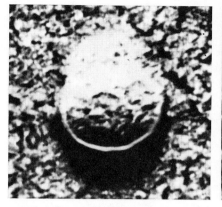

TWO CROSS SECTIONS of *Eobacterium* are seen in electron micrographs of metal-shadowed carbon replicas. In the fossil at left the outer and inner layers of the cell wall are visible. The wall, .015 micron thick, resembles the wall of living bacteria of the bacillus type.

somewhat earlier. Recently rocks lying close to the base of the Swaziland Sequence (and thus well below the level of the Fig Tree cherts) were shown to be 3.36 billion years old. In light of these age determinations it seems probable that the age of the Fig Tree cherts is in excess of 3.2 billion years.

In 1965 I collected cherts from several localities in the Fig Tree series, and samples were prepared for examination in my laboratory at Harvard University that summer. Two techniques were employed: thin sections of chert were cut for examination by reflected and transmitted white light under the microscope, and carbon replicas of etched and polished chert surfaces were made. The carbon replicas were then "shadowed" with metal and examined by transmission electron microscopy. J. William Schopf joined me in the study of the specimens.

When Schopf and I examined the thin sections under the light microscope, we could see that the rock matrix contained abundant laminations of dark-colored and virtually opaque organic matter. The

laminations were irregular but were usually aligned parallel to the strata of the chert, suggesting that they had originally been formed as part of an aqueous sedimentary deposit. No distortion was evident where the organic matter crossed the boundaries of the individual grains of chalcedony comprising the chert, which suggested to us that the process of deposition had emplaced the organic material within a silica-rich matrix before the silica was crystallized into chert. There was no evidence whatever that the silica was of secondary origin.

It was difficult to discern distinct bodies within the layers of organic matter under the light microscope. Our first success in isolating a Fig Tree organism was achieved with the carbon-replica technique; the electron microscope revealed a number of rod-shaped structures, preserved both in profile and in cross section. The rods are very small. They range in length from slightly under .5 micron to a little less than .7 micron, and in diameter from about .2 micron to a little more than .3 micron. In cross section the cell wall is sometimes seen to

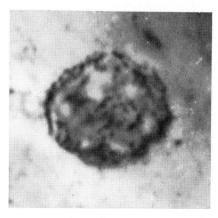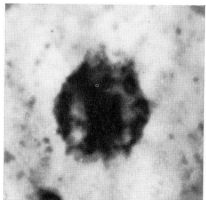

ALGA-LIKE SPHERES, seen in photomicrographs of thin sections of rock, are the other fossil organisms that are found in the Fig Tree formation. The diameter of the spheres is usually less than 20 microns. The organism is named *Archaeosphaeroides barbertonensis*.

consist of an inner and an outer layer, with a total wall thickness of .015 micron [*see top illustration on preceding page*]. This is comparable to the cell wall of many modern bacteria in both structure and dimension.

Electron microscopy also revealed the presence of organic material in the form of irregular, threadlike filaments lacking discernible structural detail. The threads are clearly native to the chert and not the products of contamination. They are as much as nine microns long and resemble decomposed plant material. Although these filaments are almost certainly of biological origin, they cannot be identified with any known type of organism. It has been suggested, probably as a result of wishful thinking, that they may be polymerized strands of abiotic organic matter from the primordial broth.

Schopf and I later succeeded in resolving larger microfossils in thin sections of Fig Tree chert under the light microscope. These fossils are spheroidal; measurements of 28 particularly well-defined specimens show that the majority are between 17 and 20 microns in diameter [*see bottom illustration on preceding page*]. Some have a darkened interior, as if cytoplasm within the spheroid had coalesced and become "coalified." Just as the rod-shaped organisms revealed by the electron microscope resemble certain modern bacteria, so the spheroids are not unlike some modern blue-green algae of the coccoid group. They may even be among the evolutionary precursors of such algae.

We have named the rods *Eobacterium isolatum*, a new genus and species. The generic name (*eo-* is the Greek root for "dawn") points to the great antiquity of the organism and to its bacterium-like appearance; the specific name defines its noncolonial, single-cell habit of growth. The spheroids we have named *Archaeosphaeroides barbertonensis*, also a new genus and species. Again the generic name refers to the organism's great age and its appearance; the specific name identifies its place of discovery. The existence of these two organisms, successful inhabitants of an aquatic environment more than three billion years ago, is evidence that the first evolutionary threshold—the transition from chemical evolution to organic evolution—had been safely crossed at some even earlier date. We now know that at least two living organisms appeared well before the first third of earth history had passed. If we accept the evidence (to which we shall return) that the alga-like Fig Tree organisms were photosynthetic, an important

geochemical event must also have occurred. With the onset of photosynthesis free oxygen would have begun to appear among the other constituents of the environment. The appearance of free oxygen was an event destined to have profound influence, both biological and geological, on the subsequent history of the earth.

Evidence of the second evolutionary threshold-crossing comes from North America. A remarkable outcropping of Precambrian rocks along the shore of Lake Superior in western Ontario shows a sequence of sediments known as the Gunflint Iron formation. The rocks at the base of the Gunflint formation include beds of black chert three to nine inches thick. The beds are exposed—in some places more or less continuously and in others as isolated outcrops—over a distance of some 115 miles in Ontario, from the vicinity of Schreiber on the east to Gunflint Lake on the west. Like the much earlier Fig Tree series, the Gunflint formation has been the subject of detailed geological investigation.

Granite underlies the Gunflint formation unconformably, that is, the basement rock had been eroded before the Gunflint sediments were deposited on its surface. Radioactive clocks have yielded two independent dates for the granite. A concentrate of its biotite contents indicates an age, in terms of the argon-potassium ratio, of 2.5 ± .75 billion years. The age of a whole-rock sample, in terms of its rubidium-strontium ratio, is 2.36 ± .70 billion years. The age of the granite thus provides a "floor" of maximum age under the Gunflint formation. The Gunflint cherts cannot be any older than these dates.

Micas separated from rocks in the upper levels of the Gunflint formation, collected near Thunder Bay, indicate an age in terms of the argon-potassium ratio of 1.60 ± .05 billion years. For technical reasons this figure is only 80 percent of the true age; it must therefore be adjusted to 1.90 ± .20 billion years. The micas thus provide a "ceiling" of minimum age for the cherts that lie at the base of the Gunflint formation. It seems reasonable to set the age of the cherts at approximately two billion years, which makes them a billion years or so younger than the Fig Tree cherts.

The only rocks in the Gunflint formation that contain microfossils are the cherts. Like the Fig Tree cherts, they are evidently the product of deposition in an aqueous environment that was rich in silica. Most of the Gunflint fossils are three-dimensional, and many show ex-

quisite anatomical details. It has been argued that the structure of these fossils has been preserved by the infiltration of silica from the surrounding sediments. In my opinion the organisms were preserved without distortion by being deposited in a siliceous solution that later crystallized into chert, much as a modern biological specimen is preserved by being embedded in plastic. The soft structure of the organism owes its preservation to the almost complete incompressibility of the silica matrix. This is as unusual as it is fortunate. In most instances of fossil preservation the matrix is composed of relatively plastic sediments that are much compressed during consolidation, with the result that any preservation of soft tissues, let alone preservation in three-dimensional form, is a rarity.

How were the Gunflint cherts deposited? The picture in Ontario is rather clearer than it is in the Transvaal. The Gunflint cherts were apparently precipitated and consolidated around an underlying complex of basement rocks, consisting of greenstone boulders and finer conglomerates, that seems to have been continuously submerged at the time. "Domes" of algae, which are visible to the unaided eye in samples of Gunflint rock, grew on the surface of the boulders. Algal "pillars" grew perpendicularly to the domes; their fossil remains consist of alternate layers of coarsely crystalline quartz and fine-grained black chert [*see bottom illustration on opposite page*].

In the 1950's I was privileged to be associated with Stanley A. Tyler of the University of Wisconsin in collecting and analyzing specimens of Gunflint chert from the Schreiber area. Only a few preliminary studies of these and other Gunflint specimens were published before Tyler died, although by then we both knew that his work had added to the fossil record an entire new group of very ancient, primitive photosynthetic organisms whose existence had not been suspected. Indeed, even though many years of study have now been devoted to the Gunflint organisms, the formation continues to yield new finds. Eight genera of primitive Gunflint plants, comprising 12 species, have been described so far, yet accompanying this article are illustrations of new forms of undetermined taxonomic status.

The most abundant of the Gunflint microfossils are filamentous structures. The majority of them are between .6 micron and 1.6 microns in diameter but a few are more than five microns across. They vary in length up to several hundred microns. Some of the filaments have inter-

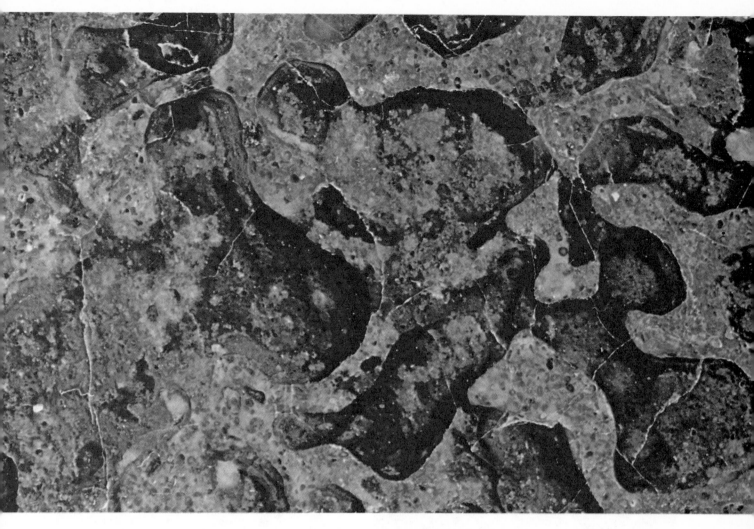

SAMPLE OF GUNFLINT CHERT of Ontario (*above*) has a knob-by surface. The knobs, exposed by weathering, are tops of pillars formed by algae in shallow water where Gunflint organisms lived.

VERTICAL SECTION through a chert sample (*below*) shows the structure of the algal pillars. Layers of quartz and of fossil-bearing black chert alternate in each pillar like sets of nesting thimbles.

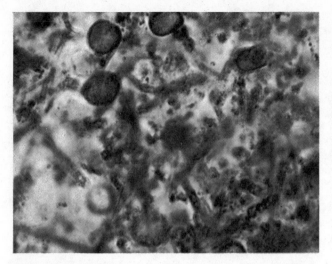

DENSE MIXTURE of organic detritus, spherical bodies and filaments is enlarged 250 diameters in this photomicrograph. All the micrographs on this page show thin sections of the Gunflint cherts.

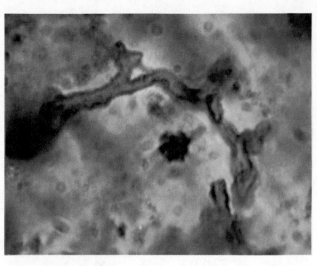

TUBULAR FILAMENT with branches and swellings is not obviously related to any living organism. Where it is not swollen it is about two microns across. It is called *Archaeorestis schreiberensis*.

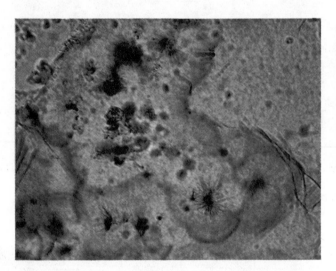

COLONY OF ALGAE is enlarged 200 diameters. The organisms, each a cluster of spikelets, have been named *Paleorivularia ontarica* because of their resemblance to the living genus *Rivularia*.

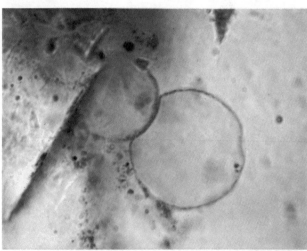

CLOSE CONTACT between a larger and a smaller cell is apparent in a chert section enlarged 1,000 diameters. Interrelations of this kind may represent a stage in the evolution of the eukaryotic cell.

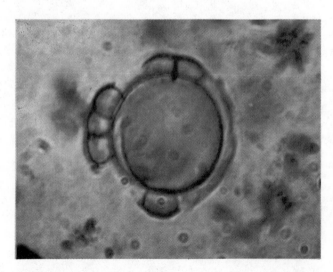

SPHERE WITHIN SPHERE, the inner one held away from the outer one by a number of smaller spheroids, is about 30 microns in diameter. The spheres comprise the extinct species *Eosphaera tyleri*.

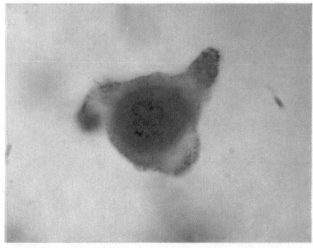

ASYMMETRICAL CELL, found in a recently prepared chert section, has not yet been formally classified. The Gunflint cherts have been studied for nearly 20 years, but they still yield new organisms.

nal walls at right angles to their length; others lack such walls. Where the walls are present, they can be broad or narrow. These primitive plants have been assigned on the basis of their morphology to four genera including five species. Among the present-day blue-green algae they resemble are the filamentous *Oscillatoria* and related forms; one of the four also shows resemblances to the modern iron-oxidizing bacterium *Crenothrix*.

Another genus of Gunflint organisms is represented by small spheroids, ranging from one micron in diameter to more than 16 microns. The walls of the spheroids vary both in thickness and in structural detail. Because of these variations they have been assigned to three separate species and placed in the genus *Huroniospora*. The spherical organisms are found in chert collected from many parts of the Gunflint formation but they are not equally abundant everywhere.

What kinds of organism are represented by the spheroids? We have only their simple morphology to judge by. Like their counterparts in the much older Fig Tree cherts, they might be noncolonial blue-green algae of the coccoid group. They might also, however, be the reproductive spores produced by the filamentous plants mentioned above, and some might even be the spores of iron bacteria. Still another possibility is that they are the fossilized bodies of free-swimming organisms whose flagella have not been preserved. Further study may lead to a choice among these alternatives.

The remaining Gunflint genera that have been intensively studied and described so far all show unusual characteristics. The organisms in one of the three genera are star-shaped and made up of radially arrayed filaments. The diameter of the "star" ranges from eight to 25 microns; in rare cases some of the filaments are branched. Although representatives of the genus are few in number and often poorly preserved, they are found throughout the Gunflint cherts.

The genus has been named *Eoastrion*—"dawn star"—to indicate the great antiquity and distinctive shape of its members. Two species have been established, one to accommodate the fossils with non-branched filaments and the other the fossils with branched filaments. There are no clear analogies between the fossils and modern organisms, although in some respects *Eoastrion* resembles the curious iron- and manganese-oxidizing organism *Metallogenium personatum*.

A most peculiar organism comprises the next of these genera. Its fossils are

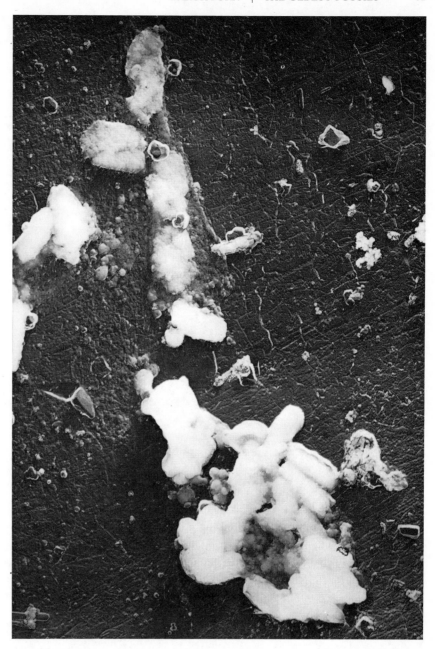

FOSSIL BACTERIA found in the Gunflint cherts are enlarged 30,000 diameters in this electron micrograph. Some two billion years old, they look like living rod-shaped species.

abundant in the cherts exposed near Kakabeka Falls, some 20 miles west of Thunder Bay. It consists of a bulb with a narrow stalk, surmounted by a structure that in some cases resembles an umbrella. The relative size of the three parts varies widely from specimen to specimen; the size of the bulb and the stalk together or of the umbrella alone ranges between 10 and 30 microns.

This odd plant has been assigned to the new genus *Kakabekia* and the species *umbellata* (which refers to the umbrella). Living organisms that superficially resemble it in form are certain multicellular polyps. All modern polyps, however, are many times larger.

The story of *Kakabekia* has had some interesting and continuing overtones. Late in 1964, quite independently (in fact, unaware) of paleontological investigations of the Gunflint fossils, Sanford M. Siegel discovered a strange new form while he was studying soil microorganisms that can survive extreme atmospheric conditions. The organism defied identification or assignment to any known taxonomic category. Siegel set it aside as being an enigma, although he kept his preparations, drawings and photomicrographs. A few months later a description of *Kakabekia* was published, and Siegel immediately noted a striking resemblance between his soil organism

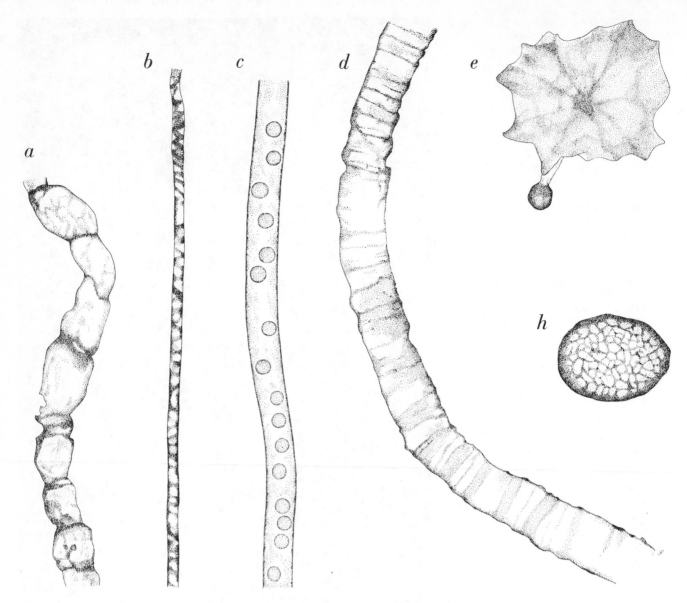

DIVERSITY OF FORM, evident among the plants fossilized in the Gunflint cherts, is indicative of evolutionary progress during the billion years that separate Fig Tree from Gunflint times. These composite drawings are based on a study of several specimens of each species and include three of the most unusual ones found. All are shown at the same scale: 2,500 times actual size. Filamentous organisms include two of the genus *Gunflintia*: *G. grandis* (*a*) and *G. minuta* (*b*). The other two are *Entosphaeroides amplus* (*c*) and *Animikiea septata* (*d*). The hydra-like *Kakabekia umbellata* (*e*) has a modern counterpart, if not a descendant, in a newfound soil

and some of the Gunflint specimens. Siegel's organism is very slow-growing, contains no chlorophyll and apparently has no nucleus; it may therefore be representative of a new group of prokaryotic microorganisms. It was first found in ammonia-rich soil collected at Harlech Castle in Wales, and it has since been recognized in soils from Alaska and Iceland and recently in soil from the slopes of the volcano Haleakala in Hawaii. Whether or not it is related to the two-billion-year-old *Kakabekia* is questionable, but the existence of two such bizarre forms is at least a remarkable evolutionary coincidence.

The eighth genus of Gunflint organisms comes from a single area near the easternmost outcropping of the chert, in the vicinity of Schreiber Beach. The organism consists of two concentric spheres; its outside diameter ranges from 28 to 32 microns. In most of the fossils the inner sphere is kept from contact with the outer one by as many as a dozen "spacers" in the form of small flattened spheroids. I have assigned this distinctive organism to the new genus *Eosphaera* ("dawn sphere") and the species *tyleri* (in honor of Tyler). No analogous living organism is known, nor has an organism resembling *Eosphaera* been found in any other Precambrian rocks. *Eosphaera* may be somewhat fancifully regarded as a "mistake" in evolution that did not survive the middle Precambrian.

Anyone who wanted to question that the organisms preserved in the Fig Tree cherts three billion years ago were photosynthetic could defend the negative position quite eloquently. Where the billion-years-younger Gunflint organisms are concerned, however, the affirmative evidence seems overwhelming. For one thing, the chemical analysis of organic material from the Gunflint cherts in several laboratories reveals the presence of the hydrocarbons pristane and phytane: two "chemical fossils" that can most reasonably be regarded as being breakdown products of chlorophyll. A second datum is the striking resemblance between the filamentous fossils that are the most abundant Gunflint forms and modern

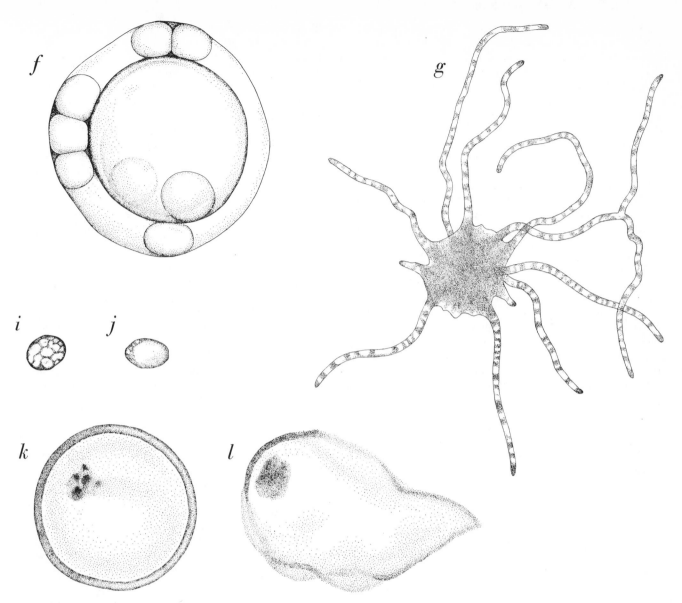

organism. The globular *Eosphaera tyleri* (*f*) evidently failed to survive Precambrian times. *Eoastrion bifurcatum* (*g*), with its array of filaments, is one of two such species. Three spherical organisms that differ chiefly in surface markings are assigned to the single genus *Huroniospora*; they are *H. microreticulata* (*h*), *H.* *macroreticulata* (*i*) and *H. psilata* (*j*). A one-celled organism that has not yet been classified (*k and l*) has internal features suggestive of a nucleus; the second specimen may have been fossilized as the process of cell division began. If these are eukaryotic cells, a key evolutionary event took place far earlier than is now thought.

photosynthetic blue-green algae. A third is the small domes and pillars in the Gunflint chert that resemble the structures formed in shallow water today by dome-building photosynthetic algae.

One may add the evidence of the relative abundance in Gunflint organic matter of the two nonradioactive carbon isotopes carbon 12 and carbon 13. The carbon in the carbon dioxide of the earth's atmosphere normally consists of about 99 percent carbon 12 and 1 percent carbon 13. In the process of photosynthesis, however, plants tend to fix slightly more carbon 12 than carbon 13, so that plant tissues are even poorer in the heavier isotope. Measurements of the ratio of the two isotopes in Gunflint organic material

were made by Thomas C. Hoering of the Geophysical Laboratory of the Carnegie Institution of Washington. The results indicate that the Gunflint material too is poor in carbon 13, and to a degree that is almost identical with the depletion in modern algae and other photosynthetic plants. The billion-years-older Fig Tree organic material shows a carbon-12-to-carbon-13 ratio that is about the same as the ratio in the Gunflint material. This result lends credence to the view that the alga-like Fig Tree organisms probably were photosynthetic too.

It seems reasonable to conclude that even if oxygen was only a trivial component of the environment in Fig Tree times, the Gunflint organisms represent

the intermediate agency that brought about the oxygen-rich environment at the end of the Precambrian. This is a development of major importance. It is not, however, the only development or even the most important development of Gunflint times. The variety of form (and therefore presumably of function) represented by the eight genera of Gunflint plants demonstrates that terrestrial life had crossed the second evolutionary threshold—the threshold of diversification—no less than two billion years ago.

The crossing of the third threshold is clearly documented in a series of late Precambrian formations, consisting primarily of limestones, sandstones and dolomites, found along the northern rim of the Amadeus basin in the Northern Territory of Australia. One member of the series is the Bitter Springs formation; a ridge in the Ross River area, about 40 miles from Alice Springs, consists of strata from its lower and middle levels. The exposed formations include isolated beds of dense black chert and laminated rocks that are associated with fossil structures built by colonial algae.

No absolute age is known for the Bitter Springs formation. Its top strata, however, lie some 4,000 feet below the lowest of the rocks that in the Ross River area are the boundary between formations of Precambrian age and those of the succeeding Cambrian period. The generally accepted date for the beginning of the Cambrian period is 600 million years ago, so that the lower position of the Bitter Springs cherts makes them considerably older. The Bitter Springs formation also underlies Precambrian sediments that are known to be some 820 million years old on the basis of their rubidium-strontium ratio. I think it is reasonable to assume that the Bitter Springs cherts are roughly a billion years old. This makes them only half as old as the Gunflint cherts and less than a third as old as the Fig Tree cherts. I collected samples of the Bitter Springs cherts in the Ross River area in April, 1965, and I added them to the Fig Tree samples for study with Schopf that summer.

A preliminary analysis of the abundant microfossils in the Bitter Springs cherts indicates that at least four general

groups of lower plants lived in the shallow seas or embayments that covered this part of central Australia in late Precambrian times. As one would anticipate, the plants included filamentous blue-green algae akin to such living forms as *Oscillatoria* and *Nostoc*. Some of these filaments are more than 75 microns long; they are thickest (about 1.4 microns) at the center and taper toward the ends to less than one micron.

Our most exciting identifications during the preliminary analysis concerned the other three groups. On the basis of the internal structures that have been preserved, all three appear to represent various green algae. The green algae, unlike the blue-green algae, are eukaryotic! Thus the Bitter Springs cherts apparently contain the earliest-known fossil evidence that an organism potentially capable of sexual reproduction, or at least possessing a nucleus, had finally evolved.

By the time Schopf had finished his detailed analysis of the Bitter Springs specimens in 1968, he had concluded that they represent a total of three bacterium-like species, some 20 certain or probable representatives of blue-green algae, two certain genera of green algae, two possible species of fungi and two problematical forms. For one of the green algae, *Glenobotrydion aenigmatis*, an unusual coincidence of fossilization has preserved several specimens at various stages of mitotic division. By arranging the specimens in order one can re-

create almost the entire sequence [*see illustration on page 56*].

With the discovery of a late Precambrian population of eukaryotic plants we approach the end of our story. The unequivocal appearance of the eukaryotic level of cellular organization at this point in earth history provides a welcome explanation for one of the principal puzzles of terrestrial evolution: Why do so many groups of higher organisms not appear in the fossil record until the span of geologic time is seven-eighths past? It is a good question, particularly in the light of what we know about the multicellular animals and higher algae early in the Paleozoic era.

Until recently the commonest explanation of the rarity of advanced organisms, multicellular animals in particular, before Cambrian times was that the supply of free oxygen in both the atmosphere and the hydrosphere had up to that point been extremely limited. Time was required to let the first prokaryotic organisms adjust to existence in an environment that had begun to contain this highly reactive element. Time was also needed for enough oxygen to leave the hydrosphere (where it was being produced by algal photosynthesis) and enter the atmosphere to establish a protective shield of ozone (O_3) between the earth's surface and the sun's harsh ultraviolet radiation. Until this shield existed neither shallow water nor bare ground would have been habitable.

An oxygen-poor environment was certainly a major obstacle in the path of

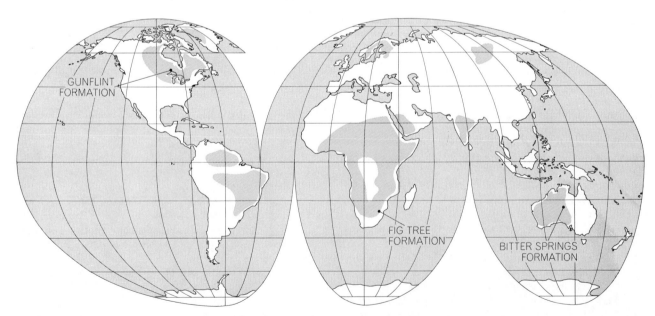

PRECAMBRIAN ROCKS occupy the old, low-lying "core" areas of continents throughout the world. Many, formed from accumulated sediments, are so altered by heat and pressure that any fossils have been obliterated. Some relatively unaltered exposures of Precambrian sediments, however, are rich in organic remains. Three such fossil-bearing Precambrian formations are shown on this map.

oxygen-dependent heterotrophs. In addition to the evidence provided by oxidized Precambrian formations, however, a number of recent studies indicate that the earth's atmosphere had actually accumulated enough oxygen to establish some kind of ozone shield considerably before the end of the Precambrian. The appearance of eukaryotic organisms late in the Precambrian, as indicated by the green algae of the Bitter Springs formation, provides a better explanation for the failure of higher organisms to appear until an even later time. The fundamental key to evolutionary progress is genetic variability. Sexual reproduction, which involves the recombination of heritable characteristics, is the highway to genetic variability and all its consequences, including the increased complexity of form and function at all levels of organization, that are thereafter apparent in the course of evolution.

This last of three thresholds, which separates the primitive world of cells without a nucleus from a world where sexual reproduction is possible, must have been crossed a long time before the formation of the Bitter Springs cherts. Only half a billion years or so later the Paleozoic seas were swarming with highly differentiated aquatic plants and animals, evolved from primitive forebears that had managed to cross all three Precambrian thresholds. Half a billion years does not seem to be evolutionary "room" enough to account for such epic progress. Moreover, there is evidence to suggest that developments in this direction may have begun in Gunflint times.

The Fig Tree, Gunflint and Bitter Springs cherts are not the only sources of Precambrian fossils, nor have Tyler, Schopf and I been the only investigators of the Precambrian fossil record. Indeed, work in the field has gone on for nearly a century. The emphasis has changed in recent years from an earlier concern with the curious "boundary" that has been traditionally accepted as separating the Precambrian from the beginning of the Paleozoic. We and our colleagues in many countries are primarily interested today in evidence that we hope will reveal even finer details of fossil cellular organization. It is a field that requires all levels of observation, from the macroscopic down to the electron microscopic. Workers in the field agree that the search for other stages and thresholds in the evolution of life must be focused on fine structures wherever these have been fortuitously preserved. The Precambrian fossil record is still meager, but significant gaps in the record are steadily

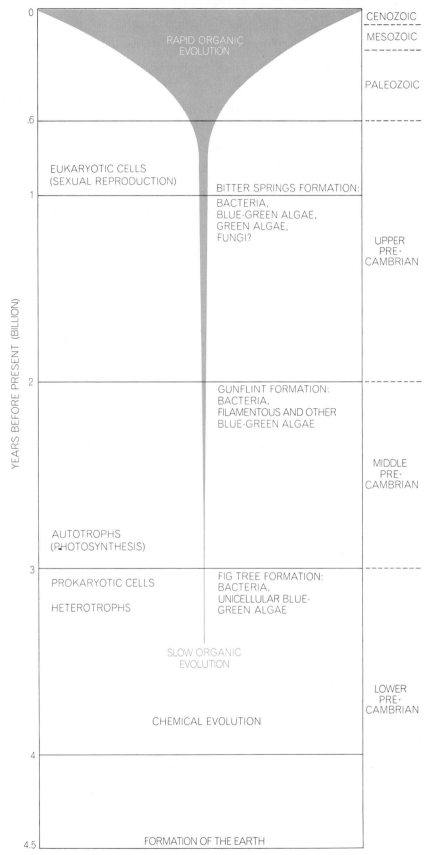

ORGANIC EVOLUTION is presented in terms of successively briefer episodes of biological advance. The Precambrian interval, the earliest and by far the longest, began when the earth was formed 4.5 billion years ago and ended 600 million years ago with the beginning of the Paleozoic era. The increasing abundance of species with the passage of time is indicated in color. Once organisms with eukaryotic, or truly nucleated, cells evolved, hastening evolutionary progress in late Precambrian times, the number of species multiplied explosively.

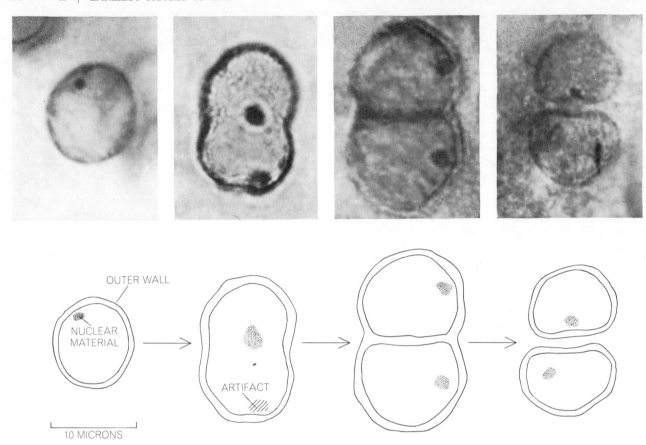

CELL DIVISION of a eukaryotic organism, a green alga of the genus *Glenobotrydion*, appears in the billion-year-old cherts of the Bitter Springs formation of Australia. J. William Schopf reconstructed the event (*drawings*), working with fossils that preserved individual organisms at different stages of division (*micrographs*). Existence of such biologically advanced algae in Bitter Springs times is proof that the evolutionary threshold leading to sexual reproduction and genetic variability had been crossed even earlier.

being filled. Indeed, the traditional concept of a clearly delimited Precambrian boundary may soon disappear from reconstructions of the history of terrestrial life.

Two final comments may be of interest to those who, like myself, are stimulated by the search for first causes no matter how often the search goes unrewarded. Some biologists now suggest that the organelles characteristic of the eukaryotic cell may once have been independent organisms that somehow came to live symbiotically inside larger host cells. It is not clear whether or not one of the host cell's responses to the presence of such a guest was a regrouping into a nucleus of the genetic material formerly scattered throughout the host's cytoplasm. If symbiosis was indeed the first step toward evolution of the eukaryotic cell, it may be that certain of the Gunflint organisms currently being studied show initial steps in this direction [*see middle illustration at right on page 50*]. A fascinating account of this transformation of various bacterial and algal organisms into components of the eukaryotic cell is given by Lynn Margulis of Boston University in a recent book. Mrs. Margulis' argument rests on biological grounds; the increasingly detailed Precambrian fossil record supports her thesis.

Astronomers and physicists have found evidence in recent years that molecules such as the hydroxyl radical (OH), carbon monoxide (CO), ammonia (NH_3), hydrogen cyanide (HCN) and formaldehyde (HCHO) are formed in the "empty" reaches of space. So far the ultimate in interstellar chemical complexity is represented by organic material that some investigators maintain they find in a peculiar class of meteorites known as carbonaceous chondrites. Until recently no carbonaceous chondrite had been recovered under circumstances that completely ruled out the possibility of accidental contamination by terrestrial organic material. As a result nagging questions about the chondrites have remained, particularly with regard to such biologically important molecules as amino acids (which are a *sine qua non* of terrestrial life). Recently, however, using stringent laboratory procedures to analyze carefully documented samples of the Murchison chondrite (which fell in southern Australia in September, 1969), Ponnamperuma and his associates at the Ames Research Center (in collaboration with Carleton B. Moore of Arizona State University and Ian R. Kaplan of the University of California at Los Angeles) seem to have proved the existence of extraterrestrial amino acids. Not only is the quantity of amino acids in the Murchison meteorite surprisingly high but also certain of them are unknown in terrestrial organisms and hence cannot be contaminants from the soil.

This finding opens up a new world of chemical evolution: a world of random synthetic processes not on the earth but in space, including the extraterrestrial formation of bodies (of which the carbonaceous chondrites seem to be fragments) that are rich in organic materials. The finding brings us back to the discussion at the beginning of this article. The chemical evolution of organic matter, the prelude to biogenesis on the earth, seems to have occurred elsewhere in the solar system or outside it. For knowledge of the earliest stages of organic evolution the biologist and paleontologist must rely on the chemist and astrophysicist.

Pre-Cambrian Animals

by Martin F. Glaessner
March 1961

*Until recently the fossils of organisms that lived earlier
than the Cambrian period of 500 to 600 million years
ago were rare. Now a wealth of such fossils has been
found in South Australia*

The successive strata of sedimentary rock laid down on the earth's crust in the course of geologic time preserve a rich record of the succession of living organisms. Fossils embedded in these rocks set apart the last 60 million years as the Cenozoic era—the age of mammals. The next lower strata contain the 150-million-year history of the Mesozoic—the age of reptiles. Before that comes the still longer record of the Paleozoic, which leads backward through the age of amphibians and the age of fishes to the age of the invertebrates. Then, suddenly and inexplicably, in the lowest layers of the Paleozoic the record of life is very nearly blotted out. The strata laid down 500 to 600 million years ago in the Cambrian period of the Paleozoic era show a diversity of primitive marine life: snails, worms, sponges and the first animals with segmented legs, the trilobites and their relatives. But the record fades at the bottom of the Cambrian. The greater part of the journey to the beginning of sedimentation, at least another 2,000 million years, still lies ahead. Yet apart from algae and a few faint traces of other forms, the Pre-Cambrian strata have yielded almost no fossils and have offered no clues to the origin of the Cambrian invertebrates.

The geological record necessarily becomes more obscure the further back it goes. The older rocks have been more deeply buried and more strongly deformed than the younger rocks. They have undergone longer exposure to the heat and pressure and the mineralizing solutions by which fossils are commonly destroyed. One can find fossils, however, in greatly deformed younger sediments, including metamorphic rocks, which have been even more thoroughly reworked by geologic processes than some older rocks. What is more, no rock-de-

forming process affects the entire surface of the globe. Some Pre-Cambrian formations have escaped extreme alteration, just as the lower Cambrian rocks are altered in some places and not in others.

The abrupt termination of the fossil record at the boundary between the Cambrian and the Pre-Cambrian has appeared to many observers as a fact or paradox of decisive importance. They have advanced many different explanations for the mystery, from cosmic catastrophes to the postulate of an interval

of time without sedimentation; from the assumption of a lifeless ocean to the thought that all Pre-Cambrian organisms may have lived at the surface of the sea and none on its bottom, or all in the deep sea and none on its shores.

The need for such speculation has at last been obviated by the discovery in the Ediacara Hills in South Australia of a rich deposit of Pre-Cambrian fossils. The first finds at this site were made in 1947 by the Australian geologist R. C. Sprigg. In sandstones that were

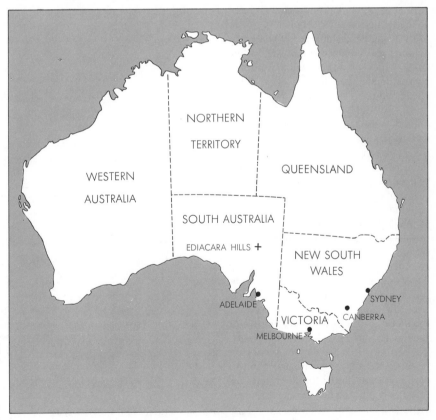

LOCATION OF PRE-CAMBRIAN FOSSIL BED is in the Ediacara Hills (*cross*), some 300 miles north of Adelaide. The geologist R. C. Sprigg made the first discoveries there in 1947.

thought to belong to the lowest strata of the Cambrian he came upon varieties of fossil jellyfish. Sprigg's find was followed up by other geologists and by students under the leadership of Sir Douglas Mawson, who found some plantlike impressions that appeared to be algae. Some time later two private collectors, Ben Flounders and Hans Mincham, brought to light not only large numbers of presumed fossil jellyfish but also segmented worms, worm tracks and the impressions of two different animals that bear no resemblance to any known organism, living or fossil. These discoveries prompted the South Australian Museum and the University of Adelaide to undertake a joint investigation of the region. Re-examination of the geology now showed that the fossil-bearing rocks lie well below the oldest Cambrian strata. This finding, taken together with the nature of the fauna represented in the fossils and their evident relationship to certain fossils discovered in South Africa before World War I and more recently in England, established that all these fossils date from the Pre-Cambrian era.

To date some 600 specimens have been collected in the Ediacara Hills. The fauna include not only jellyfish representing at least six and probably more extinct genera but also soft corals related to the living sea pens; segmented worms with strong head shields; odd bilaterally symmetrical animals resembling certain other types of living worm; and the two animals that look like no other living thing.

All the Ediacara animals were soft-bodied; none had hard shells, and their soft tissues were strengthened by nothing more than spicules: needles of calcium carbonate that served as a primitive support. All, of course, lived in the sea, some fixed to the bottom, some crawling and others free-floating or swimming. Their preservation is due to rather unusual, though not unique, conditions. The animals lived or were stranded in mud flats in shallow waters. Their impressions or their bodies were molded in the shifting sands that washed over the flats and were preserved as molds or casts in sandstone, mostly on the lower surfaces of sandstone beds. The resulting rich and varied assemblage of fossil animals gives the first glimpse of the marine life of the Pre-Cambrian era. It is a glimpse not merely of several types of animal but also of an association of creatures living together in the sea.

The soft-bodied nature of these fossils

justifies the characterization of the Pre-Cambrian as the "age of the jellyfish." The term jellyfish, however, applies to a number of highly diverse and only remotely related forms, of which the most common belong to the coelenterate phylum. These are animals that alternately take the free-swimming medusoid, or jellyfish, form and the sedentary polyp form. Sprigg concentrated on the medusoid jellyfish among his finds. He arranged some of them in two classes and four orders that have living representatives and placed the more commonly occurring specimens, which he called *Dickinsonia*, in a more problematic position with respect to living forms. But further study has indicated that none of the Pre-Cambrian medusae can be tied with any confidence to living orders, suborders or families.

Greater interest perhaps attaches to the leaf- or frondlike stalked fossils that Sprigg apparently took to be algae. The stalk is some 12 inches long and three-quarters of an inch wide. The body measures up to nine inches long and four and a half inches wide; it is characterized by transverse ridges branching off from either a tapering median field or a median zigzag groove and divided in turn by longitudinal grooves [*see bottom illustrations on page 61*]. No living algae display such structures. The true nature of these fossils appears in specimens

that show the impressions of spicules in the stalk and along the lower edges of the side branches. These suggest the spicules of otherwise soft alcyonarian corals living today and identify the fossil fronds as animals of the coelenterate phylum rather than as plants.

One group of modern corals—the sea pens (*Pennatulacea*)—has a similar arrangement of spicules, along with the stalk and side branches. Thus the fossils appear to be sea pens, which are normally rare in the geological record. The differences between the Pre-Cambrian sea pens and the modern animals are remarkably small, considering the 600 million years of evolution that separate them. In the living sea pens the frond is either deeply dissected into movable side branches or it forms an entire plate-like body. In the fossil the lateral ridges are separated by furrows and not by open slits. Coral polyps that occupy the surfaces of the fronds and stalks in modern sea pens are so small that they would not be apparent in the rather coarse sandstone casts of the fossils.

The Australian frond fossils are similar to those discovered before World War I by German geologists in Southwest Africa. Those fossils were named *Rangea* and *Pteridinium*. The Pre-Cambrian fossil discovered recently in England and named *Charnia masoni* also re-

PRE-CAMBRIAN SEASHORE AREA, reconstructed from fossils found in South Australia, supported several types of animal. Some are shown stranded in dried-up mudholes (*A*) and on sand of beach, where they were fossilized. Others appear (*lower left*) in sand and

sembles certain of the fossil Australian sea pens. The English fossil seems to possess a circular disk with concentric ribs at the end of the stalk opposite the frond. Although the connection between these two structures is uncertain, it may be that this fossil represents the two alternating coelenterate forms, that is, the free-swimming medusa and the branching colony of small polyps that remains fixed to the ocean bottom. In this case one might speculate that the Pre-Cambrian sea pens grew from free-swimming, solitary medusae. But this is as yet pure guesswork about the reproductive processes of long-dead organisms. Further discoveries may prove or disprove the connection between fronds, stalks and disks.

The most spectacular finds in the Pre-Cambrian strata of South Australia were small annelid worms named *Spriggina floundersi* after their discoverers, Sprigg and Flounders. They had a narrow, perfectly flexible body up to one and three-quarter inches long, a stout horseshoe-shaped head shield and as many as 40 pairs of lateral projections (parapodia) ending in needle-like spines. A pair of fine threads projected backward along the sides of the body from the lateral horns of the head shield, and another thread probably grew from the segment behind it [*see illustration at top right p. 60*]. Although such worms no longer

exist, they resemble the living marine *Tomopteridae*, which have similar but wider heads, transparent narrow bodies and parapodia ending in flat paddles [*see illustration, page 62*]. These modern worms, because of their special paddle adaptation to the free-swimming life, had not been considered primitive or of ancient origin. It now appears, however, that they are directly descended from extremely ancient forms. The shape of the head of the Pre-Cambrian worms suggests the possibility of a relationship between them and the arthropods, such as the now extinct trilobites, which first appear in large numbers in the Cambrian. All of these later animals represent a considerable advance over the primitive anatomical organization of the coelenterates.

The most common fossil at the Ediacara site, the *Dickinsonia*, represented by more than 100 specimens, may also be related to living worms. The fossilized bodies are quite remarkable. They are more or less elliptical in outline, bilaterally symmetrical and are covered with transverse ridges and grooves in a distinctive pattern. The size of the bodies and the number of ridges vary so much that Sprigg attempted to distinguish species by counting the ridges. One recently discovered specimen has some 20 ridges; a larger one may have had as many as 550. The animals range in

length from a quarter of an inch up to two feet. The numerous impressions of wrinkled and folded-over specimens indicate that all were soft-bodied, for there are none of the fractures that would be apparent if the creatures had possessed shells. These animals vaguely resemble certain flatworms living today. There is also one genus of annelid worm with a strikingly similar pattern of ridges formed by extensions of its parapodia. This similarity proves little or nothing, especially since no traces of eyes, legs or intestines are preserved in the fossils, but it provides some hope of finding out what these strange creatures were.

There is possibly less hope of placing in the family tree of the animal kingdom the two completely novel forms discovered in the Ediacara Hills. One had a shield- or kite-shaped body with a ridge that looked like an anchor. It was named *Parvancorina minchami* [*see illustration at top right p. 61*]. The first specimen was tiny, but others found later measure up to one inch in length. Some show faint oblique markings within the shield on both sides of the mid-ridge, as if the animal had had legs or gills underneath. Here again folded and distorted specimens occur, proving that their bodies were soft.

The other entirely new creature is even stranger. Named *Tribrachidium*, it has three equal, radiating, hooked and

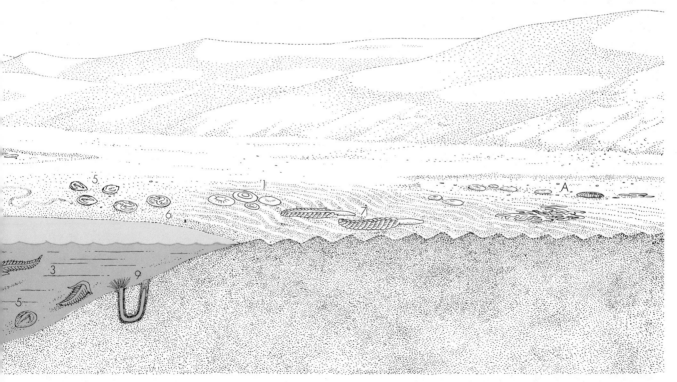

water as though seen in an aquarium. They are jellyfish-like creatures (*1*); the wormlike *Dickinsonia* (*2*); the segmented worm *Spriggina floundersi* (*3*) and worm trails (*4*); *Parvancorina* (*5*), which resembles no other known animal; *Tribrachidium* (*6*), another unknown type; the sea pens *Rangea* and *Charnia* (*7*); hypothetical algae and sponges (*8*), and a worm in a sand burrow (*9*).

tentacle-fringed arms. Nothing like it has ever been seen among the known millions of species of animals. It recalls nothing but the three bent legs forming the coat of arms of the Isle of Man.

Considered together, the South Australian fossils suggest a rough and incomplete picture of conditions in the late Pre-Cambrian. Of course, such a group of fossils constitutes no more than a small, biased sample of the life of the time. Animals buried together in slabs of sandstone did not necessarily live together. Some, if they really are medusae, were floating in the sea. Others, like the annelid worm *Spriggina,* with its numerous legs and sinuously curving body, were free-swimming. *Dickinsonia* was probably also a free-swimming form, apparently along with *Parvancorina.* Scattered miniature treelike stands of sea pens, waving their flexible

fronds, must have covered parts of the shallow sea floor. Elsewhere earthworm-like annelids, which have left only their tracks, crawled over and through the sediment, feeding on the decaying organic matter in it. Other worms inhabited the U-shaped burrows that have been found, consuming tiny creatures in the sediment and possibly also marine plankton, which left no traces in the rock. The fixed, three-rayed spread of tentacles of the strange *Tribrachidium* may be similar to the plankton-fishing structures around the mouth of the living brachiopods (lamp shells), bryozoa (lace corals) and some worms. If that is correct, *Tribrachidium* may have been a bottom dweller, possibly occupying low, conical, ridged cups, of which a few impressions have been found.

Bundles of impressions of needle-shaped spicules also occur in the Edia-

cara strata. Since spicules are characteristic of sponges, these sessile, bottom-dwelling animals may have been present. Snails and small crustaceans, as well as various protozoa (radiolarians and foraminifera), may also have existed at that time, but they would have been too small or too fragile to be preserved. Plant life likewise left no traces here.

The worm tracks are the only fossils indicating without doubt that the animals lived where their remains are found. Thus the *Spriggina* worms, the *Dickinsonia* and *Parvancorina* may have lived near or on the sedimentary beds. They are represented by individuals varying in size and growth stage, which indicates accidental death rather than transport from afar and later burial. On the other hand, the jellyfishes were probably stranded and the soft corals torn from their anchorage before they came

PRE-CAMBRIAN FOSSILS preserved in sandstone are seen in these eight photographs. This is *Dickinsonia costata,* shown actual size.

SEGMENTED WORM *Spriggina floundersi,* shown about twice actual size, resembles certain segmented worms living today.

JELLYFISH *Spriggia annulata* is one of the many types of this organism that have been found. The fossil is very slightly enlarged here.

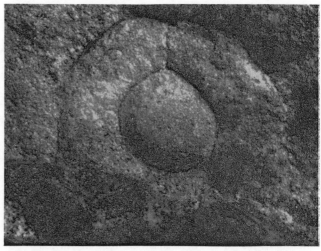

ANOTHER JELLYFISH, *Medusina mawsoni,* is shown nearly three times actual size. Jellyfish were the first fossils found.

to rest on the bottom.

The sandstone in which the fossils are found shows ripple marks and other evidences of currents, which would have had to be rather strong to transport the coarse grains of sand. Thus it is difficult at first to see how imprints of delicate, soft-bodied creatures could have been preserved. Careful study of the fossils has yielded an explanation. Only a very few of the animals came to rest on the shifting sand. Most of them came down on mud flats or on patches of fine clay that settled out of the water during calmer periods. Some of the mud patches dried out, possibly between tides, and developed deep cracks. The next high tide or shifting current covered them with a layer of sand. The lower surfaces of such sandy layers preserved the clay surfaces in the form of perfect casts,

showing the wrinkles in the clay and the cracks formed by drying as well as the shapes of the animals stuck in the clay. The sand grains were cemented by silica solutions and turned to quartzite in the transformation from soft sediment to hard rock. The clay changed to thin slatelike streaks of the mineral sericite and was compacted almost beyond recognition. Since the sericite inclusions are small and irregular, the rock does not split along their surfaces as slate would. Only the slow, natural weathering in the arid climate of South Australia can open up the rock along the vital sericitic partings where the fossils occur. Slabs of quartzite of all sizes remain in place, projecting from the hillsides until they break off. They often turn over when moving downhill and their lower surfaces become exposed to the infre-

quent rain. Then the weathering causes them to reveal their wonderful riches of Pre-Cambrian animals. But if the rocks are not collected, the fossils are ultimately worn away by the weather and by the sand drifting in on the wind from the adjoining desert plains.

The age of the fossil-containing rocks cannot be determined directly in years because it does not contain radioactive minerals suitable for dating. Fortunately in the Ediacara Hills one can follow the stratification in unbroken sequence upward until the first undoubtedly Cambrian fossils are reached in dolomitic limestone 500 feet above the Pre-Cambrian level. These fossils in the limestone are typical of the lowest Cambrian strata elsewhere and are quite unlike the strange fossil organisms in the quartzite

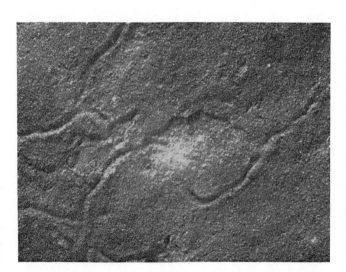

WORM TRAILS, approximately actual size, provide proof that fossilized worms lived in the area where they were preserved.

UNKNOWN TYPE OF ANIMAL, *Parvancorina minchami*, here enlarged nearly three diameters, resembles no other known organism.

SEA PEN *Rangea arborea* left this imprint, shown here twice actual size. The fossil resembles some of the living sea pens.

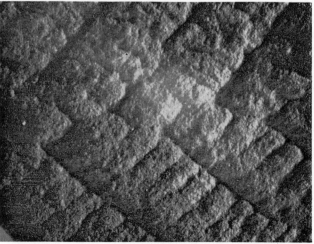

ANOTHER SEA PEN, *Charnia*, is shown actual size. Viewing photographs upside down may give fuller idea of animals' appearance.

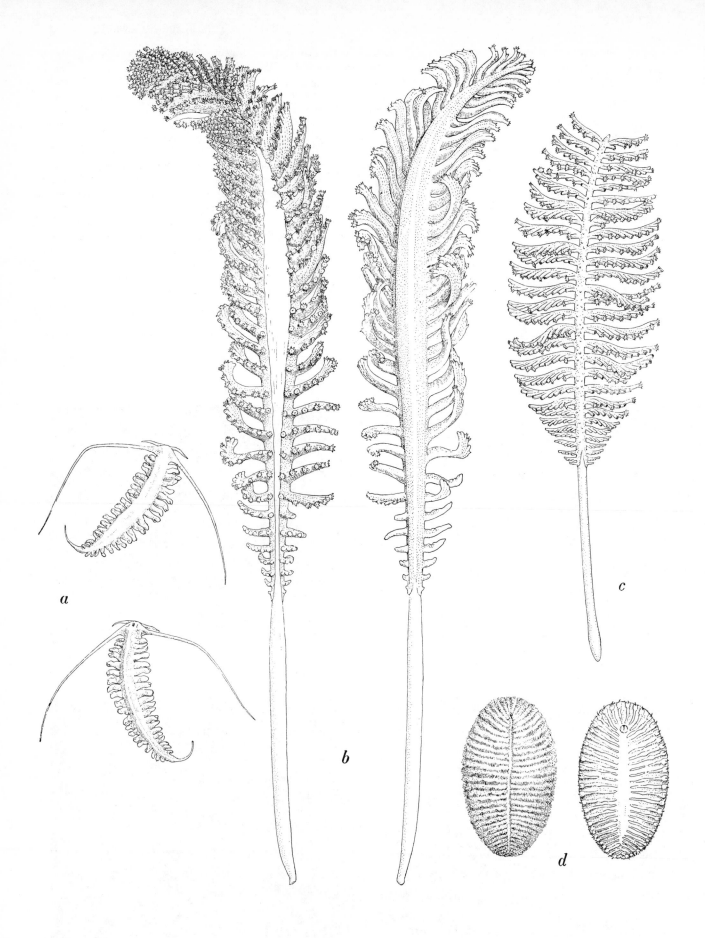

FOUR LIVING ANIMALS that resemble some of the Pre-Cambrian fossils from South Australia are a segmented worm, *Tomopteris longisetis* (*a*), seen in dorsal and ventral views; sea pens *Pennatula rubra* (*b*), shown front and back, and *Pennatula aculeata* (*c*); and the worm *Spinther citrinus* (*d*), which looks like the many specimens of *Dickinsonia* in the Pre-Cambrian rocks.

below. The quartzites higher up in the Cambrian strata do not contain any fossils of the type now known from the Pre-Cambrian, and the dolomites and limestones lower down contain no Cambrian fossils. From this distribution of fossils in the rocks it can be judged that the lack of shells and hard skeletons (other than the spicules) in the Pre-Cambrian animals was not due to any factors in the physical environment. The development of shells in the Cambrian was not a result of a sudden change in the habits or habitats of the animals. Rather, shells appeared as a step forward in biochemical evolution. Calcium metabolism underwent a change that produced hard shells and other skeletal material, providing the protection and mechanical support so important to the more advanced animals.

This is as far as the paleontologist and geologist can take the story today. The biochemist and physiologist may see in it a lead to experimentation that could well open a new chapter in the story of fundamental research in evolution.

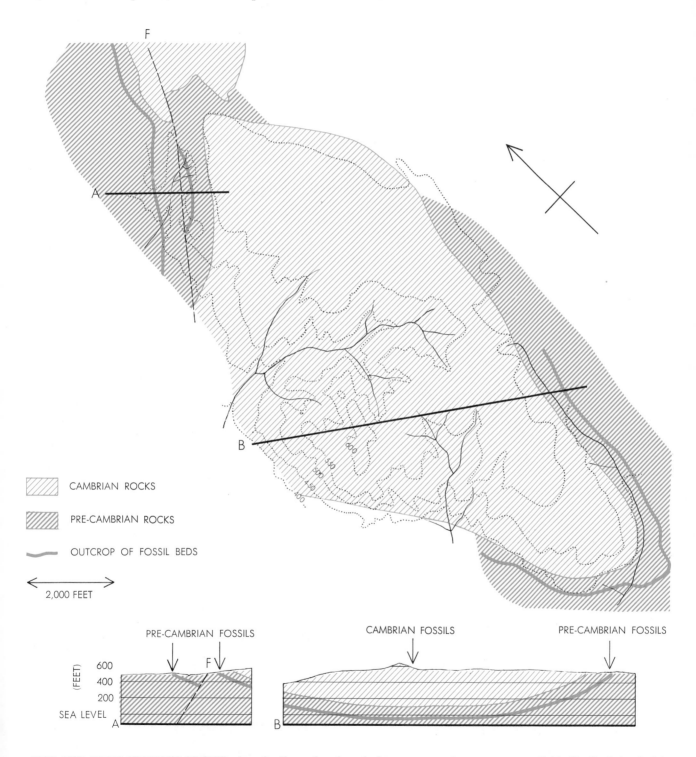

CAMBRIAN ROCKS

PRE-CAMBRIAN ROCKS

OUTCROP OF FOSSIL BEDS

2,000 FEET

PRE-CAMBRIAN FOSSILS

CAMBRIAN FOSSILS

PRE-CAMBRIAN FOSSILS

(FEET) 600 400 200 SEA LEVEL

A   F

B

MAP AND CROSS SECTIONS OF SITE where fossils are found show relative positions of Cambrian and Pre-Cambrian rocks. Lines *A* and *B* indicate locations of cross sections *A* and *B* (*below map*). Broken line *F* is a fault that has caused part of Pre-Cambrian fossil bed to move, creating two outcrops (*left*). Pre-Cambrian bed is under Cambrian rocks, except where its edges come to surface at periphery of Cambrian area. Dotted lines indicate contours. Part of the region shown here has been included in a fossil reserve.

III

HOW FOSSILS OCCUR AND WHAT THEY TELL US

III

HOW FOSSILS OCCUR
AND WHAT THEY TELL US

INTRODUCTION

The fossil record in post-Precambrian rocks that accumulated during the so-called Phanerozoic interval of geologic time is abundant and diverse in aggregate, but at any one time or place it may be quite patchy and badly biased toward one kind of organism or another. The reasons for this are rather simple. On the one hand, during the great span of time represented by the Phanerozoic eon, which includes the Paleozoic, Mesozoic, and Cenozoic eras, there have been innumerable opportunities for animals and plants to become buried and preserved in accumulating sediments. On the other hand, organisms that have mineralized skeletons and live in areas where active sedimentation occurs have a much greater potential for being fossilized than soft-bodied creatures living in places undergoing erosion. So, while it is true that the fossil record contains a marvelous variety of past life, the sampling of that life over the ages has been sporadic and incomplete. Despite this, paleontologists nevertheless can infer much about ancient organisms and their environments from their fossil remains and from the rocks in which they are imbedded.

In this part of the reader, we consider the different ways in which organisms become fossilized and the various kinds of information they provide us about the history of the earth—physical as well as biological. Some fossils, such as calcareous algae, shelly invertebrates, and vertebrates, occur as more or less unaltered hard parts. That is, after death, the organism may be buried; decomposition of the soft tissue leaves behind part or all of the mineralized skeleton. The relative resistance of the hard parts to chemical decomposition and physical disintegration permits them to survive subsequent compaction and lithification of the surrounding sediment with little or no change. Of course, percolating water in the enclosing sediment or rock can introduce chemical solutions that may dissolve and replace the original mineral matter. Yet the detailed structure of the fossil is often preserved. Unaltered or chemically replaced hard parts include such things as fish scales, crab legs, horse teeth, alligator scutes, dinosaur vertebrae, clam shells, stony corals, echinoid spines, algal crusts, or human jaw bones.

Sometimes percolating waters will deposit mineral matter—usually silica or calcium carbonate—around and within the tissues (hard or soft), thereby "embalming" them with resistant material. Bones of vertebrates, for example, are commonly found with material precipitated within the natural pore spaces. Plants, too, may be thoroughly encased in silica—not only the outer parts of the plants but even the fine interstices within.

But what about animals and plants that do not have resistant hard parts or that are not protected by later precipitated mineral matter? Organisms without any hard, mineralized skeletal parts most often just rot or wear away before being buried. Or, if such organisms are buried, microorganisms and circulating water decay and oxidize the remaining organic matter. In rare instances, however, the soft tissues may be preserved as compressed carbonized films, because the surface on which the dead organism came to rest lacked sufficient oxygen to support bacterial decomposition or oxidation. Thus, anaerobic muds may contain the soft-bodied remains of organisms that drifted down from aerated waters above, yielding rich fossiliferous layers in black shales many millions of years later.

At other times, soft-bodied organisms may be buried quickly in finer-grained sediments, which then lithify enough to mold the shape of the organism. Although later removal of the soft tissues occur, the form of the organism may be retained within the surrounding rock. Hard parts, too, can be dissolved away from the enclosing rock but leave behind a detailed mold of the original material.

Besides these direct kinds of evidence of life—unaltered and altered hard parts, compressions, and molds—we can also find indirect evidence of animals and plants: tracks, trails, burrows, and root markings. Although these so-called trace fossils are sometimes difficult to identify in terms of the parent organism that made them, they do have the advantage of being found in place. That is, whereas a shell or tooth or bone can be transported some distance from where its owner lived and died before being buried, trace fossils, by their very nature, cannot be transported. Hence, when we find them, we can be sure that that is where the organism lived. Moreover, such traces can often tell us something about the way the organism functioned or behaved. For example, some invertebrate burrows clearly indicate that the animal was moving through and feeding on the sediment; others tell us that the animal was sitting in a burrow but feeding outside it in the overlying water.

Runcorn's "Corals as Paleontological Clocks" explains how the hard, calcareous skeletons of solitary and colonial corals record daily, monthly, seasonal, and yearly growth increments. Admittedly, it is not easy to be certain which growth lines record which temporal interval. Yet it appears that certain lines occur in sets of 400 for Devonian corals, which, if daily growth increments, agrees with theoretical calculations that the number of days in the Devonian year should be about 400. The reason for more days per year in the middle Paleozoic results from the slowing of the earth's daily rotation due to the tidal friction generated by the moon. Over geologic time, the tides have slowed the earth's rotation (one rotation equals one day) so that in a course of a year, or one revolution about the sun, there are fewer and fewer rotations. As Runcorn further points out, other theories about the history of the earth-moon system as well as contending theories about an expanding, contracting, or stable earth can also be tested. Runcorn's article thus shows how paleontology can contribute to geophysical and astrophysical concepts. (See Clark, 1974, for a recent review of this subject.)

Study of growth lines in hard part-secreting invertebrates can also be used to determine if the organism lived in water shallow enough to experience tides, if there were occasional disturbances in the environment—storms, sharp changes in temperature or salinity or turbidity—causing abrupt changes in growth, or if there were regular seasonal changes that encouraged growth or restricted it. In short, because hard parts provide a direct record of an organism's growth, influences on variations in that growth might well be recorded within it. Despite the difficulty of observing just what the growth lines are and to what they might be due, this is nevertheless a promising line of paleontological research that may provide, if not definitive answers, at least contributing evidence for one hypothesis or another.

"The Petrified Forests of Yellowstone Park," by Dorf, is an example of how chemical solutions associated with volcanism precipitated silica within and around the cellulose woody tissue of Eocene trees, especially stumps, trunks, and logs. Such petrification preserved successive forests that flourished and were buried periodically by volcanic ash and breccias from nearby volcanoes. More delicate plant parts, such as leaves, are preserved as compressions and impressions in the fine-grained ash that quickly inundated the Eocene landscape during volcanism. Not only do the various kinds of fossil plants indicate what the local Eocene climate was like, but growth rings in the stumps tell us the age of the individual forests when buried by volcanic debris. Rate of volcanic sediment accumulation can also be calculated from these observations.

The article by Brues, entitled "Insects in Amber," shows us how even such delicate organisms as ants, wasps, and flies can be exquisitely preserved if the conditions are right. In particular, when the saps and resins from trees exude along their trunks and branches they can trap and engulf insects. The enclosing material, called *amber,* is fairly resistant to subsequent chemical and physical influences. The insect inside decomposes, but the finest details of the animal's body are molded by the amber and colored by the residual pigments of the chitinous body covering. Amber deposits, therefore, provide much information about the abundance and variety of past insect populations, their morphology, and their evolutionary history. Although we lack a continuous record of insect evolution, enough facts are known to observe some of the major trends, which Brues reviews in his article.

"Fossil Behavior," by Seilacher, discusses how we can learn about the way marine invertebrates behaved in their environment from their preserved traces in sedimentary rocks. Such tracks and trails record foraging for food on and within the sea floor; simple movements across the sediment, pursuing who knows what errands, or burrowing down into sands and muds either for shelter or for nutrient-rich layers. Sequences of similar traces through geologic time exhibit trends in more efficient use of sediments for their food content or in burrow construction. Sielacher also notes that certain types of traces tend to be associated with particular kinds of habitats. (See the recent volume edited by R. W. Frey, 1975, for other examples and more details about trace fossils.)

Today within paleontology, there is a whole field of "ichnology" that studies trace fossils. Much of the impetus for the development of this research in North America came from the efforts of Seilacher himself. Trace fossils not only add information about faunas, environments, and evolution in rocks that contain hard-part fossils but also extend our observations into many rocks that lack such hard parts, and so for years were thought to be "unfossiliferous." As with the Precambrian, the fossil record of the Phanerozoic has been greatly expanded as we learn to look for these very different kinds of fossils.

The last paper in Part III considers the rich fossil record of microscopic shells found in deep-sea sediments. "Micropaleontology," by Ericson and Wollin, reviews the kinds of organisms leaving such shells, including plants, such as siliceous diatoms and calcareous coccolithophorids, and animals, such as siliceous radiolarians and calcareous foraminiferans. These fossils are abundant, very small, and sufficiently diverse over the last hundred million years that they prove to be very useful in determining the geologic age of deep-sea deposits and the environmental conditions of the associated oceanic water masses.

Two very important results have come from the study of such microfossils. The first is that oceanic sediments range in age from the Jurassic to the recent and that, in general, as one proceeds toward the continents and away from midoceanic ridges where new crust is formed, ocean sediments become older and older. This observation supports the concept of sea floor spreading away from the oceanic ridges and the subduction of sea floor beneath the continents. Subduction has apparently consumed any and all pre-Jurassic oceanic sediment. The second important result is the documentation of changing global climates over the last several hundred thousands of years, change that periodically culminated in glaciation in the high latitudes. The patterns of climate change further suggest that we may experience another significant global cooling some 20,000 years from now. Rather than being at the end of the glacial cycles, we are very likely simply between cycles.

Ericson and Wollin's article illustrates well the useful but not infallible methodology of studying present-day organisms to infer past environmental conditions, another example of the "present as a key to the past." Thus, by knowing how extant species of plankton, or floating marine microorganisms, are controlled by the temperature and salinity of the surrounding waters, we can determine past oceanic conditions by examining what species are present at successively older levels of deep sediment cores. We can thereby derive climatic curves going back well into the dawn of the Pleistocene. And, if we care to, we can project these curves forward in time to estimate how the oceans might change in the future. By doing so, we use the past as a key to the future. Other earth scientists find the Pleistocene very helpful in seeing how present-day processes and phenomena become recorded in the geologic record, going from the recent to the near past and then extrapolating forward in time to predict what the near future might be like. In this way geology, a historical science, is taking on some of the attributes of a predictive science.

SUGGESTED FURTHER READING

Clark, G. R., III. 1974. "Growth Lines in Invertebrate Skeletons," *Annual Review Earth and Planetary Science*, vol. 2, pp. 77–99. A well-referenced review article explaining the different kinds of information to be had from the study of growth lines in fossils.

Cline, R. M., and Hays, J. D., ed. 1976. *Investigation of Late Quaternary Paleoceanography and Paleoclimatology*. Boulder, Col.: Geological Society of America Memoir 145. A technical—but largely understandable to the beginner—series of scientific reports about how different groups of microfossils in deep-sea cores reveal the nature and distribution of ancient climates during the last half million years or so.

Frey, R. W., ed. 1975. *The Study of Trace Fossils*. New York: Springer-Verlag. A compendium of articles describing the nature and interpretive value of a wide variety of animal and plant traces found in marine, freshwater, and terrestrial environments.

Raup, D. M., and Stanley, S. M. 1971. *Principles of Paleontology*. San Francisco: W. H. Freeman and Company. In a dozen chapters, these paleontologists cover a broad spectrum of topics from the identification and description of fossil specimens, interpretation of hard-part morphology, and stratigraphic correlation to paleoecology, evolutionary rates, and patterns of evolution. A revised edition of this basic text is due to be published soon.

7

Corals as Paleontological Clocks

by S. K. Runcorn
October 1966

Banding on certain corals evidently represents annual, monthly and daily growth. Ancient corals thus provide clues to the length of the year in past eras and to changes in the earth's rotation

The astronomers, geophysicists and other investigators whose concern is the origin and evolution of the earth are handicapped by a shortage of evidence. The events of interest to these workers occurred in times so distant that even geological records are seldom available. As a result the theories that have been advanced about such matters as the origin of the continents are largely conjectural. Moreover, as might be expected in the circumstances, the theories differ considerably and therefore are highly controversial.

An example of the kind of information that would help to overcome the handicap is a reliable measurement of the length of the day, that is, the speed of the earth's rotation on its axis. It is clear that the length of the day has increased slowly throughout geologic time; the earth's rotation has been slowed by the friction of the tides and may also have been changed slightly by internal processes. Hence the number of days in the year has decreased. Calculations on the basis of tidal friction alone indicate that the year had about 428 days at the beginning of the Cambrian period some 570 million years ago and about 400 days in the middle of the Devonian period some 370 million years ago [*see illustration on page 74*]. If a "clock" could be found that had recorded the days of ancient geological periods, it would be possible to arrive at a more precise measurement of the number of days in the year and so to obtain evidence about the earth's rotation and the factors affecting it.

Such a fossil clock may be at hand in certain corals. These organisms have long been known to have distinct bands that represent annual growth. The bands are themselves made up of narrower bands that seem to represent

monthly growth and are probably related to the tides and the monthly cycle of the moon. The intriguing possibility now under investigation is that the still finer ridges or bands found in some of the corals represent daily growth. If this is the case, a coral that could be accurately assigned to a particular geological period (by radioactive dating or the evidence of stratigraphy) would provide a measurement of the number of days in the year at that time. Corals thus hold the promise of being a powerful geophysical tool.

The corals that are of interest in this connection are those broadly known as stony because their skeleton consists of hard calcium carbonate. The portion of the coral on which the investigation focuses is the epitheca, which is the external layer on the lower part of the conical skeleton [*see illustration on page 72*]. Geologically the corals under study fall into three groups. Two are from the Paleozoic era and are known as rugose (wrinkled) and tabulate (having tabulae, or horizontal partitions). The third, which is a successor of the others, has existed from Mesozoic time to the present; it is called scleractinian, a name from the Greek word meaning "hard."

Whenever the epitheca is present, it shows fine banding, provided that the surface has not been abraded by wear or accident. There are between 20 and 60 bands per millimeter. Numerous observers have noted these bands and have regarded them as growth increments representing the periodic deposition of calcium carbonate by the coral. The first to suggest that the bands were daily growth increments was John W. Wells of Cornell University.

Wells reported in 1963 that his count

of the fine bands within the annual bands in several corals dating from the middle of the Devonian period ranged between 385 and 410 and averaged nearly 400. The average agreed closely with the number of days in the Devonian year as obtained from calculations of the effect of tidal friction. Similarly, the average of 380 bands obtained by Wells with corals from the more recent Carboniferous period showed a close correlation with the number of days in the year as calculated for that time. Wells also reported that he had counted bands on some contemporary corals, for which the annual rate of growth is fairly well known, and had found that "the number of ridges on the epitheca of the living West Indian scleractinian *Manicina areolata* hovers around 360 in the space of a year's growth."

Wells's suggestion that the ridges making up the fine banding represent daily growth gains support from some work with modern corals by Thomas F. Goreau of the University of the West Indies. Goreau showed that in these living corals the secretion of calcium carbonate varies between night and day. Curiously there has been little other work on the growth of corals. It is a field waiting to be explored.

Soon after Wells put forward his suggestion about daily growth, Colin T. Scrutton of the British Museum (Natural History) made the equally important suggestion that some corals have monthly bands. He reported that he had found what appeared to be such bands on some Middle Devonian corals from North America. Counting the number of fine bands in these larger groupings, Scrutton obtained an average of 30.6. That would represent the length of the Devonian month if the closely spaced

FOSSIL CORAL found in central New York by John W. Wells of Cornell University dates from the Middle Devonian period some 370 million years ago. The region in which the coral was found was then shallow ocean. The prominent bands visible on this cor- al, which is still in the matrix of other material in which it was found, represent annual growth. On closer view, such as that in the illustration below, closely spaced ridges or bands can be seen. Each such ridge is believed to represent one day of growth.

ANCIENT CORAL shows the closely spaced horizontal ridges that are thought to represent daily depositions of calcium carbonate by the organism when it was alive. This specimen of *Eridophyllum ar-* *chiaci* was found by Wells in central New York; it is of Middle Devonian age. Band counts on several Devonian corals gave 400 days as length of Devonian year; other methods give same figure.

bands were indeed indicative of daily growth. Dividing 30.6 into 399 (the number of days in the Devonian year according to computations of the effect of tidal friction), Scrutton obtained 13.04 as the evident number of lunar months in the year of Devonian time.

The lunar month is known to influence marine life, but the ways in which it does so are still poorly understood. There are nonetheless a few clues. For example, zooplankton rise closer to the surface of the ocean on dark nights than on moonlit nights. The lunar month that Scrutton found—the one that influences marine life—is called the synodical month: the interval between successive new moons. It is roughly two days longer than the sidereal month, in which the moon returns to the same position among the stars.

A variety of difficulties confront the investigator who hopes to use corals as paleontological clocks. For one thing, it is no easy matter to count the number of "daily" ridges in a "monthly" or "annual" band. The bands vary in distinctness, and it is often difficult to see where one band ends and another begins.

Furthermore, at present a worker in this field knows approximately how many days he should find in the Devonian month or year (or the month or year of other geological periods), and so the likelihood that this knowledge will subconsciously influence his count of bands is considerable. To meet this difficulty my colleagues and I at the University of Newcastle upon Tyne have been experimenting with a method that removes the personal factor by automating the process of counting the ridges in the bands. The technique resembles one used in X-ray crystallography. The epitheca and its hierarchy of growth bands are photographed; the negative is used as a diffraction grating to obtain spectra of the band frequencies. My colleague K. M. Creer at the university is attempting to count the bands with an analyzer that detects slight chemical variations between one band and the next.

Another major problem is the lack of certainty that the corals preserved in the geological record responded to the same environmental forces that modern corals do. Fortunately at this point the geophysicist can help to test the self-consistency of the numbers obtained by Wells and Scrutton. First Scrutton's number must be changed to the number of days in the sidereal month, so that both the Wells and the Scrutton numbers will be expressed in sidereal terms. The change gives a value of 28.4 days for the length of the sidereal month in Middle Devonian time.

Now, the laws of planetary motion or gravitational theory can be applied to the earth-moon system. By applying Kepler's first and second laws a formula can be derived giving the orbital angular momentum of the moon at any time in the past in terms of its present value. The formula, which makes use of the Wells and Scrutton numbers, shows that if the present orbital angular momentum of the moon is assigned a value of 101.6, the value in Devonian time was 100.

The tides cause the earth to lose angular momentum. Under the laws of planetary motion a loss of angular momentum by the earth can only be transferred to the orbital angular momentum of the moon. The formula using the Wells and Scrutton numbers indeed shows that the moon has picked up 1.6 percent of its orbital angular momentum since Middle Devonian time.

The formula is remarkably interesting because it relates an astronomical quantity—the orbital angular momentum of the moon—to a ratio of counts on corals as reflected in the Wells and Scrutton numbers. Moreover, by means of the ratio obtained from the formula it is possible to calculate the rate of loss of angular momentum by the earth due

TENTACLES

PHARYNX

DIGESTIVE CAVITY

SEPTUM

EPITHECA

BASAL PLATE

LIVING CORAL shows the relation of the epitheca, which is the outer portion of the coral's hard skeleton, to the growing organism. The living part of the coral is in color. The growth of the skeleton apparently occurs through daily depositions of calcium carbonate by the organism; the process produces bands on the epitheca at a spacing of 20 to 60 per millimeter.

to the tides. The figure obtained (3.9 × 10^{23} dyne per centimeter) is exactly in agreement with the figure obtained for the rate by modern astronomical calculations based on measurements of the longitude of the sun and moon over the past three centuries. This agreement between the geophysical and the astronomical calculations could be a lucky accident. That it is unlikely to be, however, is shown by the following argument.

Kepler's laws state that as the moon's angular momentum increases, the month lengthens and the moon moves farther away from the earth. The only factor other than tidal friction that could be affecting the number of days in the month (by changing the length of the day) is the earth's moment of inertia, a figure summing up the relations between the shape and size of the earth and the distribution of its internal substance. The earth's moment of inertia could have been altered by expansion or contraction of the earth or by changes in the distribution of mass in its core. Any such change would affect the length of the day. If the moment of inertia of the earth does not change, there will be a

fixed relation between the month and the day.

Since the figure obtained by Wells measures the length of the day and that obtained by Scrutton the number of days in the month, we can test if the earth's moment of inertia has changed. Using the paleontologist's counts of the bands on corals, one finds that the moment of inertia in Devonian time was very close to the modern value—it was less by about .5 percent. Thus it is evident that tidal friction has been the principal factor affecting the earth's rotation.

Although one should not assume a priori that the magnitude of the tidal slowing down remains the same over long periods, the double agreement between the geophysical and the astronomical findings about the effect of tidal friction strongly suggests that Wells and Scrutton were correct in identifying the coral bands with the day, the month and the year. The counts made by Wells and Scrutton were quite independent, and both men were unaware of the calculations I have put forward about the comparative effects of tidal friction and the moment of inertia on

the earth's rotation. Had the bands on the corals been caused by environmental forces other than the day and the lunar month, it would be quite difficult to imagine that such closely comparable findings would result from the counts.

Concluding, therefore, that certain corals are clocks of extraordinary value, it is pertinent to inquire what kind of geophysical questions the study of corals might answer. The questions fall into two broad categories. One concerns the evolution of the earth, the other the evolution of the moon.

It was once taken for granted that the earth originated as a molten object and has gradually cooled. Mountains were thought to have formed through the consequent contraction of its interior, just as the skin of an old apple wrinkles. The heat of the interior, which is revealed by volcanoes and by the rise in temperature as one penetrates the crust in mines and boreholes, was held to be original heat that had not been able to escape because of the earth's size. From the fundamental principle of the conservation of angular momentum an interesting consequence follows. Just as

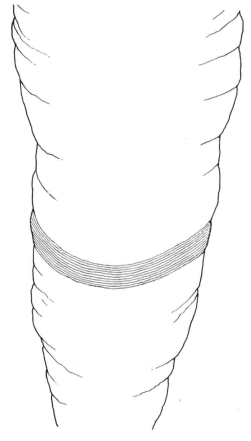

FINE GROWTH BANDS can be seen in this enlarged view of a specimen of *Holophragma calceoloides*, which dates from Middle Silurian time. The actual specimen is less than an inch long; the en-largement therefore indicates how difficult the counting of "daily" bands can be. At right a segment of 11 such bands is reproduced to show their configuration and the regularity of their spacing.

a ballet dancer spins faster as she draws in her arms, so the length of the day will shorten as the earth contracts and its mass moves toward its axis.

Since the discovery that rocks likely to be those making up the earth's interior contain radioactive elements, it has been possible to assume that the earth's heat has been generated by radioactive decay and that the earth was originally cold. Such a theory supposes that the earth originated in the accumulation of small solid fragments. This theory fits in with modern ideas on the origin of the solar system, which is thought to have started as a diffuse cloud of gas and dust that was flattened by rotation into a disk, in which the planets grew by accretion.

The earth now has an iron core with a radius about half that of the entire earth. Harold C. Urey of the University of California at San Diego has suggested that when the iron was cold, it was uniformly distributed throughout the earth; it has only slowly sunk to the center. Such an evolution would cause the day to shorten gradually, but by a much larger amount than on the basis

that the earth began hot and has cooled.

R. A. Lyttleton of the University of Cambridge has recently revived a theory, originally put forward by W. H. Ramsey of the University of the Witwatersrand, that the earth's core is not iron but rock that is in a metallic phase because of the extremely high pressures inside the earth. Lyttleton supposes this phase change is less likely to occur at low temperatures than at high ones. Assuming that the earth was originally cold, he postulates that it then had no core. The decay of disseminated radioactive elements provided energy to heat the earth, and at the center conditions eventually became right for the change of a part of the silicate rock to a denser phase, which therefore occupied a smaller volume. As the temperature continues to rise in the interior of the earth, the core grows and the earth contracts more. Lyttleton believes this theory overcomes the shortcomings of the older theories in providing for more contraction of the earth's shell.

In contrast to these theories postulating a gradually shortening day, certain other theories assume that the earth

has expanded and that the day has grown longer. An earth that has heated up will have expanded. The long ridges that run down the middle of several of the oceans offer some evidence that the earth has expanded and not contracted. The mid-Atlantic ridge and the Carlsberg ridge in the Indian Ocean have central valleys that are thought to be cracks resulting from a stretching of the crust.

Cosmological ideas have been brought into the discussion. It is conceivable that the universal force of gravity is not a constant but has been decreasing. Such an effect would cause the earth gradually to expand.

These various theories involve considerable disagreement over the length of the day in distant geological periods. Indeed, stated in terms of the length of the day in Devonian time (after allowance for the slowing of the earth by tidal friction), the theories call for a day ranging from 12 hours to 24 hours 40 minutes. Herein lies the promise of the corals. When many more corals have been studied, it should be possible to determine which of the theories is correct. Even now the boldest of the expanding-earth theories (the theory of Bruce C. Heezen of Columbia University and S. W. Carey of the University of Tasmania, who assume that the Atlantic Ocean is the result of a 40 percent expansion of the radius of the earth since Permian time, pulling apart the continents on each side of the ocean) can be disregarded. The drifting apart of the continents is now a well-established phenomenon, but its cause has been sought in ways other than expansion of the earth. One possible way is that large-scale flows in the mantle pushed the continents apart.

An account of the possible relation of corals to the origin of the moon requires some background. The height of the tides raised by the moon on the earth varies inversely as the third power of the earth-moon distance. Accordingly if the earth and moon were closer in some past era than they are now, the tidal friction would have been greater then than it is today. Assuming that the friction of the tides depends only on the earth-moon distance, and knowing the rate at which the earth loses angular momentum to the moon, one can calculate the variation of the earth-moon distance throughout geologic time.

Such a calculation was made recently by Louis B. Slichter of the University of California at Los Angeles. At the

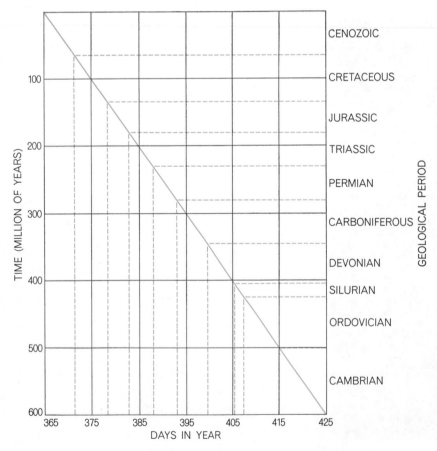

LENGTH OF YEAR in various geological periods is indicated according to computations of the effect of tidal friction, which has gradually slowed the earth's rotation and hence caused days to become longer. Counts of the bands on corals have agreed with these estimates.

CONSISTENCY OF BANDING appears in closeup of fossil corals that are widely separated in age. The specimen at left dates from the Silurian period some 415 million years ago. At right is a fossil coral dating from the Jurassic period, about 140 million years ago.

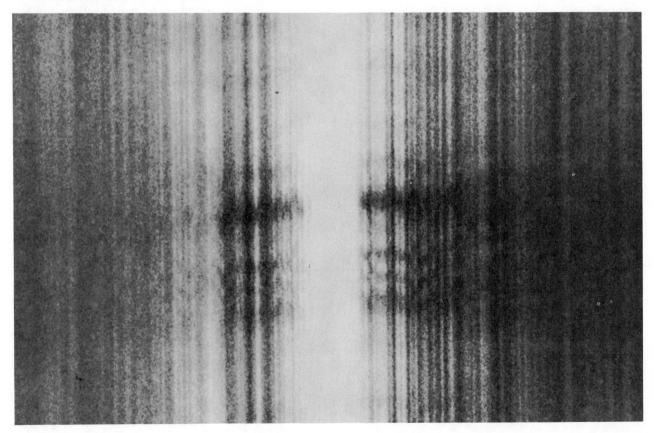

DIFFRACTION BANDS were produced by a new technique for counting bands on corals. The technique resembles X-ray crystallography. The epitheca of a coral is photographed and the negative is used as a diffraction grating to obtain spectra indicating the frequency of the bands. Thus the counting of bands is made objective, and subjective factors that could influence the count are avoided.

MODERN CORAL from the Dry Tortugas Islands of Florida is *Manicina areolata*. Its annual growth rate is fairly well known. Wells found that each increment of annual growth contained about 360 "daily" bands. The modern corals thus serve as a check of the hypothesis that the most closely spaced bands on ancient fossil corals represent daily growth.

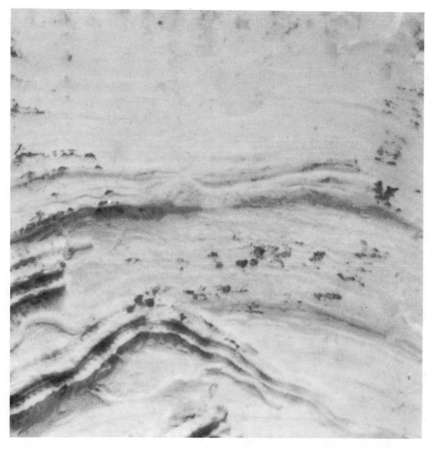

CLOSE VIEW of the specimen of *M. areolata* shows some of the fine growth bands. They are the faint horizontal lines and not the more prominent ridges, which are extraneous.

time he had only the value of the lunar tidal friction as determined from the astronomical observations of the past 2,000 years. Slichter's result showed that the earth and moon were very close together two billion years ago and were moving apart rapidly. This finding created a major puzzle about their relation before that, since the earth (and presumably the moon) has existed for some 4.5 billion years. The puzzle was not pressing then, however, because one could assume that the value of tidal friction over the past 2,000 years has been higher than normal and thus is untypical of long geological periods. Tidal friction exerts its strongest effects in shallow seas. We know that since the last ice age the melting of the polar ice caps has raised the height of the oceans, flooding many coastal areas and producing more extensive shallow seas. This phenomenon could create an abnormally high tidal friction at the present time.

The evidence from the Devonian corals, however, indicates that tidal friction was substantially the same then as it is now. Presumably, since the average value appears to have been fairly constant over many millions of years, one can use that value to extrapolate the earth-moon distance backward in time. With the loss of the varying-friction argument as an explanation of Slichter's finding, however, it becomes necessary to deal with that finding in some other manner.

If, as Slichter says, the earth and moon were close to each other two billion years ago and have separated since, where was the moon before that—in the first two to three billion years of the earth's existence? Three highly speculative answers have been given to the question. One is that the moon is a part of the earth that broke away from the region of what is now the Pacific Ocean. This idea was first put forward late in the 19th century by the British astronomer Sir George Darwin. He supposed the similarity between the density of the moon and that of the earth's mantle, together with the absence of continents in the Pacific hemisphere, might be explained by some instability of the primeval earth that increased to such an extent that the earth broke into two bodies.

Another theory is that in the first half of the earth's existence the moon was moving around the earth in a retrograde orbit, that is, in a sense opposite to its present motion. In such a situation the effect of tidal friction would be to pull the moon inward rather than drive it outward. This idea supposes the moon was drawn in from a large distance, having previously been a stray body that happened to be captured by the earth's gravitational field; finally it came into orbit very close to the earth. In time, because of asymmetries in tidal friction, the direction of the moon's orbit was reversed; hence some two billion years ago the tidal friction began to drive the moon away from the earth.

Both theories assume that the moon was quite close to the earth about two billion years ago. In both cases it is hard to see how the geological record prior to such an event could have been preserved because of the huge tides that would have swept over the earth at the time of the close approach of the moon. Yet it is possible to find rocks of about the same age as this supposed catastrophic event; these rocks still show signs of processes of formation that are similar to those that formed corresponding rocks in recent times. It is therefore difficult to conceive that the earth and moon could have been close together two billion years ago.

A third theory, which would get around the difficulty presented by the Slichter calculations, is that the moon was formed in the vicinity of the earth by an accretion process, in which a small body sweeps up debris in its path. It is difficult to decide whether or not the mechanics of such a process are feasible. If they are, and the process took a long time, it could have produced a moon that would not have begun to raise appreciable tides on the earth until perhaps two billion years ago or later.

It seems clear that more observations on corals of different geological ages will make it possible to determine the length of the day and month throughout the geological past. Those data in turn will yield important information on the early history of the earth-moon system and may, in resolving the puzzle I have been describing, provide an important clue to the origin of the moon.

Corals may not be alone in this field. There are certain fossil algae, going back in age some 600 million years, that have layers similar to those in the corals. The layers may be the result of daily, monthly and annual variations of tides, temperature and sunlight. If these and other marine organisms have recorded time in the same way as the corals, we shall indeed have factual information on the early history of the earth.

FOSSIL ALGAE may possibly be used as paleontological clocks. This stromatolite, or laminated algal structure, has been etched to bring out layering that might have been formed by daily growth. It is enlarged five diameters. Specimen grew about one billion years ago on an intertidal mud flat of the Precambrian Belt Sea in what is now northwestern Montana.

PETRIFIED TREE TRUNKS were all photographed in the vicinity, of Specimen Ridge in northeastern Yellowstone Park (*see bottom map on page 80*]. Pine stump (*left*) is on the grassy northern slope of the ridge. Sycamore (*center*) has been freed from the surrounding volcanic debris by erosion. Redwood stump (*right*) is one of the largest in the area, having a circumference of about 16 feet.

CROSS SECTION of a petrified branch of an extinct genus of pine (*Pityoxylon*) is shown slightly larger than actual size in this photograph. Colors are caused by impurities in the petrifying mineral (silica) and by residual carbon from the original wood.

8

The Petrified Forests of Yellowstone Park

by Erling Dorf
April 1964

The most extensive fossil forests of their kind in the world, they contain much information about the climate and geologic history of the Rocky Mountain region some 55 million years ago

The northeastern quadrant of Yellowstone National Park is a rugged, mountainous region lying between 6,000 and 11,000 feet above sea level. Its climate is characterized as being from cool-temperate to subarctic; its forests consist of conifers with a small admixture of hardwoods. During the Eocene epoch, which lasted from some 60 million to 40 million years ago, the same area presented a strikingly different scene. The countryside was a series of broad, flat river valleys separated by gently rolling hills. The average elevation was between 2,000 and 4,000 feet and the climate ranged from warm-temperate in the hills to subtropical in the valleys. Rainfall was probably between 50 and 60 inches a year. The composition of the dense lowland forests was roughly the reverse of what it is today, with the hardwoods dominant and the conifers in the minority.

These and many other details of the geologic, climatic and botanical conditions prevailing in the region during the Eocene have been preserved by a remarkable series of events that transformed the ancient forests into forests of stone. The first stage in this transformation was sudden and catastrophic. Volcanic eruptions to the east and northeast of the present boundaries of Yellowstone Park showered the surrounding valleys with rocks, ash and other debris, which accumulated gradually over a number of years until, at the end of the period of volcanic activity, the forests in the valleys were buried to an average depth of 10 to 15 feet. After some 200 years a new forest began to grow on top of the desolate blanket of volcanic debris. Meanwhile mineral-bearing waters below the surface had begun the long process of turning the buried tree trunks into stone. Today, many millions of years later, the volcanic matrix sur-

rounding the buried trunks has eroded away in places, leaving the petrified remains of the ancient trees standing upright exactly where the trees had been growing originally.

This remarkable accident of preservation alone would be enough to make the petrified forests of Yellowstone Park extremely valuable to the paleobotanist. The much more famous "petrified forest" in the Painted Desert region of eastern Arizona is really not a forest at all. The hundreds of huge stone logs that lie scattered in all directions in this region are far from the site where they once grew, having been carried downstream in an ancient "log drive" some 175 million years ago. Although there are a few other places in the western U.S. where petrified tree trunks still stand upright in their original position, the fossil forests of Yellowstone Park are by far the most extensive of their kind in the world, covering an area of more than 40 square miles.

What makes these forests even more extraordinary, however, is the evidence that in at least one location the whole process of burial and petrifaction took place not once but many times. On a steep bluff overlooking the Lamar River a few miles above its confluence with the Yellowstone River [*see bottom map on page 80*] no fewer than 27 distinct layers of petrified trees have been exposed by erosion. These layers, which total about 1,200 feet in depth, represent alternating periods of violent volcanic activity and quiet forest growth over a span of some 20,000 years. A detailed study of the petrified trunks, fossilized leaves and other plant remains in these layers has yielded much information about the climate and geologic history of the Rocky Mountain region during the Eocene epoch.

The "Fossil Forest" in the vicinity of the Lamar River and several other groups of petrified trees nearby were discovered in the 1870's by W. H. Holmes, an artist, explorer and geologist. In a report to the U.S. Geological Survey made in 1879 Holmes wrote: "The bleached trunks of the ancient forests...stand out on the ledges like the columns of a ruined temple." Holmes was able to count "10 or more buried forests," which he portrayed in a drawing accompanying his report. Over the past few years my students at Princeton University and I have visited the site on many occasions. Our investigations have added 17 layers of forest to Holmes's 10; our revision of his drawing appears on page 82.

The material in which the forests are buried consists chiefly of three different types of sedimentary rock: (1) conglomerates, or consolidated masses of rounded pebbles believed to have been laid down as stream deposits; (2) breccias, or similar masses composed mainly of angular fragments and probably deposited in the course of extensive mudflows or landslides, and (3) tuffs, or solidified volcanic ash deposited directly from the atmosphere, usually into lakes and streams. Of these only the tuffs contain fossilized plant remains other than petrified trunks; these remains include fossilized leaves, ferns, cones, needles and seeds [*see illustrations on page 83*]. Associated with the fossil-bearing volcanic layers are other beds—some of them 1,000 feet thick—of basalt, a dark, fine-grained, solidified lava. Since the lava was hot at the time it spread over the land, there are no petrified trunks or other plant remains in these beds.

Geologists have suspected for a long time that volcanic sediments are generally deposited much more rapidly

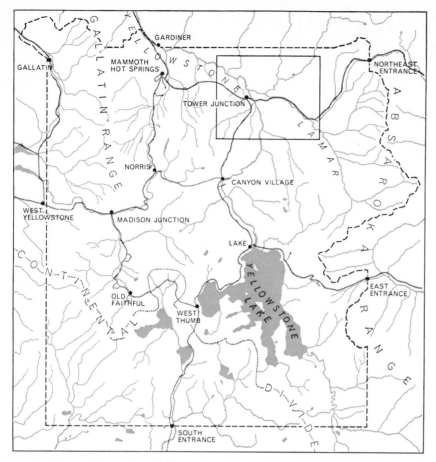

YELLOWSTONE PARK (*area inside broken black line*) is the oldest and largest national park in the U.S., occupying more than two million acres in northwestern Wyoming, Idaho and Montana. Rectangular area inside solid black line at top right contains most of the petrified trees in the park and appears in larger scale in map below. Gray lines are roads.

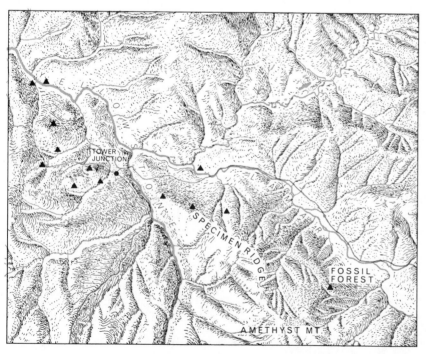

PRINCIPAL SITES where petrified trees and other fossilized plant remains have been discovered are indicated by black triangles on this map. "Fossil Forest" is at bottom right.

than other sediments. We were able to confirm this suspicion by calculating the rate of deposition of the fossil-bearing beds in the Lamar River region. We observed that each of the 27 forests in these beds was buried by a single deposit of volcanic debris. We also knew from more recent volcanic eruptions in Mexico that a new forest begins to grow on top of a volcanic sediment about 200 years after the sediment has been deposited. By counting the number of annual growth rings in the buried stumps we were also able to estimate the age of each forest at the time of its burial. In some cases we found as many as 500 of these growth rings. From these data we were able to calculate that the entire 1,200-foot "layer cake" of sediments exposed at this site must have been deposited over a period of some 20,000 years. This is equivalent to about three-quarters of an inch a year, a rate of deposition roughly 100 times more rapid than that estimated for shallow-water sand or mud sediments of a comparable age in the Gulf Coast region of southeastern North America.

The mechanisms involved in the preservation of plant fossils in volcanic sediments vary considerably. The process of petrifaction, which transforms buried stumps or logs into stone, is now known to be quite different in most cases from the old textbook explanation of a "molecule-by-molecule replacement" of plant materials by mineral matter. In 1927 Ruth N. St. John of Cornell University showed that usually the mineral matter merely fills the cavities inside the empty wood cells. During the process the tough cellular walls of the wood become surrounded almost in their original state by the petrifying mineral matter. This preserves even the most delicate microscopic details of the original wood.

In the Yellowstone Park region the petrifying mineral is almost always silica, or quartz (SiO_2), which originated in the volcanic sediments and was circulated through the buried trees by underground water, some of which was probably hot. On the polished surface of a piece of petrified wood from the region the silica can be dissolved away by carefully immersing the sample in hydrofluoric acid; this will expose a projecting residue of the embedded wood, whose original cellular structure is usually found to be very well preserved. Often the original wood retains its shape and consistency well enough to be cut with a knife or sectioned into

"FOSSIL FOREST" is the collective name given to this site on the northeastern slope of Amethyst Mountain, where the petrified remains of no fewer than 27 Eocene forests have been exposed to view by erosion. A drawing of the site appears on the following page.

EROSION of the volcanic debris under which these large upright trunks were buried during the Early Eocene epoch (some 55 million years ago) has exposed them to view today. Trunks are on northern slope of Specimen Ridge. Roots are visible at bottom right.

ALLUVIUM

BASALT

TUFF

BRECCIA AND
CONGLOMERATE

PETRIFIED TREES

BASEMENT ROCK

CUTAWAY VIEW of a cliff in the Fossil Forest region of northeastern Yellowstone Park reveals the 27 layers of volcanic sediments that contain fossilized plant remains of Eocene forests. Petrified tree trunks are in color. The fossil-bearing beds total about 1,200 feet in depth. In this idealized landscape the cliff overlooks a portion of Lamar River valley (*left background*).

slices for study in exactly the same way as ordinary wood is prepared for examination. Sections can also be cut directly from the petrified wood [*see illustration on page 84*].

The fossilized leaves and other fragile plant remains were not preserved by the process of petrifaction. Instead the rapid burial of these materials in fine volcanic ash prevented their decay, pre-serving them either in the form of com-pressions, in which some of the original plant substance is still present, or im-pressions, in which some of the original the original plant remains. (The fos-

SIX EXTINCT SPECIES of plants native to the forests in the Yellowstone region during the Eocene epoch are represented by the fossils in these photographs. At top left is the impression of a leaf from an extinct species of sycamore, the most abundant tree in the Eocene forests. At top center is a leaf from an extinct grape-vine. At top right is a fern related to today's spleenworts. Leaf at bottom left is from a tree related to the rare Chinese katsura tree. Leaf at bottom center is from a meliosma tree, whose nearest living relatives are restricted to tropical and subtropical forests. Needles at bottom right are from an ancient relative of today's red-wood tree. All the fossils except sycamore leaf contain rem-nants of original plant substance and are called compressions.

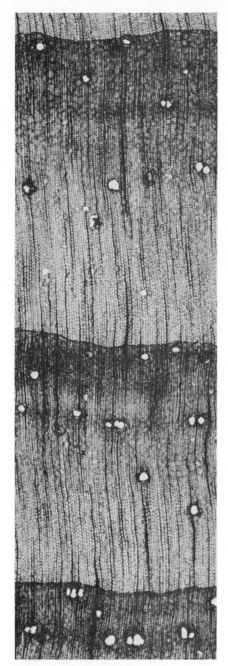

CELLULAR STRUCTURE of a petrified piece of extinct pine wood is exceptionally well preserved in this transverse section. The dark horizontal bands are annual growth rings. Section was cut directly from the petrified wood with a diamond saw.

silized remnants of animals are extremely rare in the Yellowstone volcanic deposits, probably because animals are able to migrate out of a volcanic region as soon as the proportion of ash and dust in the air makes breathing difficult.)

So far we have been able to identify more than 100 different species of plants in the fossil beds along the Lamar River valley. Of these about a fourth are conifers, ferns and other nonflowering plants. The remaining three-fourths are various flowering species, predominantly hardwoods. The most numerous hardwood species in the Eocene forests were apparently large-leaved sycamores, walnuts, magnolias, chestnuts, oaks, redwoods, maples and dogwoods. Most of these species have their nearest living relatives in today's warm-temperate to subtropical forests, such as those found in the southeastern U.S. Also fairly common were species of figs, laurels and bays, whose nearest living relatives are found chiefly in today's tropical forests. Exotic species, whose descendants have since been completely exterminated in North America, included ancient relatives of the Oriental katsura tree and the Asiatic chinquapin and breadfruit trees. Among the less abundant species were climbing ferns, pines, soapberries, hickories, bayberries, elms and willows.

The simple assumption on which our conclusions about the climatic conditions prevailing in this region during the Eocene epoch are based is that these fossil plant species must have had the same general climatic requirements as their nearest living relatives. Thus we conclude that the Eocene climate in the vicinity of what is now Yellowstone Park was essentially the same as that which now prevails in the Gulf Coast region of southeastern North America. The change from a humid, nearly subtropical climate to the present cool-temperate to subarctic conditions was probably the result of a general world-wide cooling accompanied by a gradual uplifting of the entire Rocky Mountain

area by as much as 7,000 feet over the past 40 million years.

Our study of the fossil plant species from Yellowstone's volcanic rocks has also enabled us to determine more accurately than ever before the age of these rocks. By comparing them with known fossil species from different parts of the world we have found that they date from either the latest part of the Early Eocene epoch or from the early part of the Middle Eocene (roughly 55 million years ago). The span of time covered by such a range is probably less than a million years.

Is volcanic activity of the type that produced these fossil forests likely to begin again in the Yellowstone Park region? New evidence has recently been presented by F. R. Boyd of the Carnegie Institution indicating that the last outpouring of lava in this region occurred during the Pleistocene epoch, probably less than 100,000 years ago. In more recent years the continuing activity of geysers, hot springs and other hydrothermal phenomena as well as recurring earthquakes strongly suggest that volcanic activity is merely dormant in the Yellowstone Park region and may well resume at any time.

Insects in Amber

by Charles T. Brues
November 1951

In which the insect species that were trapped in the gum of pine trees 30 to 90 million years ago are compared with·those caught in flypaper under similar conditions today

AMONG all the living creatures on earth the most abundant, the most varied and the most highly specialized are the insects. Many people think our era should be called the Age of Insects rather than the Age of Man. There are those who argue that the insects will some day dominate the living world. Is there any evidence to support this prediction?

We need not pay too much attention to the naïve fear that the insects may kill off the human species by their activities in destroying crops and spreading disease; man himself, by his prodigious consumption of natural resources and multiplication of population, is contributing to the downfall of his species on a far larger scale than the insects could. But what of the insects themselves? Are

they still on the ascendancy—an ever-increasing horde destined to take over the earth? Or have they begun to go downhill toward extinction, like the giant dinosaurs and other vanished groups that once ruled the planet?

The only place where we can look for an answer to this question is in the fossil history; perhaps by tracing the record of the rise and evolution of insects through the long geological ages we may get some hints as to their trend. Unfortunately the record is woefully incomplete. Most of the vast parade of life that has passed across our planet is lost forever in decay. All we have left to read are the remains of an occasional organism that was accidentally trapped and turned to stone before it could disintegrate. In the case of the insects the rock

record is particularly barren, because these delicate creatures have no hard, bony skeleton and are not readily fossilized.

Yet there is one deposit vault where we *can* find ancient insects, more beautifully preserved than any fossil ever disinterred from the rocks. This reservoir is amber: an ancient tree-sap which trapped insects like flypaper and then hardened to preserve the insects intact for millions of years. Found in several parts of the world, these pieces of amber provide us with collections of insects

A GALLERY of insects in amber is shown in the photographs on these four pages. The specimens are from the collection of the Museum of Comparative Zoology at Harvard.

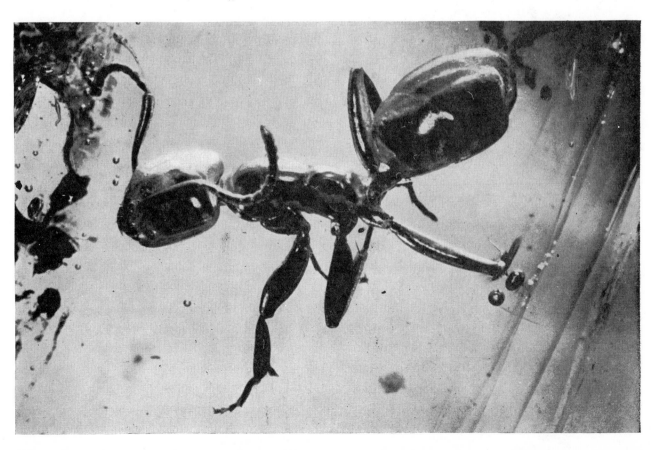

from 30 to 90 million years old, so well preserved that they can be examined in almost as great detail as modern species that have been caught alive and mounted by careful laboratory methods.

THE richest tomb of ancient insects yet discovered is Baltic amber found in Germany. Most of us are acquainted with Baltic amber; it is commonly cut into small ornaments and beads, which used to be worn as necklaces by superstitious Victorian ladies to ward off the discomforts of hay fever, asthma and other allergic afflictions. Such beads often contain small insects. Baltic amber is the fossilized resin of an extinct species of pine that grew in the Baltic region during the geological period known as the Oligocene, some 70 million years ago. When first exuded from the living trees, the resin trapped many small insects on its sticky surface. Engulfed in this viscous substance, the insects were preserved without damage just as they were. As the resin solidified into amber they remained like delicate trinkets imbedded in transparent plastic. After the trees had died and rotted away, the amber, unaffected by the wood-destroying fungi and bacteria, was left in small chunks in the soil. Eventually erosion washed it with the soil into the sea. Since amber is only slightly heavier than sea water, the waves carried it about and cast it up on the shores of the Baltic. The pieces of crude amber now found vary greatly in size; some weigh a pound or

more, but most are very much smaller. Fortunately the pieces are often accumulated in pockets in the earth, where they may be dug out or mined.

The insects preserved in amber of course are not entirely whole, for there is nothing to prevent the decay of their internal organs. But all their external details, even to the most minute bristles and hairs, are faithfully preserved, due to the fact that their outer covering is made of a tough, horny material called chitin (from the Greek word for coat of mail). When we examine one of these specimens, what we really see is its mold in the amber, lined with a pigment composed of metamorphosed and carbonized material from the chitinous skeleton. Any attempt to free the fossil insect by dissolving away the amber is doomed to failure, for once the supporting amber has been removed, the specimen crumbles into dustlike fragments. Hence we must study it just as it lies in its amber tomb.

Clear amber is a transparent material of a yellowish or brownish color. Usually, however, the amber is clouded with trapped mold, vegetable matter, tiny air globules and bubbles of water vapor exuded from the insects. This may obscure the insects, but the specimens can often be salvaged for examination by careful cutting and polishing of the amber into small blocks or slabs.

Many thousands of insects in amber, ranging from a partial to a nearly perfect condition of preservation, have been col-

lected. A number of entomologists, including the writer, have made technical studies of these insects and have been able to classify them with such exactitude that it is now possible to compare the insect life of 70 million years ago with that of today.

TO THE best of our knowledge insects first came into existence on this planet about 250 million years ago. Their rise and evolution roughly coincides with that of the air-breathing vertebrates. Among the earliest insects were some winged forms very different from any now living, and some hardy types, such as the cockroaches, that still exist in much the same form in the warmer parts of the globe. The evolution of insects progressed rapidly toward great variation and specialization. By the dawn of the Age of Mammals, some 70 million years ago, they were present in numbers and variety closely comparable to the picture that they present today.

The insects of that period, as preserved in the Baltic amber, were very similar to those that now inhabit the temperate regions of Europe and North America. To be sure, very few of the species that lived then exist in exactly the same form today. But most of the genera and almost all of the families of that time still survive in the form of modern variants of the ancient types.

Most remarkable among the Baltic amber insects are the ants. These greatly specialized social insects, today a dom-

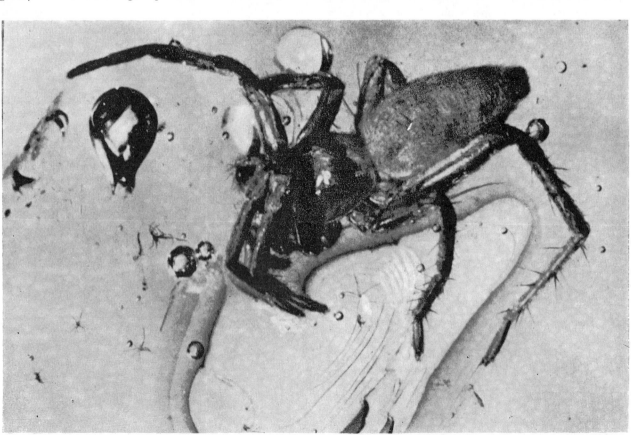

inant group, especially in the tropics, were even more abundant 70 million years ago than they are at present. Moreover, they included a wide variety of types. Some of those types are now extinct or have disappeared from the Baltic and live only in other parts of the world. For example, one genus of ants that was first discovered in Baltic amber was later found to have modern descendants living in Malaya. There is a family of parasitic wasps in the Baltic amber which now survives only in Australia and South Africa. (Wasps are members of the same order of insects as the ants.) On the other hand, the most abundant ant in Baltic amber is hardly distinguishable from the mound-building black ant (*Formica fusca*) that now ranks among the commonest of ants in Europe and North America. It is clear that the speed of evolutionary change has varied greatly among different kinds of insects. Some have evolved rapidly into new types; others have undergone no visible change at all. On the whole, however, considering the ants, beetles, flies, wasps, bugs and other types of insects that are found in amber, it looks as if the insect fauna existing at that time was not too unlike that of the present day.

Yet we cannot be sure that the insects preserved in amber are really representative of the insect population of that era. Obviously there are many kinds of insects in a forest that will not be trapped in resin on pine trees. Some are too large and powerful to be caught; many, large and small, are not in the habit of visiting pine trees. Consequently any statistical comparison of the modern insect population with that in the amber is necessarily open to gross error.

SOME years ago we took a census of insects in a present-day forest by a method simulating that of the Baltic amber, in order to get a statistically comparable population. The method we used was to tack sheets of sticky "tanglefoot" flypaper on the trunks of large pine trees. Insects flying about or crawling on the tree trunks were caught in the tanglefoot just as they had been in the exuding resin of the pines in ages past. The numerous insects and other tiny fauna trapped on the paper were soaked off in alcohol and were recovered in good condition for examination. Our collections were made in primeval sections of the Harvard Forest in Massachusetts, where the tree types and other conditions are very similar to those that existed in the Oligocene forest. We obtained a statistically valid sample of more than 21,000 insects.

Just as we expected, we found that most of our captives were of the smaller varieties of insects in the forest, proving that the scarcity of large species in the Baltic amber does not mean such species were absent from the ancient forests. In our count of the trapped insects we encountered a surprise. While working in the moist, shaded forest, we were constantly annoyed by mosquitoes, black flies and deer flies, which appeared in swarms and bit us unmercifully. We also saw several mosquitoes caught while the papers were being put on the trees. Yet in our final tally of the collections we found we had captured only 26 mosquitoes, 18 black flies and 3 deer flies. Evidently such insects constitute only a very minor proportion of a forest's total insect population. The reason we are so impressed with their numbers is that they seek out the company of warm-blooded animals, and we notice them more than other insects. I must admit that the comparatively low density of the mosquito population was an eye opener for me as an entomologist.

INSECTS are classified into three major groups. The first group are primitive, wingless types that mature into adults without much change. (This group does not include fleas or lice, which are descended from ancestors that once had wings but lost them as they evolved into parasites.) The second group undergo a partial metamorphosis as they mature and acquire wings. The third group go through three distinct stages of growth—the larva, the pupa and finally the winged adult. These three major groups represent the evolutionary path of the insects, proceeding from the most primitive to the most highly developed. In the early ages, 200 million years ago, the first group was dominant; today the third group is by far the most abundant.

What does the comparison of the population on our flypaper with that in the Baltic amber show as to the trends during the past 70 million years? One thing it shows is that so far as numbers go, the more specialized types of insects continue to gain at the expense of the more primitive. Consider, for instance, the largest and most specialized order of insects: the flies. Flies account for a considerably larger proportion of the total insect population now than they did then. In the amber they made up 54 per cent of all the insects trapped; in our flypaper census their proportion has risen to 72 per cent. In particular the group known as *Muscoidea,* which includes the common house-fly, fruit flies and their allies shows a great relative increase. Yet on the whole the flies, notwithstanding their rise in numbers, cannot be said to have exhibited any great surge of evolutionary enterprise or invention during these recent millions of years. Some of the insects in amber were just as specialized and peculiar as any found today.

Moreover, not all of the highly developed types of insects have grown in numbers. A striking example are the ants: according to our flypaper census they are only about one ninth as abundant today as they were in the ancient forests.

In general the picture we get from these two comparable censuses, spanning the past 70 million years, is that of a decline in primitive types of insects, a gain in the relative abundance of the specialized orders and some substantial changes in the ratio of certain groups to the total population. But by and large the insect population of today remains remarkably similar to that of the earlier age. All the major orders of insects now living were represented in the ancient Oligocene forest. Some of the specific types have persisted throughout the 70 million years since then with little or no change, indicating a pronounced fixity that gives little promise of adaptive change in the future. Furthermore, the insects of that age already showed great variety; indeed, in some groups that we have been able to compare in detail we find a greater diversity in the Oligocene insect fauna than in the present one.

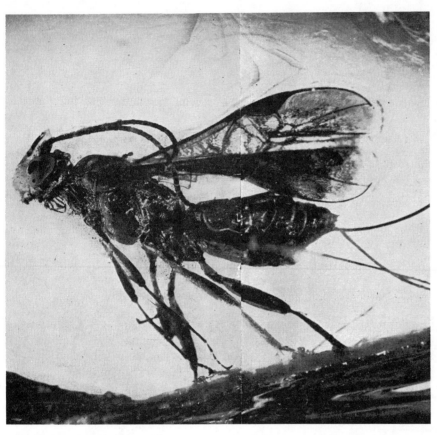

CAN we still be sure, then, that this is the Age of Insects? Are insects still increasing in abundance and variety, or have they passed their heyday? We have no way of comparing their total numbers then and now, but our sample censuses suggest at least that many abundant groups of insects have passed their prime. Although we may still be in the Age of Insects, it seems safe to say that they are not now coming into bloom.

What path they may follow in the future it is impossible to predict, because man is rapidly breaking down their environment. The march of civilization, with its attendant mechanical monstrosities, is remaking the whole face of animate nature—for better or for worse.

Fossil Behavior

by Adolf Seilacher
August 1967

Some fossils represent the tracks or burrows of ancient animals. Such fossils can seldom be identified with a particular animal, but they do show how the animal behaved and something of how behavior evolved

Most of what is known about the evolution of plants and animals has been learned from fossils. One might think that this information is limited to the anatomical changes in living organisms, but such is not the case. There is a class of fossils that provides evidence on animal behavior. These fossils consist not of animal remains but of fossilized animal tracks and burrows.

The great majority of such fossils are markings made in the soft sediments of the ocean floor by ancient invertebrates: ancestral marine worms, starfishes and sea snails, extinct arthropods such as the trilobites, and the like. The tubes and tunnels, trails and feeding marks left by these animals are preserved as raised or depressed forms in layers of sediment that gradually became rock. Paleontologists call such forms "trace fossils"; they classify them in taxonomic arrays and assign them names. Geologists find them useful as indicators of age in formations of sedimentary rock that do not contain the usual kind of fossil. Trace fossils are also clues to the interrelations of organisms and environments—the ecology—in the ancient oceans. This article, however, is concerned with what trace fossils reveal about the behavior of the animals that produced them and how such behavior evolved over periods of millions of years.

The most obvious question to ask about trace fossils is: What animal is responsible for a given track or burrow? This question is one of the most difficult to answer. Except in the case of some trilobites and a few other arthropods, clear-cut "fingerprints" preserved in tracks and burrows are rare or difficult to recognize. As far as the identity of the animal that made them is concerned, many trace fossils may remain a mystery forever. Such fossils can nevertheless be

sorted out and classified according to the behavior that gave rise to them. Many of their differences are functional; for example, a hole that was dug as a shelter would differ from a hole made by a sediment-feeding animal in the course of a meal. Similarly, an animal that feeds on the surface of the ocean bottom produces one kind of trail when it is foraging and a distinctly different kind when it is trying to elude a predator.

I have found it useful to put trace fossils in groups representing five activities [*see illustrations on pages 91 and 92*]. The first group consists of crawling tracks, indicative of nothing more than simple motion. The second is made up of foraging tracks, marks left by animals that moved along the ocean bottom (or just below it) in the course of feeding. The third consists of feeding burrows, as distinct from foraging tracks, made by animals that tunneled well into the bottom sediments. The fourth contains "resting" tracks made by animals that took temporary refuge by burying themselves in sandy bottoms. The last group is composed of dwelling burrows, the permanent shelters of animals such as marine worms that lived and fed without moving from place to place but gathered their food from outside the burrow.

One way of reconstructing fossil behavior is to devise a model of it in the form of a program of commands such as might be written for a computer. The

validity of the model can be tested by determining if the sequence of commands will produce actions that are compatible with the fossil evidence. A simple example is provided by "pipe-rocks," seacoast sediments that have now become sandstone and are named for the abundance of vertical dwelling burrows they contain. The pipe-rock burrows, known in trace-fossil taxonomy as *Scolithos*, appear to be the work of animals with a simple pattern of behavior. A model program for the *Scolithos* animals might consist of two commands. The first would be "Dig down vertically for n times your length," the second "Avoid crossing other burrows." Behavior responsive to these two commands would suffice to produce pipe-rocks.

The trace fossils left by animals that feed on sediments, either by grazing on the bottom or by tunneling into it, show particularly prominent behavior patterns. The nutrients in a given area or volume of sediment are best extracted by orderly movements rather than random ones. Efficient sediment-feeders produce regular winding or branching tracks in which repetition of the same kind of turn produces an intricate pattern. Some 40 years ago the German paleontologist Rudolf Richter pointed out the significance of these patterns in a pioneering behavioral analysis of the trace fossil *Helminthoida labyrinthica*.

The animal responsible for the *H. lab-*

⟶

FOSSIL TRACKS on the following page suggest how the behavior of animals that lived on the ocean floor evolved in the Paleozoic era (from 600 million to 230 million years ago). At left are primitive "scribbles." The track at top was made during the Ordovician period by a wormlike animal; the track in the middle, during the Cambrian period, possibly by a snaillike animal; the track at bottom, during the Cambrian by a trilobite. At right are more advanced spirals and meanders. The one at top was made during the Mississippian period by a wormlike animal; the one in the middle, during the Pennsylvanian period, possibly by a snaillike animal; the one at bottom, during the Ordovician by a trilobite.

yrinthica markings was a particularly efficient sediment-feeder. Its tunnels are found in fine-grained sedimentary rocks in the Alps and in Alaska, usually silts and marls of Cretaceous and Eocene age (between 135 million and 36 million years ago). To rephrase Richter's analysis as a model program, it appears that the animal obeyed only four commands. The first command was "Move horizontally, keeping within a single stratum of sediment." Obedience to this command is evident in the fact that the animal's tunnels lie within the horizontal laminations of sediment. The second command was "After advancing one unit of length make a U-turn." Here obedience to the command is apparent in the "homostrophy," or uniformity of turning, that is characteristic of the animal's tunnels. The third command was "Always keep in touch with your own or some other tunnel." Biologists call obedience to such a command "thigmotaxis," meaning an involuntary approaching movement in response to the stimulus of contact with some object. That this command was followed is evident in the closeness of the tunnels to one another. The fourth command was "Never come closer to any other tunnel than the given distance *d.*" Obedience to this command, which is termed "phobotaxis," would have prevented the animal from digging across another tunnel; crossed tunnels are in fact absent from the fossils.

The model program does not, of course, specify the sensory responses that would have allowed the animal to follow its orders. It is possible, however, to guess what these responses were. For three of the commands nothing more complex seems to be needed than positive or negative chemotaxis: either approach to a chemical stimulus or avoidance of it. Each of the laminations in the sediment would probably have had, so to speak, a characteristic flavor that

would have served to guide the animal's horizontal movements and thus would have resulted in obedience to the first command. Obedience to the third and fourth commands is somewhat harder to understand until one considers that as the animal moves it churns up sediment along the sides of its tunnel. Such areas of disturbance are visible in some of the fossils. It seems likely that the animal could chemically distinguish between the disturbed sediment of an adjacent tunnel and the undisturbed sediment in the opposite direction.

The second command, on which the animal's turning maneuver depends, seems to require something other than a chemical stimulus. Its successful execution must be related to the fact that the *H. labyrinthica* animal was shaped like a worm. The length of its body could thus serve as a measuring rod. When it had dug a straight section of tunnel far enough forward, its tail would emerge from the last U-turn and straighten out. Tail-straightening was probably the only cue required for the animal's head to start its next U-turn.

One fact of real life that is neglected in model command programs is that a need to disobey commands must frequently arise. If the tunnel being dug by the *H. labyrinthica* animal had penetrated between two other tunnels, for example, obedience to the third and fourth commands would have left it trapped if it had not violated the first command and moved upward or downward. In such circumstances disobedience is required for survival.

Evidence that the commands were not inflexible is found in the way the animal's meandering turns were executed. The length of the individual "lobes"—the straight passages between the U-turns—are not uniform; some are shorter than average and some are longer. One can

assume that a short lobe was formed when contact with some unidentified object triggered obedience to the fourth, or "avoid," command before the entire "forward march" portion of the second command had been executed. A long lobe could have been formed if some accidental bend in the straight part of the tunnel misled the animal into delaying its next U-turn. There is no way to test the hypothesis concerning short lobes. Examination of many fossils shows, however, that in a majority of cases long lobes are associated with secondary bends.

The same behavioral program operating in different animals can be expected to give rise to a wide variety of patterns because of the differences in the animals' modes of locomotion, feeding, responses to commands and even simpler differences such as body length and turning ability. The *H. labyrinthica* animal, for instance, could make very tight turns. The animal that produced the trace fossil *Helminthoida crassa*, however, was apparently less agile: its turns tend to be shaped like a teardrop. Judging by these loose meanders, the *H. crassa* animal was also less sensitive to thigmotactic commands.

A further example of such meandering trace fossils should be mentioned because it demonstrates how a relatively small change in a command program can result in a highly complicated meander. In the Alps and Alaska and other regions of folded sedimentary rocks there are series of a particular kind of sandstones called graywacke. On the bottom of graywacke layers one finds the complicated trails of *Spirorhaphe*. If one draws a loose inward spiral, makes a U-turn at the center and spirals outward again in the space between the inward whorls, one has traced the path taken by the *Spirorhaphe* animal.

I thought at first that it would be very

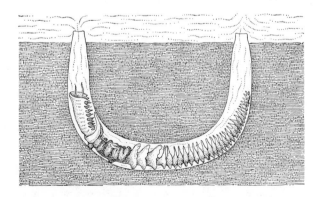

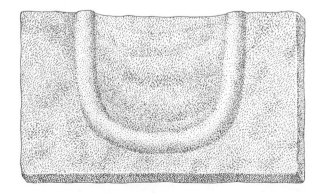

DWELLING BURROW differs from other trace fossils because it represents the shelter of an animal that is stationary and may even be anchored. The animal illustrated here is a polychaete worm that feeds by filtering particles of food from the water around it.

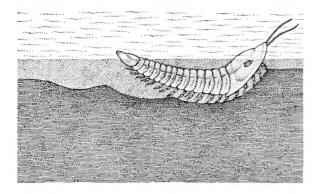

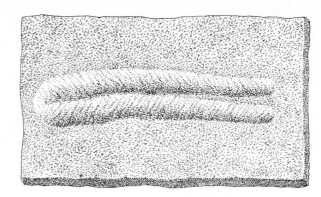

CRAWLING TRACKS comprise one of the five main categories of "trace fossils," which record the various activities but seldom

the identity of the animals that made them. The examples illustrated on this page, however, are all attributable to trilobites.

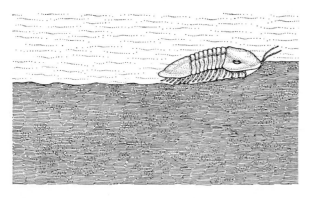

FORAGING TRACKS differ from crawling tracks by recording not simple motion but an animal's search for food over the surface

of the ocean floor. The illustrations of fossil tracks and burrows on these pages and on pages 93 and 94 were made by Thomas Prentiss.

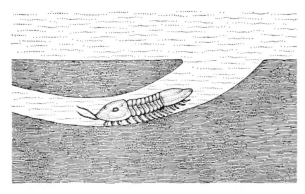

FEEDING BURROW is made by an animal that tunnels into the sediments of the ocean floor and eats the bits of organic matter it

extracts from the silt. The burrow also provides shelter. The meander and the spiral illustrated on page 93 are feeding burrows.

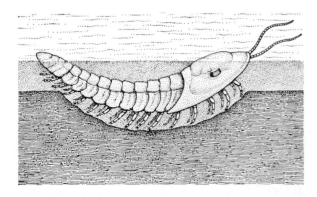

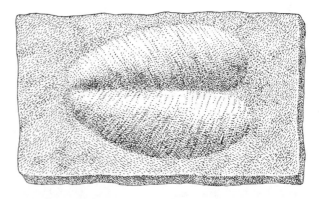

RESTING TRACKS are the marks produced when an animal, often a scavenger, makes a shallow, temporary hiding place just below

the surface of the ocean floor, usually where the bottom is sandy. Their shape generally corresponds to the outline of the animal.

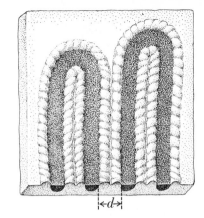
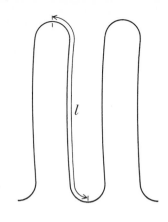

HORIZONTAL TRAIL (*left*), composed of tight meanders, was made by an unknown wormlike animal. The diagram in perspective shows how the animal's passage through the sediments on which it fed churned up the silt on each side of its tunnel. The width of this churned area established the critical distance ("*d*" *in perspective diagram*) that separates adjacent trails. The average distance between one of the animal's U-turns and the next ("*l*" *in plan view*) seems to have been determined by the length of its body.

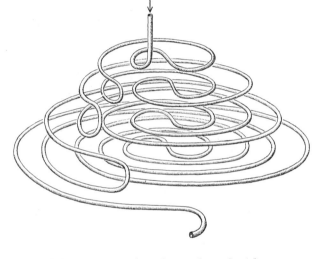

SPIRALING TRAIL (*left*) is actually a double spiral made by a wormlike animal that circled inward, made a U-turn at the center and circled outward again in the empty area between its entering whorls. "Mistakes" in the position of the U-turns in many specimens and the presence of short alien trails nearby (*short segment in center at left*) made author aware that the fossils preserved only a single, horizontal segment of what had been an elaborate, three-dimensional burrow in the silt (*hypothetical outline at right*).

difficult to devise a command program for the motions required to make such a double spiral. In examining a number of *Spirorhaphe* specimens, however, I noted small "mistakes" in several of them. In each case the animal had been forced to reverse its course too soon because the center of its spiral was already occupied by an alien U-turn (for which no inward and outward spirals could be seen). The meaning of these mistakes soon became apparent: each *Spirorhaphe* fossil preserves only one horizontal layer of what had been a multifloored, three-dimensional tunnel [*see lower illustration above*]. The "alien" U-turns are the earlier work of the same *Spirorhaphe* animal tunneling down from above. The spiral portions of the earlier whorls, which lay at a higher level, have not been preserved.

Although the three-dimensional tunnels are complex in appearance, the program required to produce them is simpler than the one required for the one-story pattern. Only two commands need be added to the four that produce horizontal meanders. The first additional command is "After spiraling inward make a U-turn and keep in contact with the adjacent tunnel on the horizontal." The second is "After spiraling outward turn down and keep in contact with the adjacent tunnel vertically." The wormlike *Spirorhaphe* animal could distinguish between inward and outward spirals by means of a cue similar to the one responsible for homostrophy among the animals that meander horizontally. As the animal spiraled inward its head end would be more tightly curved than its tail end. As it spiraled outward after

making its U-turn the tail end would be more tightly curved.

What are the ways in which the evolution of behavior among trace-fossil animals can be detected? One is the comparative study of groups of fossils that appear to be both closely related and closely associated in space and time. One such group is the Graphoglypt family; two of the trace fossils I have mentioned, *Spirorhaphe* and *Helminthoida crassa*, belong to this group. Graphoglypt trace fossils are alike in their time range, being found mainly in formations of Cretaceous to Tertiary age (between 135 million and some two million years ago). They are found in the same kinds of rock and are preserved in the same way. It seems probable, in view of the factors they have in common, that they were made by closely related animals, al-

CHANGE IN BEHAVIOR over 150 million years is apparent in the trace fossil *Dictyodora,* the trail left by an animal that seems to have possessed a siphon. In Cambrian times it fed by meandering a few millimeters down in the sediment (*a*). The path produced during its foraging is at left; a three-dimensional reconstruction of the entire fossil track is at right. Later the animal left the upper sediment, eating its way deeper along a corkscrew-like path and then meandering in a restricted manner ("*b*" *and* "*c*"; *path at left, reconstruction at right*). By middle Mississippian times the animal no longer meandered (*d*) but only corkscrewed deep into the silt.

though what animals they were we do not know. When the group is examined as a whole, the various Graphoglypt genera and species can be arranged in a few lines of descent that have one trend in common. In terms of behavioral programming, the lines of descent show a gradual shift from rigid behavior patterns to patterns that gave the animals freedom to adapt to local circumstances. This trend toward flexibility enabled the animals to forage more efficiently, a factor that is of obvious value in terms of natural selection.

A second example of behavioral evolution is apparent when trace fossils are examined in a more general way to see what changes have taken place over millions of years in behavior that deals with one specific biological task. The obvious task to study is foraging, and we quickly find that several methods of foraging other than meandering exist. The simplest of them resembles the scribbling of children: the tracks form a series of circles with slightly offset centers. The program for scribbling has one simple command: "Keep heading to one side but don't stay in the same track." Scribbling, however, covers an area much less efficiently than meandering does.

A somewhat more complex method of foraging, and one more efficient than scribbling, consists of moving in an outward spiral. The tighter the spiral, the more efficient the coverage. The program for tight spiraling has two commands. One is "Keep circling in one direction"; the other is "Keep in touch with the spiral whorl made earlier." This program is simpler than the four-command program for meandering, but spiraling is less efficient than meandering. The strips between the spirals remain unexploited.

Several unrelated groups of trace fossils provide evidence that the progress from simple to complex forms of foraging is an evolutionary one [*see illustration on page 90*]. In Cambrian times, at the beginning of the Paleozoic era some 600 million years ago, none of the animals of the ocean floor had evolved the meandering method. Scribbling was practiced by some trilobites and by what were probably varieties of marine snails. In a heterogeneous group of trails attributed to various unknown worms good scribbles appear just after the end of the Cambrian, in Ordovician times (between 500 million and 425 million years ago). Eventually, however, scribbling disappears altogether and worms, snails and trilobites all begin to forage by meandering, with spiraling as an occasional alternative method. Complex meanders and dense double spirals are features of

ANTLER-SHAPED BURROWS are the work of an anchored animal of the Cretaceous period, roughly 100 million years ago. The lobes in the sediment that the animal excavated for their food content were narrow and much of the adjacent sediment was left unexploited.

SKIRTLIKE BURROW is the work of the same kind of animal some 40 million years later, in the Tertiary period. One large excavation is more efficient than the series of lobes made in Cretaceous times. Early in the life of the animal, however, its burrow is Cretaceous in design (*bottom left*), showing that behavioral ontogeny can recapitulate phylogeny.

comparatively late geologic history, not appearing until Cretaceous times, at the end of the Mesozoic era (between 135 million and 63 million years ago). They indicate further progress in the foraging efficiency of sediment-feeders.

There is a well-documented instance in which an anatomical change in a trace-fossil animal was accompanied by a change in its behavior. The fossil *Dictyodora* is the work of an unknown sediment-feeder that tunneled into the ocean floor and filled its tunnel behind it. The *Dictyodora* animal, however, seems to have been equipped with a long, thin siphon that allowed it to maintain contact with the water above it. As the animal traveled in loose meanders through the sediments, its siphon, dragging behind it, left its own marks in the ooze.

From Cambrian to Devonian times, between 600 million and 350 million years ago, *Dictyodora* animals foraged only a few millimeters down in the sediment. Their siphons were short. This particular ecological niche, however, was within reach of many smaller competitors. By Mississippian times (beginning about 350 million years ago) the *Dictyodora* animals had become adapted to feeding at a deeper and less crowded level. Their siphons had become longer and their meandering trails were well below the reach of their sediment-feeding contemporaries.

As the *Dictyodora* animals evolved anatomically they also altered their behavior. In earlier millenniums they had apparently not begun to feed until they had worked their way down to a specific foraging level. The Mississippian animals, however, ate their way deep into the sediments, leaving a corkscrew trail behind them, before they began their horizontal meanders. Moreover, the first several meanders no longer followed the older loose pattern. Instead they curled concentrically around the initial corkscrew.

As time passed the *Dictyodora* animals' behavior became simpler. Trace fossils from later Mississippian formations in East Germany show a high percentage of *Dictyodora* burrows in which loose meanders are altogether absent, concentric meanders taking their place. Trace fossils from southern Austria, which are possibly even younger, show a further behavioral change. Most of the burrows have no meanders at all; the central corkscrew trail has simply been drilled deeper. It is doubtful that the *Dictyodora* animals remained in the same burrow all their lives. During their reproductive cycle, for example, the animals probably rose to the surface of the sediment. Thereafter they would have had to dig new burrows, if they did not begin a new mode of life altogether.

Trace fossils include the burrows of a few animals that led a sedentary existence. These burrows, of course, are a record of the animals' entire life. One such fossil is *Zoophycos;* it is the burrow of an unknown wormlike animal that foraged through the sediments by constantly shifting the lobes of its U-shaped tunnel. The two openings of the tunnel remained fixed in the sediment, but thin concentric layers inside the lobes record the tunnel's shifts of position. *Zoophycos* burrows have been found in sedimentary rocks as old as the Ordovician. In Cretaceous and Tertiary times (from 120 million to some two million years ago) the *Zoophycos* animal seems to have taken over the *Dictyodora* animal's ecological niche in the deep sediments that have since become Alpine rocks.

When the burrows of a *Zoophycos* animal of Cretaceous and Tertiary times are compared, an evolution in behavior is readily apparent. The earlier burrow consists of narrow lobes that join to form an antler-like pattern [*see top illustration on preceding page*]. The sediment that lay between the lobes remained exploited. In Tertiary *Zoophycos* burrows, however, the individual lobes have fused into a continuous, skirtlike pattern inside which all the sediment has been explored [*see bottom illustration on preceding page*]. A much more effective coverage of a food-bearing area has resulted from what was probably a relatively small change in the *Zoophycos* animal's behavioral program.

Surprisingly this more efficient behavior is found only in the portions of the burrow occupied by the adult Tertiary animal. The behavior of the young animal remained like that of its Cretaceous ancestor, producing the same narrow lobes. Thus we see that the old adage about ontogeny recapitulating phylogeny can be applied to an animal's behavior as well as to its anatomy.

Scanty though the trace-fossil material now available is, it gives one hope that the early evolution of behavior patterns will become as valid an area of study as the evolution of anatomical structures. Further insights into fossil behavior should not only add a new aspect to paleontological research but also help to counter the belief that paleontologists are concerned only with dead bodies and have no real comprehension of ancient life.

Micropaleontology

by David B. Ericson and Goesta Wollin
July 1962

*Some fossils are so small they can be identified only
with the aid of a magnifying glass or a microscope.
They label sedimentary layers and provide excellent
clues to ancient changes in climate*

The words "fossil" and "paleontology" usually evoke pictures of
dinosaur bones or other good-sized
pieces of vertebrate skeleton. This article
is concerned with micropaleontology,
which deals with fossils of entirely different magnitude. They are the shells, or,
more properly, the skeletons, of minute
aquatic animals. None of these microfossils can be recognized without the
help of a strong magnifying glass; some
must actually be examined in the
electron microscope. Their minuteness
makes them especially useful for geological research. They can be brought
up unbroken—and in enormous numbers
—by a narrow coring pipe or even by an
oil-well drill. Found both in the ocean
floor and in dry-land formations that
were once covered by water, microfossils
have long served oil prospectors as stratigraphic markers. More recently they
have begun to furnish important information about processes of change in the
structure of the earth's surface and in the
earth's climate.

To meet the needs of the micropaleontologist an organism must have characteristics other than small size. First, of
course, the organism must build a skeleton, or some hard part durable enough
to fossilize under normal conditions of
sedimentation. The fossils in a group
should be distinguished as to genus and
preferably species. This implies some
complexity of organization. Species that
flourished during the shortest periods of
time, and over the greatest geographical
area, are the best geological indicators.
They make it possible to differentiate
sharply among layers of sedimentary
rock of different age and to match strata
in widely separated parts of the earth.

In order to obtain evidence of past
climate and other environmental conditions the investigator must think of fossils as once living animals with specialized adaptations to their particular surroundings. He then tries to reconstruct
those surroundings by analogy with the
ecological requirements of near relatives
of the ancient organisms that are alive
today. As might be expected, the method
becomes more difficult as the evolutionary distance between fossils and living organisms increases. A paleoecologist
must be a good detective and find meaning in all sorts of apparently trivial and
irrelevant observations. Certain potentially useful microfossils are just beginning to attract serious study. Previously
they were ignored because they are so
very small. The coccoliths, disk-shaped
plates of calcium carbonate, are one
example. They are planktonic, which
means that the ocean currents in which
they float have carried them to many
quarters of the globe and that they have
been settling to the bottom for some
500 million years. One cubic centimeter
of sediment can contain 800 million of
them; they must be enlarged 800 diameters to be seen at all. Because they were
found in the topmost layer of sediment
on the ocean floor, it was realized that
the organism depositing them must be
extant. For a long time what the organism was remained a mystery; the finest
collecting nets could not catch it. Eventually it was located not in a man-made
trap but in the filtering apparatus of the
common marine animal called *Salpa*.
Recently the electron microscope has revealed that coccoliths possess an astonishingly complex structure. When they
have been more thoroughly described
and classified, they should be helpful in
correlating sedimentary strata from continent to continent as well as in the
oceans.

Another group of tiny organisms, now
extinct, were the star-shaped discoasters.
A little larger than coccoliths, their
world-wide distribution in deeper sedi-
ments suggests that they too were planktonic. Although known for almost 100
years, they have not been applied in
stratigraphy because of the mistaken
notion that all known species lived continuously from 60 million years ago to
the time of their disappearance. In reality they evolved fairly rapidly, and some
species make excellent guide fossils. The
earliest discoaster stars had as many as
24 rays, or points. By 20 million years
ago the number of rays had declined to
six, and the last forms had only five slender rays. The exact time of their extinction is unknown, but if, as is suspected, it was just before the Pleistocene
(the geological period of the last ice
age), five-rayed discoasters will occupy
an important position as reliable indicators for this important boundary in
geologic time.

In their day-to-day work micropaleontologists deal chiefly with diatoms,
radiolaria, conodonts, ostracodes and
foraminifera. All of these except the
diatoms are large enough to be studied
in a low-power microscope at a magnification of between 30 and 100 diameters—an important consideration when
hundreds of samples must be examined
each day, as is the practice in some oil-company laboratories.

Among the microfossils the diatoms
and the radiolaria compete for the first
place in beauty. Both secrete shells of
opal (silica combined with some water).
In the lacy skeletons of radiolaria the
opal looks like clear spun glass; in diatom
shells it takes on a jewel-like quality,
often displaying a many-colored fire that
must be seen to be appreciated. Many
species of diatoms live exclusively in
fresh water, whereas all radiolaria live
in salt water. In consequence diatom
fossils are somewhat more informative;
from the species found in a sedimentary

FORAMINIFERA are the most important microfossils in the study of earth history. Various planktonic species produced the fossil shells shown here at a magnification of about 20 diameters.

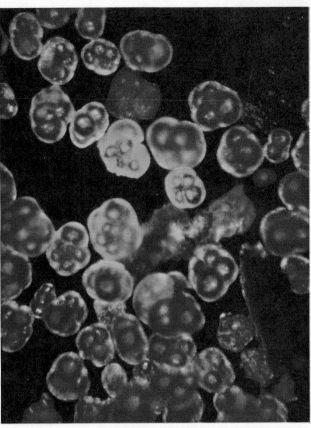

GLOBIGERINA PACHYDERMA, a foraminifer, lives in the Arctic Ocean. In this photomicrograph light from below shows details of chambers in the shells. Magnification here is some 60 diameters.

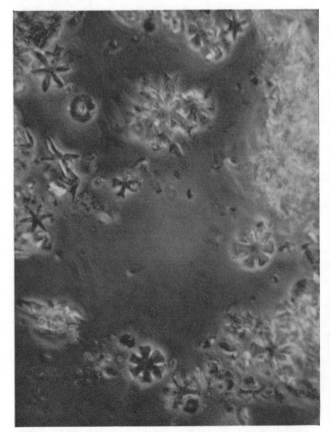

DISCOASTERS, extinct perhaps a million years, are useful as guide fossils. Forms with six rays shown here lived about 40 million years ago. They are magnified approximately 1,000 diameters.

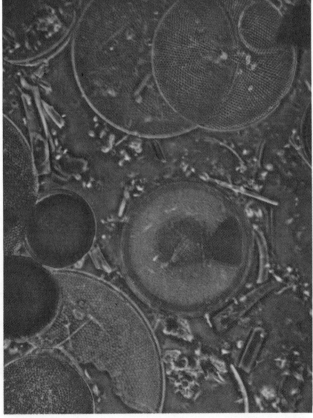

DIATOMS are sometimes abundant in sediments and sometimes missing. These were illuminated by polarized light. Magnification is 200 diameters. Photomicrographs were made by Roman Vishniac.

layer it is usually possible to tell whether the sediment was formed in a lake or in the sea. There are a great many species of diatoms and radiolaria; diatoms can occur locally in such fantastic numbers that they produce fairly thick layers of sediment, known as diatomite, consisting almost entirely of their fossils. The opal skeletons of both diatoms and radiolaria, however, tend to dissolve in water, and they cannot be counted on to show up in a particular region. Often they are missing from just those places where an index fossil is most needed.

Conodonts are small tooth-shaped or platelike objects with one or more points. Like vertebrate teeth, they are made of calcium phosphate. Although many "genera" and "species" have been described since they were discovered more than a century ago, no one knows what kind of animal produced them. Whatever it was, it became extinct some 240 million years ago during the Triassic period. The variation in the form of conodonts from level to level in older sediments makes them particularly use-

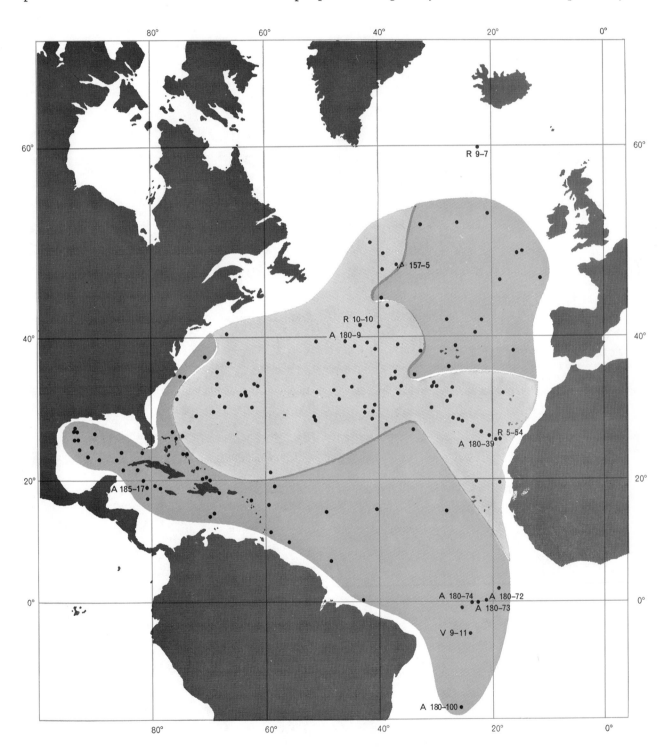

COILING DIRECTIONS of living *Globorotalia truncatulinoides,* a planktonic foraminifer, define three provinces in the Atlantic Ocean. Most of the spiral-shaped shells of this species found on the sea floor coil to the left in the gray region and to the right in the colored areas. The dots mark sites where cores have been taken. The letters and numbers identify the cores diagramed on the following pages. Core V 9–11, brought up just south of the Equator, carries the Pleistocene record back at least 600,000 years.

ful to the petroleum geologist. Because of their small size they often come up undamaged in rock cuttings from oil wells or exploratory holes.

The ostracodes present no mysteries. These odd little relatives of crabs and lobsters are very much alive today and flourish wherever there is enough water, whether it is fresh, brackish or salty.

They are the only crustaceans that have two valves, or shells, which makes them look like tiny clams. The valves vary in length from half a millimeter (a fiftieth of an inch) to three or four millimeters. They first appear in the geological column in sediments deposited at least 450 million years ago in the Cambrian period of the early Paleozoic era.

In the course of evolution the shells have varied greatly in shape and ornamentation, and many species have lived only for short intervals of geologic time. Knowledge of the present distribution of living genera in open salt water, sounds, estuaries, lagoons and lakes helps in analyzing past conditions when similar genera show up in ancient sediments.

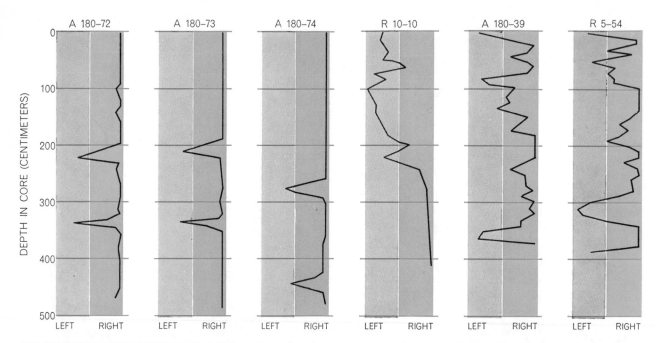

CHANGES IN COILING DIRECTION of *Globorotalia truncatulinoides* with depth in cores make it possible to correlate cores from different locations. The correlation is readily apparent in the three cores A 180–72, –73 and –74. Cores A 180–39 and R 5–54 also show correlation. Samples of shells were studied every 10 centimeters or so from top of cores to bottom. Variation in each diagram is from 100 per cent left-coiling at left to equal ratio in the middle to 100 per cent right-coiling at right. The older shells are deeper.

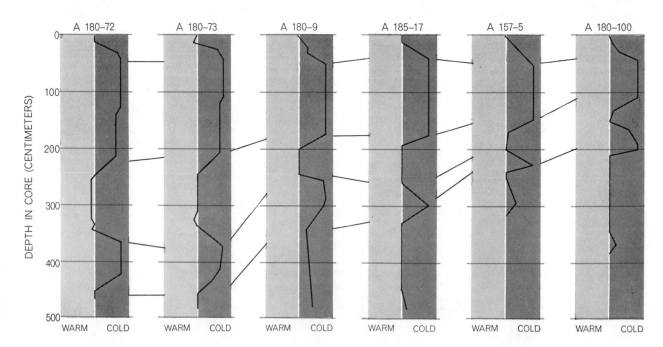

CLIMATE CURVES are based on relative number of warm- and cold-water forms of planktonic foraminifera found in each of six deep-sea sediment cores. "Warm" and "Cold" indicate warm and cold climates compared with present-day climate, which is vertical line in center of each diagram. Thin lines connect faunal changes believed to have occurred at the same time in the various locations. Obviously rates of sedimentation have differed widely. Such curves provide the data for establishing a chronology of the ice ages.

The ostracodes have only one drawback: they are not nearly so numerous as the universal favorites among microfossils, the foraminifera.

The very popularity of foraminifera greatly enhances their usefulness. Over the years a mountain of information of all sorts has accumulated, and most of it is readily available. Descriptions of genera and species have been brought together in an enormous catalogue published by the American Museum of Natural History, which now includes 69 volumes and is growing constantly as descriptions of new species are added.

The foraminifera are protozoa, or single-celled animals, that build tests (shells) of various materials. On the basis of the mode of construction the animals are divided into two large groups. Calcareous species construct tests of calcium carbonate precipitated directly from sea water; arenaceous species build their shells of sand grains, flakes of mica, sponge spicules or even small discarded calcareous tests of other species, any of which they can cement together with secretions of calcium carbonate or iron salts. The size of the shell varies widely with the species. Some long-extinct giant foraminifera exceeded 15 centimeters (six inches) in diameter. The majority of fossil foraminifera range from .2 millimeter to two millimeters.

The architectural unit of the test is the chamber. A few species have only one chamber, but most build anywhere from two to several hundred. On this basic principle of structure the foraminifera improvise endlessly. To duplicate all the strangely shaped chambers and their intricate arrangements would tax the ingenuity of a topologist. Geometric versatility has made possible all the thousands of different species that have come and gone during the past 500 million years.

Almost all the foraminifera species live on the ocean bottom. Although some attach themselves permanently to rocks, most of them potter along at a few millimeters per hour, pushing themselves by means of pseudopodia extruded through minute pores in the shells. Clearly this way of life does not make for wide distribution. Beginning some 100 million years ago in the Upper Cretaceous period, however, a few types became planktonic. Although they constitute only about 1 per cent of the known species, the enormous volume of living space open to them has permitted great proliferation: individuals of the planktonic forms make up almost 99 per cent of the

fossils found in ocean sediments. In some places accumulation of the shells has produced thick deposits of chalk. The white cliffs of Dover and Normandy are such deposits, now uplifted and partly eroded. Today large areas of the bottom of all the oceans are receiving a slow but constant rain of discarded tests of planktonic foraminifera, which make up on the average 30 to 50 per cent of all bottom sediments.

Shells of the important planktonic species have fairly simple forms. The dominant theme is a series of chambers arranged in a spiral. As the animal grows it adds chambers of steadily increasing size. In most species the general form resembles that of a small shell. Like snail shells, some tests coil to the left and others coil to the right, the two kinds being mirror images of each other.

Because of the rapid succession of distinctive species throughout successive geological ages, foraminifera are ideally suited to the needs of the petroleum

geologist, who must deal with many kinds of thick sedimentary rocks, some of them heavily folded and faulted. If one were asked to invent a class of ideal fossils for identifying and matching strata, it would be hard to improve on the foraminifera. It is small wonder that most of the hundreds of paleontologists working for oil companies devote full time to the study of these microfossils.

Foraminifera are a good deal more than tags for sedimentary rocks. To workers in pure geology, and particularly to those in the hybrid branch known as marine geology, they furnish invaluable keys to the remote past. So many samples of ocean sediment have been examined by now and their microfossils classified that it is possible to chart the approximate distribution of the most common species of planktonic foraminifera. The charts show that some species live only in low latitudes, others are most abundant in middle latitudes and

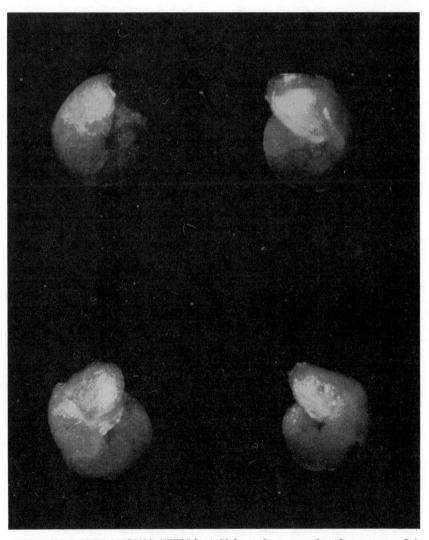

LEFT- AND RIGHT-COILING SHELLS of *Globorotalia truncatulinoides* appear at left and right respectively. Coiling direction is apparently associated with climatic conditions.

still others live in high latitudes. (One species, *Globigerina pachyderma,* ranges up to the North Pole.) Evidently water temperature plays an important part in the distribution of the various species.

If so, the animals must live near the surface; only there does the temperature vary significantly with latitude. The fact that living foraminifera are caught in plankton nets towed through the photic zone (the sunlit top 100 meters of the ocean) seems to bear this out. But the surface samples long presented something of a problem. Their shells are thin-walled and transparent, whereas shells of the same species taken from the bottom are almost all heavily encrusted with calcium carbonate or calcite.

Until recently it was supposed that the material precipitates on the empty tests after they settle to the bottom. This hypothesis had fatal flaws, however. For one thing, oceanographers generally agree that at depths of several thousand meters in the ocean calcium carbonate dissolves far more rapidly than it precipitates. Our laboratory at the Lamont Geological Observatory of Columbia University finally undertook a close examination of the distribution of calcite on individual shells, which furnished the clue to the correct explanation. We discovered that calcite is thickest on the earliest formed chambers and diminishes steadily in the chambers farther out on the growth spiral. This can only mean that the living foraminifera precipitate the calcite and that the thin-walled specimens in the photic zone are immature.

Recently Allan W. H. Bé of Lamont has caught heavily encrusted tests containing living foraminifera by towing plankton nets at depths of more than 500 meters. They provide the final proof that at least some species mature and reproduce well below the photic zone. Evidently the embryos rise to the photic zone, fatten on diatoms and other photosynthetic organisms there and then sink to a lower level, where they complete their life cycle. The pattern does not conflict with the idea that temperature variations in the upper layers of water determine the geographical distribution of various species; each individual passes a critical period of its life in the photic zone. On the other

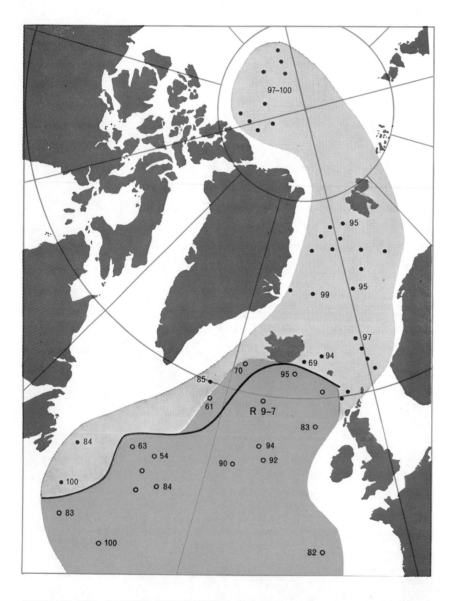

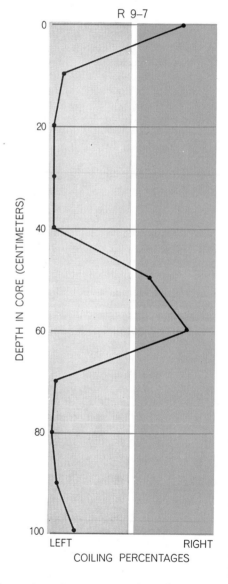

CLIMATE AND COILING DIRECTION of *Globigerina pachyderma* seem to be closely correlated in northernmost Atlantic and adjacent seas. Right-coiling of living pachyderma (*colored area on map*) is associated with warmer water, left-coiling (*gray area on map*) with coldest water. Position of the isotherm marking an average surface temperature of 7.2 degrees centigrade in April (*black line on map*) closely follows border of right-left provinces. Open circles indicate dominance of right-coiling at tops of cores; closed circles mean left-coiling. Percentages of shells coiling in dominant direction are shown for some cores. Curves at right

hand, it suggests a possible explanation for a rather puzzling distribution of the species *Globorotalia menardii.* Fossils of this group are extremely abundant in sediments south and southwest of the Canary Islands, whereas they are entirely absent from the region to the north and northeast. Yet the Canaries current flows southwestward through the area and would sweep all *G. menardii* out of the region south of the islands if they spent their entire lives in the photic zone. We believe that the animals sink below the Canaries current as they approach maturity and enter a deep countercurrent that returns them and the new generation of embryos to the northeast. As yet no one has attempted to detect the countercurrent directly, but mathe-

matical analysis of deep circulation in the Atlantic suggests it should be there.

The discovery that foraminifera live part of their lives deeper than 500 meters has an important bearing on recent attempts to estimate ancient water temperatures from measurements of oxygen isotopes in fossil shells. The technique is based on the fact that in warmer waters there is a slight preponderance of heavier isotopes in the shells and in cooler waters a slight preponderance of lighter isotopes. If the relative abundance of the isotopes is to be significant, the depth at which the shells incorporated their oxygen must be known. Deep waters at every latitude are quite cold.

Since warm-water and cold-water species have not changed greatly in the past million years or so, the micropaleontologist can use the fossils to obtain a quite objective picture of changing climatic conditions during the period. Cores of sediment from ocean bottoms furnish a continuous record of geological and climatic events in contrast with the garbled record available from the distorted layers of rock on dry land. From our study of planktonic foraminifera in more than 1,000 cores in our laboratory, we feel that we have been able to arrive at the first accurate set of dates for the most recent glacial and interglacial periods. Carbon-14 analysis of the fossils shows that the last ice age ended 11,000 years ago instead of 20,000 years ago, as had previously been thought. The new date, now generally accepted, may change some ideas about the rate of human evolution. Judging from the average rate of sedimentation, the last part of the last glaciation began approximately 60,000 years ago, after an interstadial period (an interglacial of short duration) of 30,000 years [*see illustration on page 105*]. The preceding glacial period lasted only 20,000 years, whereas the interglacial before it appears to have gone on for 110,000 years. This is as far back as reliable core data go, although a single core from the equatorial Atlantic appears to carry the Pleistocene record back at least 600,000 years.

Since the last glaciation ended 11,000 years ago, and since the minimum interglacial period seems to have been 30,000 years, it would seem that man can look forward to at least 20,000 more years of the present mild climate, if not to even warmer weather. If a warmer climate should melt the glaciers that remain today, the sea level would rise, but probably no more than 10 meters or so. This would be a considerable nuisance—it would put much of New York City under

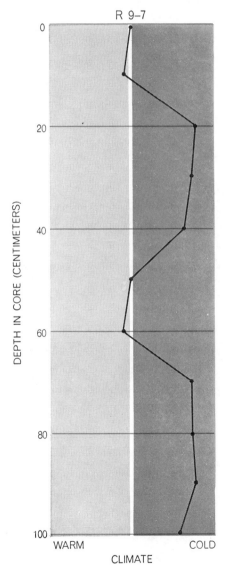

R 9-7

DEPTH IN CORE (CENTIMETERS)

WARM   COLD

CLIMATE

show (*left curve*) relative percentages of left- and right-coiling *G. pachyderma* at 10-centimeter intervals in core R 9–7, and abundance of cold and warm forms of all other planktonic foraminifera in core (*right*).

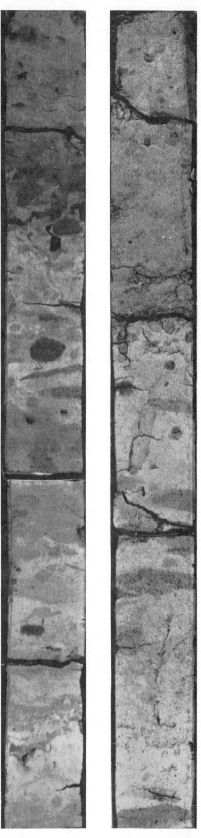

DEEP-SEA CORES represent a vertical cross section of ocean-bottom sediments. These are small parts of cores A 180–72 (*left*) and A 180–73 (*right*). Sometimes layers are quite apparent in cores; at other times no layers can be seen with the naked eye, but the microscope reveals sharp changes in the fossil contents of the undersea sediments.

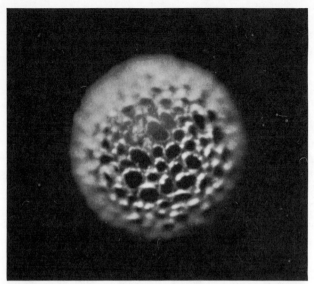

RADIOLARIA produce lacy skeleton that looks like clear spun glass. It is made of opal, which tends to dissolve in water.

OSTRACODES are related to crabs and lobsters but have two valves, or shells, that closely resemble shells of tiny clams.

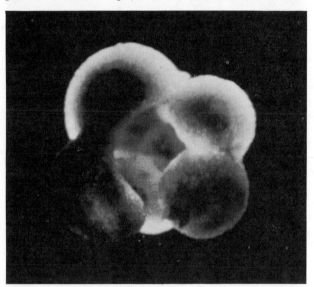

GLOBIGERINA BULLOIDES is found in cool to cold water. This and forms that follow are all species of planktonic foraminifera.

GLOBIGERINA INFLATA is a cool-water form found in middle latitudes. It occurs only in sediments of the Pleistocene epoch.

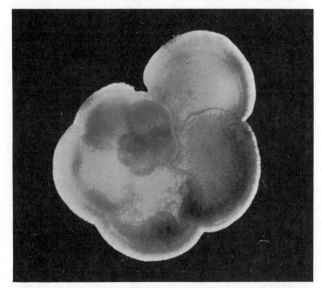

GLOBOROTALIA MENARDII is found in waters of mid-latitudes and tropics. All Pleistocene forms of this species coiled to left.

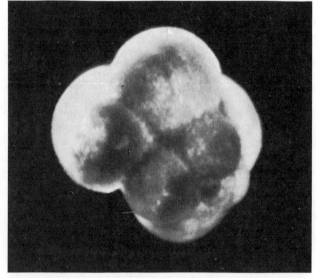

GLOBIGERINA PACHYDERMA is a climate indicator for northern seas. Shells on this page are enlarged 60 to 120 diameters.

water, for example—but it would hardly threaten the existence of mankind.

Arriving at accurate dates for the late Pleistocene would have been far easier if it were true, as geologists formerly believed, that marine sediments accumulate everywhere at the rate of one centimeter per 1,000 years. Then each sample core would represent the same time scale. Our studies of hundreds of cores show that the rate of accumulation can be that show, but in many places it is as much as 50, 100 and even 250 centimeters in 1,000 years, depending largely

on the underwater topography. "Turbidity" currents, consisting of silt-laden water denser than surrounding water, frequently run down even gentle ocean-bottom slopes, depositing several meters of mud in a few hours in some places; in other places they may scour away sediments accumulated over thousands of years [see "The Origin of Submarine Canyons," by Bruce C. Heezen; SCIENTIFIC AMERICAN, August, 1956]. Another phenomenon responsible for large variations is slumping, in which the sediments simply slide away down a

steep slope. The removal of upper layers of sediments by these processes is not all bad, however; it has brought 100-million-year-old fossils near enough to the sea floor to be reached by our coring tubes. So far we have not brought up a core in which the entire upper part of the Pleistocene record has been removed, leaving the older ice-age sediments intact. Several cores of this type, or longer cores from places already sampled, would carry our time scale back to the beginnings of human evolution.

The coiling direction of shells of *Globorotalia truncatulinoides* is proving extremely useful in matching sediments from various locations. In our laboratory we have discovered that the ratio between right-coiling and left-coiling shells varies from place to place in such a way as to define three distinct geographical provinces in the North Atlantic [see upper illustration on page 100]. Carbon-14 dating of samples from cores shows that the present pattern of distribution has persisted for about 10,000 years. Evidently some environmental factor has maintained the pattern in spite of the mixing effect of general ocean circulation. Going back in time by determining coiling ratios in fossil samples taken every 10 centimeters in cores, we find that the pattern of distribution changed rather suddenly from time to time during the late Pleistocene, presumably in response to changing currents or shifting water masses. Although we cannot yet say just what the changes were, we can match layers in different cores by means of the coiling-direction ratios. Petroleum geologists in Europe and India have applied this same method to other species to match strata penetrated by oil wells.

In the case of *Globigerina pachyderma*, the dominant direction of coiling follows closely the temperature of surface water in the northernmost Atlantic. Again we find evidence of shifts in the boundary between the left-coiling (cold water) and right-coiling (warm water) populations. Here we believe the shifts were determined directly by temperature changes in the late Pleistocene [see illustration on pages 102 and 103]. The coiling changes in a typical core show a period of cold climate preceded by a time of mild climate during which right-coiling was as strongly dominant as it is today. In this lower zone of the core and at the top of the core we find abundant fossils of various other warm-water species; these are absent in the intervening zone of left-coiling. (A direct causal relation between coiling direction and

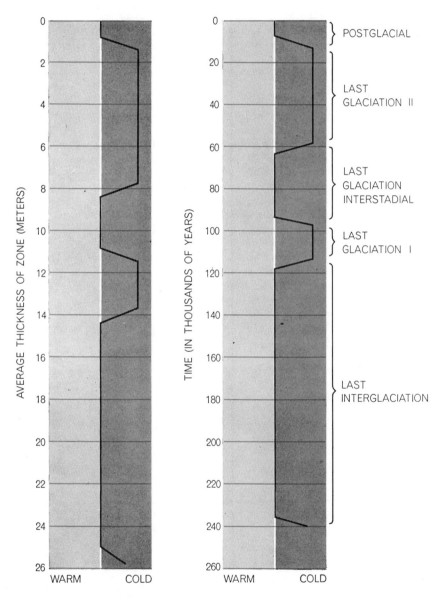

CHRONOLOGY FOR LATE PLEISTOCENE (right) is drawn from micropaleontological studies of planktonic foraminifera in 108 deep-sea sediment cores and on extrapolations of the rate of sedimentation from 37 carbon-14 dates in 11 of the cores. The authors consider this the most accurate timetable available for the past 240,000 years. Present-day climate is at center line of diagrams. The average-thickness curve (left) shows greater rate of sedimentation during glaciations than during interglacial or interstadial periods. This is caused by lowering of seas, which exposes continental shelf. Rivers thus carry material from continents out to edge of shelf and dump land sediments into deep sea, whereas in warmer times sea is higher and continental sediments are deposited on shelf.

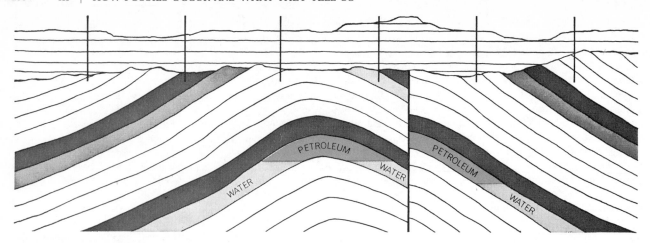

CROSS SECTION OF OIL POOL shows "slim holes," or exploratory bore holes (*vertical lines*), passing through a series of nearly horizontal sediments and into folded, faulted and partly eroded beds of an earlier era. Two beds of same dark gray tone are both shale, which is likely to lie over oil. Microfossils from bottoms of slim holes show relative ages of beds and thereby reveal existence of fold and fault. The deeper shale bed is of course much older. Although slim holes have not struck oil, they have shown presence of a dome that in this case contains oil, as well as the fault that has moved part of the oil-bearing stratum upward.

temperature is hardly conceivable. If both the direction of coiling and the temperature tolerance are determined genetically, the two characteristics may be associated because their genes are linked.)

Such changes in coiling direction of *Globigerina pachyderma* at lower levels in sediment cores provide insight into oceanographic conditions in the North Atlantic during the late Pleistocene. For example, left-coiling dominates from top to bottom of all cores taken in the area of present-day left-coiling. This indicates that the inflow of relatively warm Atlantic water into the Norwegian Sea was never greater in the late Pleistocene than it is today. Again, the change to left-coiling directly below the tops of cores in the area of present-day right-coiling implies that the inflow of warm

Atlantic water decreased during the last glaciation. As yet we cannot tell whether this resulted from a decrease in energy of the general circulation of the North Atlantic or from the lower sea level that accompanied the glaciations. The latter would make the submarine ridge between Iceland and Scotland a more effective barrier to influx. We hope that further study of the foraminifera in sediment cores will answer this and similar questions. Since ocean circulation, particularly that of the North Atlantic, must have had a powerful influence on Pleistocene climates, a better understanding of this circulation may yet provide the basis for a satisfactory theory of the cause of ice ages.

Finally, our microfossil studies have thrown new light on the origin of the Atlantic Ocean. No cores from any ocean

have yielded fossils older than the late Cretaceous period, about 100 million years ago. It is particularly suggestive that the rather thorough sampling of the bottom of the Atlantic during the past 15 years has yielded no older fossils. Can we conclude from this that the Atlantic basin came into existence in its present form at some time during the Cretaceous period? To admit this possibility is to question the widely held belief in the permanence of continents and ocean basins. For a definitive answer we shall probably have to wait until samples have been raised from one or more deep borings in the Atlantic. In the meantime the accumulating evidence suggests that a drastic reorganization of that part of the earth's surface now occupied by the ocean basins took place about 100 million years ago.

IV

REEFS, DINOSAURS, MAMMALS, AND HUMANS

REEFS, DINOSAURS, MAMMALS, AND HUMANS

IV

INTRODUCTION

In this part, we want to consider several different evolutionary histories as revealed by the fossil record. In Part I, we discussed the complementary aspects of theory and history that we find in a historical science such as paleontology. Thus, Part I informed us about evolutionary theory, Part II reviewed the evidence for the earliest chapters in the history of life, and Part III provided selected examples of how past life becomes preserved in the stratigraphic record and examples of the various kinds of information that can be derived from the resulting fossils. We now want to examine the fossil record more closely to trace out some interesting biological lineages. Only by such specific consideration of the fossil record can we determine what *in fact* happened in the history of life, rather than what *might* have happened.

Each of the readings in Part IV looks at different elements of this history—different not so much in time but in the scale or level of events that are unfolding. We begin by tracing the evolution of the reef ecosystem through time, seeing how different organisms at various epochs contributed to the formation of these great shallow water structures. Next we consider the reptiles as a class and the evolutionary role they played in the history of vertebrates. The third paper reviews the fossils that lie intermediate between the reptiles and mammals, and thereby provide the structural link between these two vertebrate classes that would otherwise seem quite separate, judging from living members of each (e.g., alligators vs. shrews). The following article demonstrates how the habits and habitats of an extinct organism such as the cave bear can be reconstructed from bones and teeth. In this instance, the life history of the animal, as well as its evolutionary history, are well documented by fossils. The last two articles discuss the human fossil record, tracing first the history of *Homo erectus*, the species that gave rise to our own species, and then the evolution of that species, *Homo sapiens*, into three different geographic races. Thus in Part IV we look at historical events: the evolution of an ecosystem, reptiles into mammals, the short and happy life of cave bears, and the rise and spread of that most dominant species of all, the human one. (At least, it's the only one that studies its own fossil record!)

"The Evolution of Reefs," by Newell, explains the idea of an ecosystem, namely an interdependent association of plants and animals who continuously cycle energy and matter among themselves. The particular ecosystem in question is that of reefs that thrive in shallow tropical waters. Newell shows how this ecosystem became established during life's great expansion in the late Precambrian and early Cambrian and has stayed with us ever since. Although there have been many comings and goings of different kinds of organisms throughout the Phanerozoic, the basic organization of a "reef" has remained intact. Thus, while there has been considerable turnover in the actors, the scenario has not changed much during the long run of this evolutionary play. However, there have been times in the past when reef communities have been greatly restricted, even collapsed. Newell attributes these times of stress to the dramatic changes in climate and size of shallow seas that accompanied the fragmentation and reassembly of the continents through plate tectonics. (See the article by Valentine and Moores in Part V for more about the role of plate tectonics in the history of life.) Although reefs are only a part of the total marine environment, their history does provide a good summary of the evolutionary highlights of calcareous algae and shelly invertebrates throughout the Phanerozoic.

"Dinosaur Renaissance," by Bakker, nicely illustrates the devious ways that paleontologists must pursue to reconstruct the life of extinct organisms. Using such diverse evidence as bone microstructure, predator-prey ratios, and geographic distribution of living and fossil reptiles, Bakker comes to the startling conclusion that dinosaurs were "warm-blooded"; that is, unlike modern reptiles, they maintained a high internal body temperature. Dinosaurs survive today through their direct descendants, the birds. Still another line of reptiles, the so-called mammal-like reptiles, which led to the mammals, is also considered by Bakker to have been warm-blooded. Not only are these ideas important on their own, but Bakker's paper also shows paleontological reasoning at its best: weaving together different lines of independent evidence from the recent and the past into a coherent, internally consistent hypothesis. There is no *direct* way of telling from their scattered remains, hundreds of millions of years after the fact, if dinosaurs were indeed warm-blooded. As with most fossils, paleontologists must rely on indirect inferences to build their case for one interpretation or another. Just as art is often an original recombination of commonplace materials and images, so too is good historical science that brings together in an imaginative way seemingly unrelated observations and concepts. We should also note that, as in the Newell article, reconstruction of ancient continental configurations plays an important part in the Bakker hypothesis. The development of global plate tectonics in the 1960s provides the theory that makes possible such continental reshuffling and that thereby contributes to fundamental paleontologic theory.

Colbert's article, "The Ancestors of Mammals," describes the fossil evidence for demonstrating the evolutionary link between the reptiles and the mammals. If we just look at living reptiles—alligators, lizards, and turtles, among others—and compare them with living mammals—cats, dogs, bats, and whales and the rest—we do not see the evolutionary connection very easily. However, if we go backward in time to the late Paleozoic and early Mesozoic, we do find fossils of reptiles that approach the mammals. Appropriately enough, these fossils are called *mammal-like reptiles*.

One might assume that the reptiles that led to mammals were in some way rather advanced reptiles—that mammals only appeared at the culmination of reptile history to open a new chapter of vertebrate evolution. Such an assumption unfortunately is quite wrong. In fact, reptiles ancestral to mammals are relatively close to the late Paleozoic root stock, and these mammal-like reptiles appear early during the general reptilian radiation that was to lead to the dinosaurs, pterodactyls, and ichthyosaurs. More often than not, this is typical in evolutionary history: Advanced later descendants that become a major lineage usually evolve from the ancestral primitive stock rather than from the later, more highly specialized forms. Consider hominids: We humans evolved within the order of primates that is evolutionarily close to the ancestral stock of all mammals, the order of the insectivores. Humans did not evolve from later, highly specialized mammals such as bats or whales or elephants.

Many of the diagnostic differences between living mammals and reptiles are not fossilizable: hair, mammary glands, reproductive organs, high internal body temperature, and so on. Thus it is not always easy to differentiate fossil reptiles and mammals that are near the borderline between the two classes. Colbert points out that there is one useful discriminating hard-part character, namely the nature of the articulation between the upper and lower jaw and a related feature, the number of ear ossicles in the middle ear. Because of the loss of information that occurs during fossilization, paleontologists sometimes have to rely on such arbitrary criteria in their identification and classification of fossil materials. Yet, while not perfect, if such features are thoughtfully chosen—as in the case of mammals—they can be very useful.

"The Cave Bear," by Kurtén, provides an instructive case history of the relatively brief evolutionary life of this animal during the glacial age. Because these organisms spent the harsh winters in caves, opportunities for preservation of their bones and teeth were excellent. So much so, in fact, that literally thousands of fossils of the cave bear have been found across Europe. Moreover, these remains provide evidence about mortality rates, age structure of the populations, diet, and even the presence of such diseases as gout. Kurtén discusses the probable adaptive niche occupied by these mammals and the reasons for their eventual extinction. Compare this kind of "normal" extinction—predictable evolutionary turnover among species—with that described by Newell (Parts IV and V), whereby many different kinds of organisms die off, apparently owing to some large-scale, global environmental change. Kurtén's article thus informs us how a once-living, complex, biological system such as the cave bear can be reconstructed from just the animal's scattered, although abundant, hard parts.

The last two articles by Howells examine the fossil evidence for the evolution of our own human lineage. As we noted in Part I, when Darwin wrote the *Origin of Species* virtually the only human fossil known was the Neanderthal skull, now attributable to an archaic form of *Homo sapiens*. In the late nineteenth and early twentieth centuries, important discoveries of hominid fossils were made in South Africa, Java, China, and Europe. More recently, in the 1960s and especially in the 1970s, still more crucial finds have occurred. Indeed, the field of human paleontology is currently one of the most active and exciting in the realm of scientific exploration. We can expect that significant new theories about human prehistory over the next decade will make many of our present ideas seem quite naive and incomplete.

"Homo erectus," by Howells, reviews the relatively slow, step-by-step accumulation of the evidence for human evolutionary history. Because of the importance of individual discoveries, each specimen was often given a separate species, or even generic, classification. Before long, a most confusing terminology resulted, which badly obscured the real evolutionary affinities of the fossils with each other and with modern humans. As Howells shows, this confusion ran rampant within those specimens now all considered to be within the species *Homo erectus*, which presumably is immediately ancestral to our own, *Homo sapiens*.

Until recently it was assumed that *Homo erectus* descended directly from the so-called African ape-men, or the Australopithecines. Recent fossil finds in East Africa, however, indicate that *Homo erectus* extends backward to almost two million years and was contemporary with *Australopithecus*. Although these discoveries were made after Howells wrote this article, he did suggest that this might well be the case, noting that the "transition from an australopithecine level to an *erectus* level about a million years ago . . . seems almost too late."

The special importance of *Homo erectus* lies in the fact that this was the first really widespread hominid—earlier species being restricted mainly to Africa—that ranged across tropical and temperate regions of the Eastern Hemisphere. The cultural achievements, too, of this species were quite advanced and included the controlled use of fire, reasonably elaborate making and using of tools, communal hunting of large game, and at least rudimentary language. In short, it appears that *Homo erectus* represents the evolutionary stage of humans that radiated far and wide from their earlier, more restricted African savanna habitat, a stage increasingly able to cope with the surrounding environment. Whereas earlier hominids might be thought of as an initial testing of the evolutionary waters, *Homo erectus* is the first full-blown realization of what it means to be human.

In "The Distribution of Man," Howells continues the story of hominid evolution and discusses the rise of geographic variants, or races, that is due to the very wide distribution of *Homo sapiens*. Although discussions about race are too often sociologically loaded, it is legitimate to postulate that at least some of the geographic variation seen among humans is simply the result of natural selection, producing strains most adaptable for the given environment. We can think of these racial differences as delicate adjustments to the local habitat. Such differences today have little evolutionary significance, because cultural adaptations to the environment are more important and because ability to interbreed indicates that the genetic bases for these differences are trivial. As Howells points out, some of these differences might merely be the spread of mutations by genetic drift within small populations. In any event, we should marvel at the richness of human diversity, both in physical traits and cultural achievements, rather than let these differences become sources of suspicion, antagonism, or fear.

SUGGESTED FURTHER READING

Bakker, R. T. 1975. "Experimental and Fossil Evidence of the Evolution of Tetrapod Bioenergetics," in *Perspectives of Biophysical Ecology*, D. Gates and R. Schmerl, ed. New York: Springer Verlag. A somewhat difficult, but important, article that examines various levels of metabolism and thermoregulation in spiders, amphibians, reptiles, birds, and mammals. Bakker's approach is ingenious and thoughtful, nicely tying together how the different organisms exploit their energy resources.

Beerbower, J. R. 1968. *Search for the Past*. Englewood Cliffs, N.J.: Prentice-Hall. An introductory paleontology text that not only discusses basic principles, but also provides clear and succinct histories of major invertebrate and vertebrate groups.

Heckel, P. 1974. "Carbonate Buildups in the Geologic Record," *Society of Economic Paleontologists and Mineralogists Special Publication*, vol. 18, pp. 90–154, Tulsa, Okla. An in-depth treatment of the geological history of reefs in terms of their stratigraphy and sedimentology, contributing organisms, and environmental controls; abundant comprehensive references.

Isaac, G. Ll. and McCown, E. R. 1976. *Human Origins: Louis Leakey and the East African Evidence*. Menlo Park, Calif.: W. A. Benjamin. Rapid advances in human paleontology have made even recent texts somewhat obsolete. This volume presents many of the new paleontological and archaeological data relating to human origins. (If you want to really keep up on this field, keep your eye on *Science* and *Nature*, two periodical magazines that regularly announce latest discoveries and ideas.)

The Evolution of Reefs

by Norman D. Newell
June 1972

*The community of plants and animals that builds
tropical reefs is descended from an ecosystem of two
billion years ago. The changes in this community
reflect major events in the history of the earth*

To a mariner a reef is a hazard to navigation. To a skin diver it is a richly populated underwater maze. To a naturalist a reef is a living thing, a complex association of plants and animals that build and maintain their own special environment and are themselves responsible for the massive accumulation of limestone that gives the reef its body. The principal plants of the reef community are lime-secreting algae of many kinds, including some whose stony growths can easily be mistaken for corals. The chief animal reef-builders today are the corals, but many other marine invertebrates are important members of the reef community.

This association of plants and animals in the tropical waters of the world is the most complex of all ocean ecological systems; as we shall see, it is also the oldest ecosystem in earth history. Its closest terrestrial counterpart, in terms of organization and diversity, is the tropical rain forest. Both settings evoke an image of exceptional fertility and exuberant biomass. Both are dependent on light in much the same way; the sunlight filters down through a stratified canopy, and the associations at each successive level consist of organisms whose needs match the available illumination and prevailing conditions of shelter. There is even a parallel between the birds of the rain forest and the crabs and fishes of the reef. Both play the part of lords and tenants, yet their true role in the history and destiny of the community is essentially passive.

It is a common belief that a reef consists principally of a rigid framework composed of the cemented skeletons of corals and algae. In reality more than nine-tenths of a typical reef consists of fine, sandy detritus, stabilized by the plants and animals cemented or otherwise anchored to its surface. Physical and biochemical processes that are little understood quickly convert this stabilized detritus into limestone. The remains of dead reef organisms make a substantial contribution to the detritus. This major component of the reef, however, has a fabric quite different from the upward-growing lattice of stony algal deposits and intertwined coral skeletons that forms the reef core.

The Reef Community

The interaction of growth and erosion gives the reef an open and cavernous fabric that in an ecological sense is almost infinitely stratified and subdivided. In the dimly lit bottom waters at the reef margin, rarely more than 200 feet below the surface, caves and overhanging ledges provide shelter for the plants and animals that thrive at low levels of illumination. From the bottom to the surface is a succession of reef-borers, cavern dwellers, predators and detritus-feeders, each living at its preferred or obligatory depth, that includes representatives of nearly every animal phylum. Near and at the surface the sunlit, oxygen-rich and turbulent waters provide an environment that contributes to a high rate of calcium metabolism among the myriads of reef-builders active there.

The most familiar of the reef animals, the corals, are minute polyps that belong to the phylum Coelenterata. The polyps live in symbiosis with zooxanthellae, microscopic one-celled plants embedded in the animals' tissue, where they are nourished by nitrogenous animal wastes and, through photosynthesis, add oxygen to the surrounding water. Experiments show that the zooxanthellae promote the calcium metabolism of the corals. The corals themselves are carnivores; they feed mainly on small crustaceans and the larvae of other reef animals.

The limestone-secreting algae—blue-green, green and red—are the principal food base of the reef community, just as the plant life ashore nourishes terrestrial herbivores. The algae are distributed across the reef in both horizontal and vertical zones. The blue-green algae are most common in the shallows of the tidal flat, an area where red algae are absent. The green algae are predominantly back-reef organisms and the reds are mostly reef and fore-reef inhabitants [*see illustration on page 123*].

The other important members of the reef community are all animals. Next in significance to the corals as reef-builders are several limestone-secreting families of sponges, members of the phylum Porifera. The phylum Protozoa is represented by a host of foraminifera species whose small limy skeletons add to the deposits in and around reefs. Several species of microscopic colonial animals of the phylum Bryozoa also contribute their limestone secretions, as do the spiny sea urchins and elegant sea lilies of the phylum Echinodermata, the bivalves of the phylum Brachiopoda and such representatives of the phylum Mollusca as clams and oysters, all of whose accumulated skeletons and shells contribute to the reef limestones.

Many organisms in the community do not add significantly to the reef structure; some burrowers and borers are even destructive. The marine worms that inhabit the reef are soft-bodied and thus incapable of contributing to the reef mass. The hard parts of such reef dwellers as crabs and fishes are systematically consumed by scavengers. A few fragments may escape, but except for such passive and minor contributions to the reef detritus these organisms are not reef-builders.

The reef community is adapted to a low-stress environment characterized by the absence of significant seasonal change. The mean winter temperature of the water where reefs grow is between 27 and 29 degrees Celsius and the difference between summer and winter monthly mean temperatures is three degrees C. or less. The water is clear (so that the penetration of light is at a maximum), agitated (so that it is rich in oxygen) and of normal salinity. Even under these ideal circumstances many reef organisms (for example corals) do not grow at depths greater than 65 feet. This adaptation to freedom from stress makes the community remarkably sensitive to environmental change.

The fossil record documents hundreds of episodes of sweeping mass extinction, some continent-wide and others worldwide. These times of ecological disruption have simultaneously affected such disparate organisms as ammonites at sea and dinosaurs ashore, the plants on land and the protozoans afloat among the

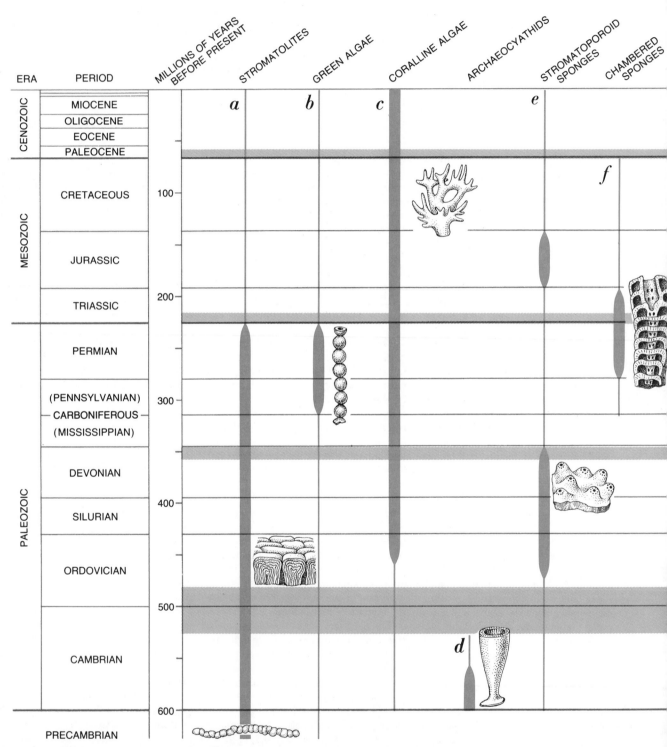

FOUR COLLAPSES (*bands of color*) have altered the composition of the reef community since the initial association between plant and animal reef-builders was established nearly 600 million years ago. This was when a group of spongelike animals, the archaeocyathids (*d*), appeared among the very much older reef-forming algal stromatolites (*a*) at the start of the Paleozoic era. In less than 70 million years the archaeocyathids became extinct; their demise marks the first community collapse. A successor community arose in mid-Ordovician times. Its members included coralline algae (*c*); the first corals, tabulate (*h*) and rugose (*i*); stromatoporoid sponges (*e*), and communal bryozoans (*m*). The group flourished almost to the end of the Devonian period, some 350 million years ago, the

oceanic plankton. The causative phenomena underlying the disruptions must therefore be unlike the ordinary, Darwinian causes of extinction—natural selection and unequal competition—that tend to affect species individually and not en masse.

For generations, in a reaction to the biblical doctrine of catastrophism that dominated 18th-century geology, scholars have viewed the apparent lack of continuity in fossil successions with skepticism. The breaks in the record, they proposed, were attributable to inadequate collections or to accidents of fossil preservation. At the same time, certain pioneers—T. C. Chamberlin and A. W. Grabau in the U.S. and Hans Stille in Germany—saw the breaks in fossil continuity as reflections of real events and sought a logical explanation for them. These men were eloquent proponents of a theory of rhythmic pulsations within the earth: diastrophic move-

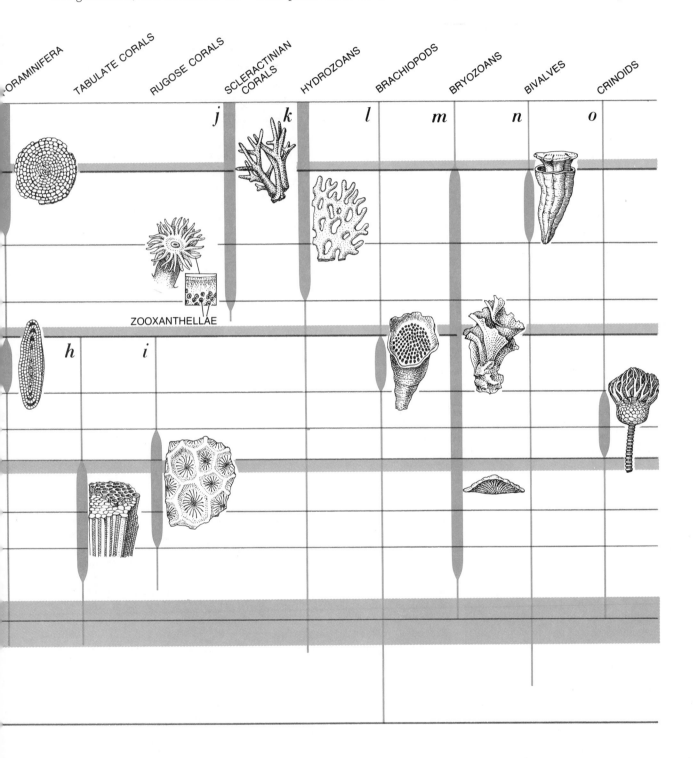

time of the second collapse. Its successor, some 13 million years later, contained a new sponge group (f) and increased numbers of green algae (b), foraminifera (g), brachiopods (l) and crinoids (o). These reef-builders flourished until the end of the Paleozoic era and the third collapse. The next resurgence occupied most of the Mesozoic era. It was marked by the appearance of modern cor-als (j) and a dramatic upsurge of a mollusk group, the rudists (n), that became extinct at the time of the fourth community collapse some 65 million years ago. Draining of shallow seas during the Cenozoic era produced cooler climates and led to formation of the Antarctic ice cap. Both developments have been factors in restricting the successor community in diversity and distribution.

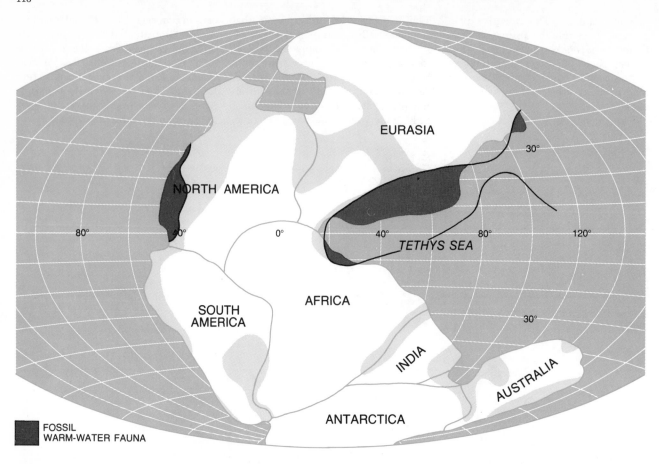

EURASIA

30°

NORTH AMERICA

80°   40°   0°   40°   80°   120°

TETHYS SEA

SOUTH
AMERICA

AFRICA

30°

INDIA

AUSTRALIA

ANTARCTICA

FOSSIL
WARM-WATER FAUNA

TWO FACTORS that have affected world geography and climate are the movement of the continental plates and their greater or lesser invasion by shallow seas. The status of both factors during three key intervals in earth history is shown schematically here and on the opposite page. Near the end of the Paleozoic era (*a*) the continental plates had gathered into a single land area. Many of the reef-building species were then pantropical in distribution. By the end of the Mesozoic era (*b*) sea-floor spreading had separated the continental plates. The Atlantic Ocean had become enough of a barrier to the migration of reef organisms between the Old and New World to allow the evolution of species unique to each region. At the end of the Mesozoic era the shallow seas encroaching on the continents were drained completely. Early in the Cenozoic era (*c*) the shallow seas had been reestablished in certain zones, but the total continental area above water was larger than in the Mesozoic. The change resulted in a trend toward greater seasonal extremes of climate; the distribution of tropical and subtropical organisms, however, remained quite broad, as palm-tree fossils show.

ments that had been accompanied by significant fluctuations in sea level and consequent disruptive changes in climate and environment.

The lack of any demonstrable physical mechanism that might have produced such simultaneous worldwide geological revolutions kept a majority of geologists and paleontologists from accepting the proposed theory of the origins of environmental cycles. Today, however, we have in plate tectonics a demonstrated mechanism for changes in sea level and shifts in the relative extent of land and sea such as Chamberlin and his colleagues were unable to provide [see the article "Plate Tectonics," by John F. Dewey; Offprint 900]. Significant changes in the volume of water contained in the major ocean basins, produced by such plate-tectonic phenomena as alterations in the rate of

lava welling up along the deep-ocean ridges or in the rate of set-floor spreading, have resulted sometimes in the emergence and sometimes in the flooding of vast continental areas. The changing proportions of land and sea meant that global weather patterns alternated between mild maritime climates and harsh continental ones.

Now, reef-building first began in the earth's tropical seas at least two billion years ago. As we have seen, the modern reef community is so narrowly adapted to its environment as to be very sensitive to change. It seems only logical to expect that the same was true in the past, and that changes in reef communities of earlier times would faithfully reflect the various rearrangements of the earth's land masses and ocean basins that students of plate tectonics are now documenting. The expectation is justified;

whereas many details of earth history will be clarified only by future geological and paleontological research, the record of fossil reef communities accurately delineates a number of the main catastrophic episodes.

As one might expect, the oldest of all types of reef is the simplest. Algae alone, without associated animals, are responsible for limestone reef deposits, billions of years old, that were formed in the seas of middle and late Precambrian times.

The Precambrian algae produced extensive accumulations of a distinctively laminated limestone that are found, flanked by aprons of reef debris, in rock formations around the world. Geologists call these characteristic limestone masses stromatolites, from the Greek for "flat" and "stone." The microscopic organisms that built the stromatolites are rarely

b

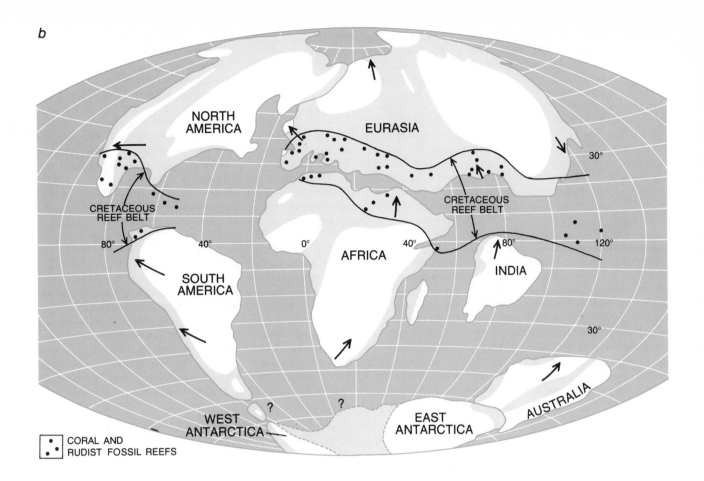

NORTH AMERICA

EURASIA

CRETACEOUS REEF BELT

80°

40°

0°

AFRICA

40°

CRETACEOUS REEF BELT

30°

80°

120°

SOUTH AMERICA

INDIA

30°

?

?

WEST ANTARCTICA

EAST ANTARCTICA

AUSTRALIA

CORAL AND
RUDIST FOSSIL REEFS

c

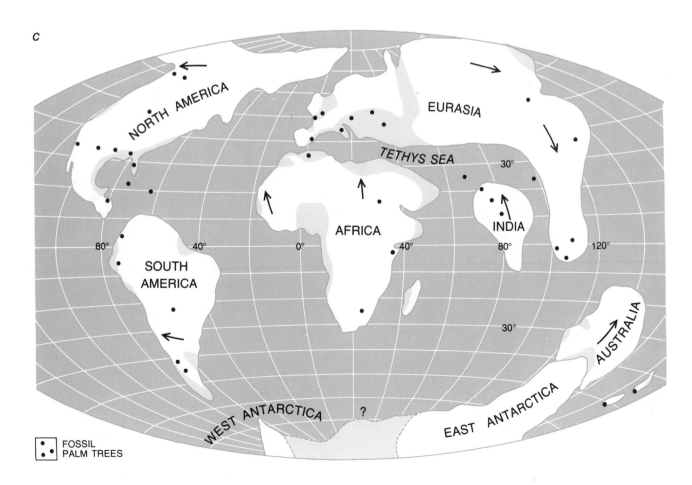

NORTH AMERICA

EURASIA

TETHYS SEA

30°

80°

40°

0°

AFRICA

40°

INDIA

80°

120°

SOUTH AMERICA

30°

WEST ANTARCTICA

?

EAST ANTARCTICA

AUSTRALIA

FOSSIL
PALM TREES

preserved as fossils, but they must surely have been similar to the filamentous blue-green algae that form similar masses of limestone today.

The accomplishments of these Precambrian algal reef-builders were not inconsiderable: individual colonies grew upward for tens of feet. They did so by trapping detrital grains of calcium carbonate and perhaps by precipitating some of the lime themselves. The resulting fossil bodies take the form of trunklike columns or hemispherical mounds.

As I outline the evolution of the reef community the reader will note that I often speak of first and last appearances. This does not mean, of course, that the organism being discussed was either instantly created or instantly destroyed. Each had an extensive evolutionary heritage behind it when some chance circumstance provided it with an appropriate ecological niche. Similarly, the decline and extinction of many major groups of organisms can be traced over

periods of millions of years, although in numerous instances the time involved was too short to be measured by the methods now available.

Enter the Animals

The long Precambrian interval ended some 600 million years ago. The opening period of the Paleozoic era, the Cambrian, saw the first establishment of a reef community. The stromatolites' first partners were a diverse group of stony, spongelike animals named archaeocyathids (from the Greek for "ancient" and "cup"). In early Cambrian times these stony animals rooted themselves along the stromatolite reefs, grouped together in low thickets or scattered like shrubs in a meadow. It is not hard to imagine that the vacant spaces within and between these colonies provided shelter for the numerous bottom-feeding trilobites that inhabited the Cambrian seas. Not every reef harbored archaeocyathids, however; some reefs of early and

middle Cambrian times are composed only of stromatolites.

By the end of the middle Cambrian, some 540 million years ago, the archaeocyathids had vanished. No single cause for this extinction, the first of the four major disasters to overtake the reef community, can be identified. One imaginable cause—competition from another reef-building animal—can be ruled out completely. The seas remained empty of any kind of reef-building animal throughout the balance of the Cambrian and until the middle of the Ordovician period, some 60 million years later. All reefs built during this long interval were the work of blue-green algae alone.

Fossil formations in the Lake Champlain area of New York, a region that lay under tropical seas in middle Ordovician times some 480 million years ago, contain the first evidence of a renewed association between reef-building plants and animals. The community that now arose was a rather complex one. Stromatolites continued to flourish and a second kind

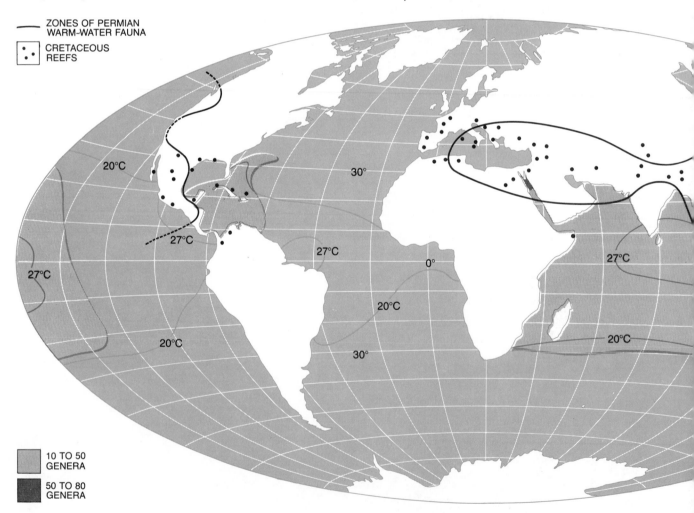

ZONES OF PERMIAN WARM-WATER FAUNA

CRETACEOUS REEFS

10 TO 50 GENERA

50 TO 80 GENERA

REEF-BUILDERS TODAY are confined to a narrow belt, mainly between 30 degrees north and 30 degrees south latitude (*light gray*), and even within this belt the greatest number of species are found where the minimum average water temperature is 27 degrees Celsius. Paleozoic and Mesozoic reef fossils, however, are found in areas far outside today's limits (*black lines and dots*). This suggests that the true Equator in those times lay well to the north of the equators shown on the maps on the preceding pages. The asym-

of plant life appeared: the coralline red alga *Solenopora,* a direct progenitor of the modern coralline algae. Colonial bryozoans, previously insignificant in the fossil record, now assumed an important role in the expanding reef community. Animal newcomers included a group of stony sponges, the stromatoporoids, some shaped like encrusting plates and others hemispherical or shrublike in form. These calcareous sponges were to play a major role in the community for millions of years. The most significant new animals, however, in light of subsequent developments, were certain stony coelenterates: the first of the corals. The intimate collaboration between algae and corals, apparently unknown before middle Ordovician times, has continued (albeit with notable fluctuations) to the present day.

These new arrivals and the other corals that appeared during the Paleozoic era were mainly of two types. In one type the successive stages of each polyp's upward growth were recorded as a

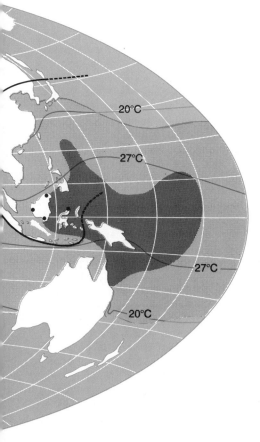

metry of the fossil-reef belt with respect to the present belt suggests that much of a once wider array has been engulfed by subduction along continental-plate boundaries.

series of parallel floors that subdivide the stony tube sheltering the animal; these organisms are called tabulate corals. The conical or cylindrical stone tube that sheltered the second type has conspicuous external growth wrinkles on its surface; these corals are called rugose.

A little more than 350 million years ago, near the end of the Devonian period, worldwide environmental changes caused a number of mass extinctions. Among the victims were many previously prominent marine organisms, including several groups in the reef community. That community now underwent a major retrenchment. Up to this point the tripartite association between algae, sponges and corals that first appeared in Ordovician times had proliferated for 130 million years without significant disturbance. The environmental alterations that nearly wiped out the reef community at the end of that long and successful period of radiation remain unidentified, although one can conjecture that a change from a mild maritime climate to a harsh continental one probably played a part. In any event the episode was severe enough so that only scarce and greatly impoverished reef communities, consisting in the main of algal stromatolites, survived during the next 13 million years. Not until well after the beginning of the Carboniferous period was there a community resurgence.

Some 115 million years passed between the revival of the reef community in Carboniferous times and the end of the Paleozoic era; the interval includes most of the Mississippian and all of the Pennsylvanian (the two subdivisions of the Carboniferous) and the closing period of the Paleozoic, the Permian. The revitalized assemblage that radiated in the tropical seas during this time continued to include stromatolites, numerous bryozoans and brachiopods and a dwindling number of rugose corals. Except for these organisms, however, the community bore no great resemblance to its predecessor of the middle Paleozoic. Both the stromatoporoid sponges and the tabulate corals are either absent from Carboniferous and Permian reef deposits or are present only in insignificant numbers.

Two new groups of calcareous green algae, the dasycladaceans and codiaceans, now attained quantitative importance in the reef assemblage. As if to match the decline of the stromatoporoid sponges, a second poriferan group—the calcareous, chambered sphinctozoan sponges—entered the fossil record. At the same time a group of echinoderms—the crinoids, or sea lilies—assumed a

larger role in the reef community. As the Paleozoic era drew to a close, the crinoids and the brachiopods achieved their greatest diversity; their skeletons preserved in Permian reef formations number in the thousands of species.

The Third Collapse

Half of the known taxonomic families of animals, both terrestrial and marine, and a large number of terrestrial plants suffered extinction at the end of the Paleozoic era. The alteration in environment that occurred then, some 225 million years ago, had consequences far severer than those of Devonian times. In the reef community the second successful radiation—based principally on a new tripartite association involving algae, bryozoans and sphinctozoan sponges—came to an end; reefs are unknown anywhere in the world for the first 10 million years of the Mesozoic era.

What was the cause of this vast debacle? There is little enough concrete information, but analogy with later and better-understood events encourages the conjecture that once again unfavorable changes in climate and habitat were major factors. In late Paleozoic times all the continents had come together to form a single vast land mass: Pangaea. Continental ice sheets appeared in the southern part of this supercontinent, a region known as Gondwana, in Carboniferous and early Permian times. These glaciers are concrete evidence of cooler climates; paleomagnetic studies show that the glaciated areas were then all located near the South Pole. Any relation between these early Permian ice sheets and the widespread late-Permian extinctions, however, is not yet evident. What is probably more significant is evidence that, at least during a brief interval, all the shallow seas that had invaded continental areas were completely drained at the end of the Paleozoic era. Serious climatic consequences must have resulted from the disappearance of a mild, primarily maritime environment.

In late Paleozoic times a wide tropical seaway, the Tethys, almost circled the globe. The only barrier to the Tethys Sea was formed by the combined land masses of North America and western Europe, which were then connected. The tongue of the Tethys that eventually became the western Mediterranean constituted one end of the seaway. The opposite end invaded the west coast of North America so deeply that great Permian reefs arose in what is now Texas.

The Mediterranean extremity of the Tethys Sea was the setting for a significant development when, after 10 million years of eclipse, the reef community was once again revitalized. There, in mid-Triassic times, a new group of corals, the scleractinians, made its appearance. The scleractinians were the progenitors of the more than 20 families of corals living in the reef community today. At first the new coral families, six in all, were represented in only a few scattered reef patches found today in Germany, in the southern Alps and in Corsica and Sicily. Even by late Triassic times, some 200 million years ago, the new corals were still subordinate as reef-builders to the calcareous algae.

The Mesozoic Community

During the 130 million years or so of Jurassic and Cretaceous times the reef community once again thrived in many parts of the world. The stromatoporoid sponges, all but extinct since the community collapse of Devonian times, returned to a position of some importance during the Jurassic. The new coral families steadily increased in diversity and reached an all-time peak in the waters bordering Mediterranean Europe during the Cretaceous period. In that one region there flourished approximately 100 genera of scleractinians; this is a greater number than can be found worldwide today. The reef community in this extremity of the Tethys Sea was also rich in other reef organisms. It included the two groups of sponges and such reef-builders as sea urchins, foraminifera and various mollusks. In addition a hitherto minor group of coralline red algae, the lithothamnions, now began to play an increasingly important role. By this time the stromatolites, important reef-builders throughout the Paleozoic, were no longer conspicuous members of the reef community.

Early in the Cretaceous period, some 135 million years ago, there was an unfavorable interval: reefs are unknown in the fossil record for some 20 million years thereafter. This pause merely set the stage, however, for a further efflorescence. Both in the Mediterranean area and in the waters of the tropical New World, hitherto unknown or insignificant coral families appeared. The present Atlantic Ocean was just beginning to form. Regional differences between reef communities in the Old World and the New that appeared at this time testify to the growing effectiveness of the Atlantic deeps as a barrier to the ready migration of reef organisms.

An extraordinary evolutionary performance was now played. Certain previously obscure molluscan members of the reef community, bivalves known as rudists, abruptly came into prominence as primary reef-builders. The next 60 million years saw a phenomenal rudist radiation that brought these bivalves to the point of challenging the corals as the dominant reef animals. Along the sheltered landward margins of many fringing barrier reefs rudists largely supplanted the corals. Their cylindrical and conical shells were cemented into tightly packed aggregates that physically resembled corals, and many of the aggregates grew upward in imitation of the coral growth pattern. Before the end of the Cretaceous period some 65 million years ago the rudists had attained major status in the reef community. Then at the close of the Cretaceous they quite abruptly died out everywhere.

Ever since the pioneer days of geology in the 18th century the end of the Cretaceous period has been known as a great period of extinctions. Nearly a third of all the families of animals known in late Cretaceous times were no longer alive at the beginning of the Cenozoic era. The reef community was not exempt; in addition to the rudists, two-thirds of the known genera of corals died out at this time. Forms of marine life other than reef organisms were also hard hit. The ammonites, long a major molluscan group, had suffered a decline during the final 10 million years of the Cretaceous; by the end of the period they were all extinguished. The belemnites, another major group of mollusks, declined sharply, as did the inoceramids, a diverse and abundant group of clams that had previously flourished worldwide. Among the foraminifera, the free-floating, planktonic groups suffered in particular.

The environmental changes that devastated life at sea also took their toll of land animals. Perhaps the most spectacular instance of extinction ashore involved the group that had been the dominant higher animals during most of the Mesozoic era: the dinosaurs. Of the 115-odd genera of dinosaurs found in late Cretaceous fossil deposits, none survived the end of the period. The concurrent breakdown of so many varied communities of organisms clearly suggests a single common cause.

What was the source of this biological crisis? We have now come sufficiently close to the present to have a better grasp of the evidence. Throughout almost all of Mesozoic times life both on land and in the sea seems to have been remarkably cosmopolitan. A broad belt

of equable climate extended widely in both directions from the Equator and there was no very evident segregation of organisms into climatic zones. The earth has always been predominantly a water world; the total land area may never have exceeded the present 30 percent of the planet's surface and has often been as little as 18 percent. In the late Cretaceous almost two-thirds of today's land area was submerged under shallow, continent-invading seas. It appears that during times of extensive inundation such as this one there was nothing like the blustery global circulation of air and strong ocean currents of today.

Paleoclimatology reveals the contrast between contemporary and Cretaceous conditions. By measuring the proportions of different isotopes of oxygen present in the carbonate of fossil foraminifera and mollusks it is possible to calculate the temperature of the water when the animals were alive. Today the temperature of deep-ocean water is about three degrees C. Cesare Emiliani of the University of Miami has shown that early in the Miocene epoch, some 20 million years ago, the bottom temperature of the deep ocean was about seven degrees C. He finds that in the Oligocene, 10 million years earlier, the temperature was about 11 degrees and that in late Cretaceous times, 75 million years ago, it was 14 degrees. He suggests that the onset of cooling may have been lethal to dinosaurs. In any event it is clear that cold bottom water has been accumulating in the deep-ocean basins since early in the Cenozoic era.

The accentuated seasonal oscillations in temperature and rainfall at the end of the Cretaceous period, a trend that evidently started when the shallow continental seas began to drain away into the deepening ocean basins, have been credited by Emiliani and also by Daniel I. Axelrod of the University of California at Davis and Harry P. Bailey of the University of California at Riverside with the simultaneous reduction in numbers or outright extinction of many other animal and plant species at that time. Genetically adapted to a remarkably equable world climate over millions of years, many of these Mesozoic organisms were ill-prepared for the extensive emergence of land from under the warm, shallow continental seas and the climate of accentuated seasonality that followed. Perhaps we should be surprised not that so many Cretaceous organisms died but that so many managed to adapt to the new conditions and thus survived.

Near the end of the Paleocene epoch, some 10 million years after the great

FOSSIL BARRIER REEF was built in upper Devonian times by a community of marine plants and animals living in the warm seas that covered part of Australia more than 350 million years ago. Exposed by later uplift and erosion, the reef now forms a belt of jagged highlands, known as the Napier Range, in Western Australia. A stream has cut a canyon (*foreground*) through the reef rock.

YOUNGER FOSSIL REEF, built some 250 million years ago in the Permian period, forms a rim of rock 400 miles long surrounding the Delaware Basin (*right*) on the border between Texas and New Mexico. Most of the reef is buried under later deposits, but one exposure forms this 40-mile stretch of the Guadalupe Mountains. El Capitan (*foreground*), a part of the reef front, is 4,000 feet high.

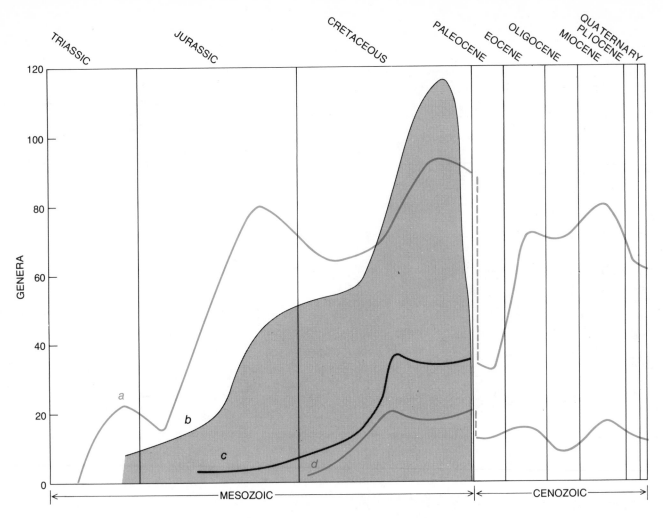

MESOZOIC INCREASE in the diversity of ocean and land animal genera reached a peak in the Cretaceous period. It was followed both by extinctions and by severe reductions in the number of genera that chanced to survive. At sea the explosively successful rudists (*c*) and ashore the long-dominant dinosaurs (*b*) became extinct. Corals (*a*) and globigerine foraminifera (*d*) fell in numbers.

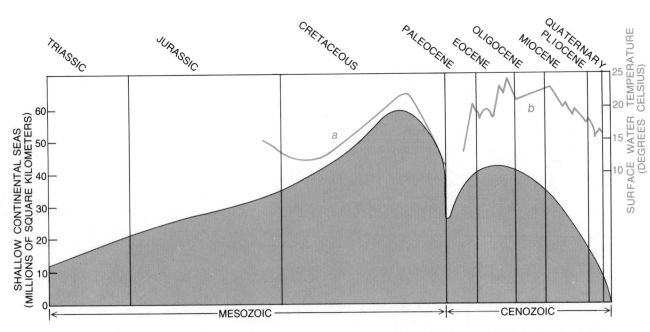

SHRINKING OF SEAS that had invaded large continental areas (*gray*) began near the end of the Mesozoic era and accelerated in the Miocene epoch. Oxygen isotopes in belemnite (*a*) and forami- nifera (*b*) skeletons show that the ocean has become progressively cooler since early in Miocene times (*color*). A relation seems to exist between a larger land area and greater seasonal extremes in climate.

Cretaceous collapse, a reef community sans rudists reappeared in the tropical seas. The following epoch, the Eocene, saw a new radiation of the scleractinian corals. Several genera that now appeared worldwide were unknown earlier in the fossil record; many of them are still living today.

The Cenozoic Decline

A sharp reduction in coral diversity that began in late Eocene times and lasted throughout the Oligocene epoch seems to reflect a continued increase in seasonality of climate and a substantial lowering of mean temperatures over large areas of the globe. Nonetheless, communities built around a bipartite association of corals and coralline algae continued to build extensive reefs in the Gulf of Mexico and the Caribbean area, in southern Europe and in Southeast Asia. Continued sea-floor spreading and deepening of the Atlantic basin, which enhanced the effectiveness of the Atlantic as a barrier to migrating corals, are evidenced in increasing differences between Caribbean and European coral species in late Oligocene fossil deposits.

By now the earth had been free from continental glaciers for almost 200 million years, but a change in the making was to have profound consequences for world climate: the Antarctic ice cap was becoming established. Fossil plant remains and foraminifera from nearby deep-sea deposits both testify that even in early Miocene times the climate in much of the Antarctic continent was not much different from the blander days of

the early Cenozoic, when palm trees grew from Alaska to Patagonia. At this time, moreover, Antarctica was some distance away from its present polar position. Nonetheless, mountain glaciers had begun to appear there millions of years earlier, in Eocene and Oligocene times. Sands of that age, produced by glacial streams and then rafted out to sea on shelf ice, are found in offshore deep-sea cores. The cooling trend was well established before the end of the Miocene epoch. In the Jones Mountains of western Antarctica lava flows of late Miocene age overlie consolidated glacial deposits and extensive areas of glacially scoured bedrock.

With the formation of the Antarctic ice cap some 15 to 20 million years ago a factor came into being that strongly influences world weather patterns to this day. The ice cap energizes the world weather system. In the broad reaches of open ocean surrounding Antarctica the surface water is aerated and cooled until it is too heavy to remain at the surface. The cold water sinks and spreads out along the sea floor, following the topography of the bottom. The result is a gravity circulation of cold water from Antarctica into the world's ocean basins, with a consequent lowering of the mean ocean temperature and cooling of the overlying atmosphere. The energetic interactions of the atmosphere above with the ocean and land below, in turn, strongly influence global wind patterns and worldwide weather. Today's climate is the product of a long cooling trend, marked by ever greater seasonal extremes; the trend became strongly ac-

centuated when the Antarctic ice cap came into being late in the Miocene epoch.

This event and others in the Cenozoic era are faithfully recorded in terms of changes in the reef community. For example, in spite of the development of new barriers to the migration of reef organisms during the Mesozoic era, such as the Atlantic deep, the reef community had remained predominantly cosmopolitan up to the close of the Cretaceous. By Miocene times, however, what had once been essentially a single pantropical community was effectively divided into two distinct biogeographic provinces: the Indo-Pacific province in the Old World and the Atlantic province in the New World.

In the Old World the increasingly unfavorable climate had eliminated the reef community in European waters. It was during Miocene times, when Australia reached its present tropical position, that the Old World reef-builders first began to colonize the shallows of the Australian shelf; in terms of maximum diversity, the headquarters of the Indo-Pacific province today lies in the Australasian region where seasonal contrasts in water temperature are minimal.

Like the Atlantic deep, the deep waters of the eastern Pacific formed a generally effective barrier to the migration of reef organisms from the Indo-Pacific province into the hospitable tropical waters along the west coast of the Americas, principally around Panama. In Miocene times this Pacific coastal pocket was still connected to the reef-rich Caribbean, the headquarters of the At-

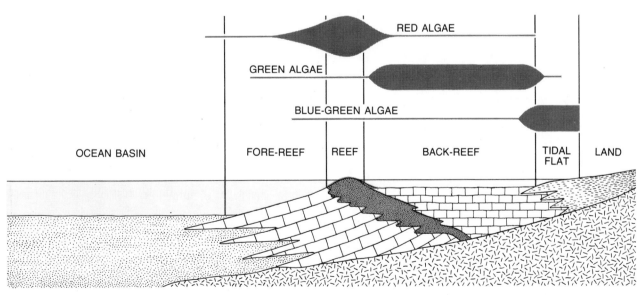

REEF ANATOMY centers on a rigid core (*color*) composed of the cemented skeletons of algae and corals. In this diagram growth of the core began at right and continued upward and outward. Most of the bulk of the reef consists of wide areas of stabilized detritus that are continuously being converted into the limestone that comprises the fore-reef and back-reef. The sandlike detritus is stabilized by the growth of the animals and plants in the reef community. The zones occupied by different algae are indicated here.

lantic province. Contact between the Atlantic province and the small Pacific enclave continued until the two areas were separated during the Pliocene epoch by the uplifting of the Isthmus of Panama.

The Pliocene saw a further contraction of the world tropics. The community of reef-builders gradually retreated to its present limits, generally south of 35 degrees north latitude and north of 32 degrees south latitude. Rather than serving as a center for new radiations, this tropical belt essentially became a haven. The epoch that followed, the Pleistocene, was marked by wide fluctuations in sea level and sharp alternations in climate that accompanied a protracted series of glacial advances and retreats. Oddly enough, the reef community was scarcely affected by such ups and downs; the main reasons for this

apparent paradox seem to be that neither the total area of the deep tropical seas nor their surface water temperature underwent much change during the Pleistocene ice ages.

Is a fifth collapse in store for the most complex of ocean ecosystems? It would be foolhardy to respond with a flat yes or no. If, however, the past is prologue, the answer to a collateral question seems clear. The question is: Will a reef community in any event survive? The most significant lesson that geological history teaches about this complex association of organisms is that, in spite of the narrowness of its adaptation, it is remarkably hardy. At the end of the long interlude that followed each of four successive collapses, the reef community entered a new period of vigorous expansion. Moreover, without exception each revitalized aggregate of reef-builders in-

cluded newcomers in its ranks.

The conclusion is inescapable that even during the times most unfavorable to the reef ecosystem the world's tropical oceans must have had substantial refuge areas. In these safe havens many of the threatened organisms managed to adapt and survive while others seem to have crossed some evolutionary threshold that had prevented their earlier appearance in the community. Today the Atlantic and Indo-Pacific provinces are two such refuge areas.

As long as the cooling trend that began in Cretaceous times does not entirely destroy these tropical refuge areas, one long-range conclusion seems firm. Any collapse of the present reef community will surely be followed by an eventual recovery. The oldest and most durable of the earth's ecosystems cannot easily be extirpated.

Dinosaur Renaissance

by Robert T. Bakker
April 1975

The dinosaurs were not obsolescent reptiles but were a novel group of "warm-blooded" animals. And the birds are their descendants

The dinosaur is for most people the epitome of extinctness, the prototype of an animal so maladapted to a changing environment that it dies out, leaving fossils but no descendants. Dinosaurs have a bad public image as symbols of obsolescence and hulking inefficiency; in political cartoons they are know-nothing conservatives that plod through miasmic swamps to inevitable extinction. Most contemporary paleontologists have had little interest in dinosaurs; the creatures were an evolutionary novelty, to be sure, and some were very big, but they did not appear to merit much serious study because they did not seem to go anywhere: no modern vertebrate groups were descended from them.

Recent research is rewriting the dinosaur dossier. It appears that they were more interesting creatures, better adapted to a wide range of environments and immensely more sophisticated in their bioenergetic machinery than had been thought. In this article I shall be presenting some of the evidence that has led to reevaluation of the dinosaurs' role in animal evolution. The evidence suggests, in fact, that the dinosaurs never died out completely. One group still lives. We call them birds.

Ectothermy and Endothermy

Dinosaurs are usually portrayed as "cold-blooded" animals, with a physiology like that of living lizards or crocodiles. Modern land ecosystems clearly show that in large animals "cold-bloodedness" (ectothermy) is competitively inferior to "warm-bloodedness" (endothermy), the bioenergetic system of birds and mammals. Small reptiles and amphibians are common and diverse, particularly in the Tropics, but in nearly all habitats the overwhelming majority of land vertebrates with an adult weight of 10 kilograms or more are endothermic birds and mammals. Why?

The term "cold-bloodedness" is a bit misleading: on a sunny day a lizard's body temperature may be higher than a man's. The key distinction between ectothermy and endothermy is the rate of body-heat production and long-term temperature stability. The resting metabolic heat production of living reptiles is too low to affect body temperature significantly in most situations, and reptiles of today must use external heat sources to raise their body temperature above the air temperature—which is why they bask in the sun or on warm rocks. Once big lizards, big crocodiles or turtles in a warm climate achieve a high body temperature they can maintain it for days because large size retards heat loss, but they are still vulnerable to sudden heat drain during cloudy weather or cool nights or after a rainstorm, and so they cannot match the performance of endothermic birds and mammals.

The key to avian and mammalian endothermy is high basal metabolism: the level of heat-producing chemical activity in each cell is about four times higher in an endotherm than in an ectotherm of the same weight at the same body temperature. Additional heat is produced as it is needed, by shivering and some other special forms of thermogenesis. Except for some large tropical endotherms (elephants and ostriches, for example), birds and mammals also have a layer of hair or feathers that cuts the rate of thermal loss. By adopting high heat production and insulation endotherms have purchased the ability to maintain more nearly constant high body temperatures than their ectothermic competitors can. A guarantee of high, constant body temperature is a powerful adaptation because the rate of work output from muscle tissue, heart and lungs is greater at high temperatures than at low temperatures, and the endothermic animal's biochemistry can be finely tuned to operate within a narrow thermal range.

The adaptation carries a large bioenergetic price, however. The total energy budget per year of a population of endothermic birds or mammals is from 10 to 30 times higher than the energy budget of an ectothermic population of the same size and adult body weight. The price is nonetheless justified. Mammals and birds have been the dominant large and medium-sized land vertebrates for 60 million years in nearly all habitats.

In view of the advantage of endothermy the remarkable success of the dinosaurs seems puzzling. The first land-

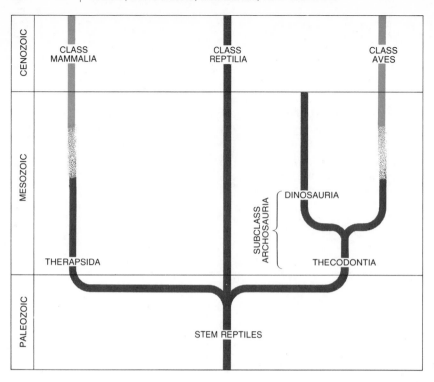

USUAL CLASSIFICATION of land vertebrates (excluding the Amphibia) is diagrammed here in a highly simplified form. The classes are all descended from the original stem reptiles. Birds (class Aves) were considered descendants of early thecodonts, not of dinosaurs, and endothermy (*color*) was thought to have appeared gradually, late in the development of mammals and birds. The author proposes a reclassification (*see illustration, page 140*).

HAIRY THERAPSIDS, mammal-like reptiles of the late Permian period some 250 million years ago, confront one another in the snows of southern Gondwanaland, at a site that is now in South Africa. *Anteosau-*

vertebrate communities, in the Carboniferous and early Permian periods, were composed of reptiles and amphibians generally considered to be primitive and ectothermic. Replacing this first ectothermic dynasty were the mammal-like reptiles (therapsids), which eventually produced the first true mammals near the end of the next period, the Triassic, about when the dinosaurs were originating. One might expect that mammals would have taken over the land-vertebrate communities immediately, but they did not. From their appearance in the Triassic until the end of the Cretaceous, a span of 140 million years, mammals remained small and inconspicuous while all the ecological roles of large terrestrial herbivores and carnivores were monopolized by dinosaurs; mammals did not begin to radiate and produce large species until after the dinosaurs had already become extinct at the end of the Cretaceous. One is forced to conclude that dinosaurs were competitively superior to mammals as large land vertebrates. And that would be baffling if dinosaurs were "cold-blooded." Perhaps they were not.

Measuring Fossil Metabolism

In order to rethink traditional ideas

about Permian and Mesozoic vertebrates one needs bioenergetic data for dinosaurs, therapsids and early mammals. How does one measure a fossil animal's metabolism? Surprising as it may seem, recent research provides three independent methods of extracting quantitative metabolic information from the fossil record. The first is bone histology. Bone is an active tissue that contributes to the formation of blood cells and participates in maintaining the calcium-phosphate balance, vital to the proper functioning of muscles and nerves. The low rate of energy flow in ectotherms places little demand on the bone compartment of the blood and calcium-phosphate system, and so the compact bone of living reptiles has a characteristic "low activity" pattern: a low density of blood vessels and few Haversian canals, which are the site of rapid calcium-phosphate exchange. Moreover, in strongly seasonal climates, where drought or winter cold forces ectotherms to become dormant, growth rings appear in the outer layers of compact bone, much like the rings in the wood of trees in similar environments. The endothermic bone of birds and mammals is dramatically different. It almost never shows growth rings, even in severe climates, and it is rich in blood vessels and frequently in Haversian

FEATHERED DINOSAUR, *Syntarsus*, pursues a gliding lizard across the sand dunes of Rhodesia in the early Jurassic period some 180 million years ago. This small dino-

rus (*right*), weighing about 600 kilograms, had bony ridges on the snout and brow for head-to-head contact in sexual or territorial behavior. Pristerognathids (*left*) weighed about 50 kilograms and represent a group that included the direct ancestors of mammals. The reconstructions were made by the author on the basis of fossils and the knowledge, from several kinds of data, that therapsids were endothermic, or "warm-blooded"; those adapted to cold would have had hairy insulation. The advent of endothermy, competitively superior to the ectothermy ("cold-bloodedness") of typical reptiles, is the basis of author's new classification of land vertebrates.

saur (adult weight about 30 kilograms) and others were restored by Michael Raath of the Queen Victoria Museum in Rhodesia and the author on the basis of evidence that some thecodonts, ancestors of the dinosaurs, had insulation and on the basis of close anatomi- cal similarities between dinosaurs and early birds. Dinosaurs, it appears, were endothermic, and the smaller species required insulation. Feathers would have conserved metabolic heat in cold environments and reflected the heat of the sun in hot climates such as this.

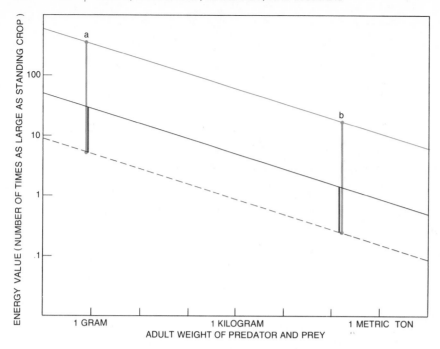

PREDATOR-PREY RATIO remains about constant regardless of the size of the animals involved because of the scaling relations in predator-prey energy flow. The yearly energy budget, or the amount of meat required per year per kilogram of predator, decreases with increasing weight for endotherms (*colored curve*) and for ectotherms (*solid black curve*). The energy value of carcasses provided per kilogram by a prey population decreases with increasing weight at the same rate (*broken black curve*). The vertical lines are proportional to the size of the prey "standing crop" required to support one unit of predator standing crop: about an order of magnitude greater for endothermic predators (*color*) than for ectothermic ones (*gray*), whether for a lizard-size system (*a*) or a lion-size one (*b*).

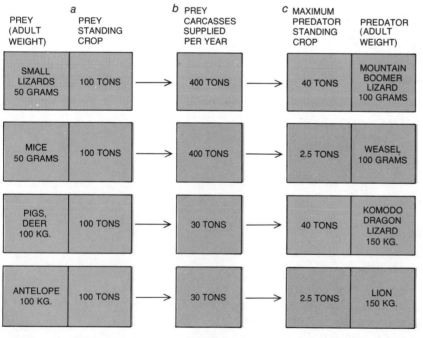

ENERGY FLOW and predator-prey relation are illustrated for predator-prey systems of various sizes. Standing crop is the biomass of a population (or the potential energy value contained in the tissue) averaged over a year. For a given adult size, ectothermic prey (*gray*) produce as much meat (*b*) per unit standing crop (*a*) as endotherms (*color*). Endothermic predators, however, require an order of magnitude more meat (*b*) per unit standing crop (*c*). The maximum predator-prey biomass ratio (*c/a*) is therefore about an order of magnitude greater in an endothermic system than it is in an ectothermic one.

canals. Fossilization often faithfully preserves the structure of bone, even in specimens 300 million years old; thus it provides one window through which to look back at the physiology of ancient animals.

The second analytic tool of paleobioenergetics is latitudinal zonation. The present continental masses have floated across the surface of the globe on lithospheric rafts, sometimes colliding and pushing up mountain ranges, sometimes pulling apart along rift zones such as those of the mid-Atlantic or East Africa. Paleomagnetic data make it possible to reconstruct the ancient positions of the continents to within about five degrees of latitude, and sedimentary indicators such as glacial beds and salt deposits show the severity of the latitudinal temperature gradient from the Equator to the poles in past epochs. Given the latitude and the gradient, one can plot temperature zones, and such zones should separate endotherms from ectotherms. Large reptiles with a lizardlike physiology cannot survive cool winters because they cannot warm up to an optimal body temperature during the short winter day and they are too large to find safe hiding places for winter hibernation. That is why small lizards are found today as far north as Alberta, where they hibernate underground during the winter, but crocodiles and big lizards do not get much farther north than the northern coast of the Gulf of Mexico.

The third meter of heat production in extinct vertebrates is the predator-prey ratio: the relation of the "standing crop" of a predatory animal to that of its prey. The ratio is a constant that is a characteristic of the metabolism of the predator, regardless of the body size of the animals of the predator-prey system. The reasoning is as follows: The energy budget of an endothermic population is an order of magnitude larger than that of an ectothermic population of the same size and adult weight, but the productivity—the yield of prey tissue available to predators—is about the same for both an endothermic and an ectothermic population. In a steady-state population the yearly gain in weight and energy value from growth and reproduction equals the weight and energy value of the carcasses of the animals that die during the year; the loss of biomass and energy through death is balanced by additions. The maximum energy value of all the carcasses a steady-state population of lizards can provide its predators is about the same as that provided by a prey population of birds or mammals of about the same numbers and adult body size.

Therefore a given prey population, either ectotherms or endotherms, can support an order of magnitude greater biomass of ectothermic predators than of endothermic predators, because of the endotherms' higher energy needs. The term standing crop refers to the biomass, or the energy value contained in the biomass, of a population. In both ectotherms and endotherms the energy value of carcasses produced per unit of standing crop decreases with increasing adult weight of prey animals: a herd of zebra yields from about a fourth to a third of its weight in prey carcasses a year, but a "herd" of mice can produce up to six times its weight because of its rapid turnover, reflected in a short life span and high metabolism per unit weight.

Now, the energy budget per unit of predator standing crop also decreases

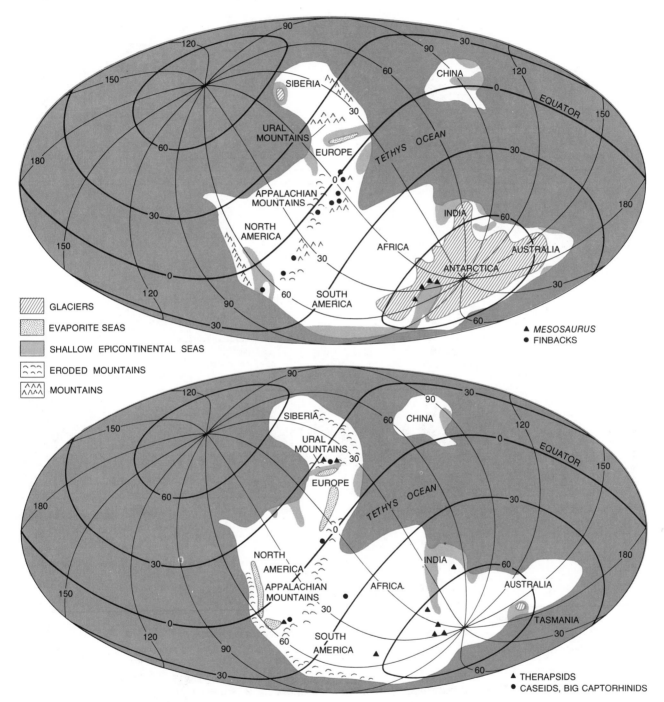

GLACIERS

EVAPORITE SEAS

SHALLOW EPICONTINENTAL SEAS

ERODED MOUNTAINS

MOUNTAINS

▲ MESOSAURUS
● FINBACKS

▲ THERAPSIDS
● CASEIDS, BIG CAPTORHINIDS

PERMIAN WORLD is reconstructed here (on an oblique Mollweide projection, which minimizes distortion of area) on the basis of paleomagnetic and other geophysical data. All major land masses except China were welded into a supercontinent, Pangaea. In the early Permian (*top*) the Gondwana glaciation was at its maximum, covering much of the southern part of the continent. Big finback pelycosaurs and contemporary large reptiles and amphibians were confined to the Tropics; they had ectothermic bone and high predator-prey ratios. The only reptile in cold southern Gondwanaland was the little *Mesosaurus,* which apparently hibernated in the mud during the winters. The late Permian world (*bottom*) was less glaciated, but the south was still cold and the latitudinal temperature gradient was still steep. Big reptiles with ectothermic bone, caseids and captorhinids, were restricted to the hot Tropics, as large reptiles had been in the early Permian. By now, however, many early therapsids, all with endothermic bone and low predator-prey ratios, had invaded southern Gondwanaland. They must have acquired high heat production and some insulation.

with increasing weight: lions require more than 10 times their own weight in meat per year, whereas shrews need 100 times their weight. These two bioenergetic scaling factors cancel each other, so that if the adult size of the predator is roughly the same as that of the prey (and in land-vertebrate ecosystems it usually is), the maximum ratio of predator standing crop to prey standing crop in a steady-state community is a constant independent of the adult body size in the predator-prey system [see top illustration page 128]. For example, spiders are ectotherms, and the ratio of a spider population's standing crop to its prey standing crop reaches a maximum of about 40 percent. Mountain boomer lizards, about 100 grams in adult weight, feeding on other lizards would reach a similar maximum ratio. So would the giant Komodo dragon lizards (up to 150 kilograms in body weight) preying on deer, pigs and monkeys. Endothermic

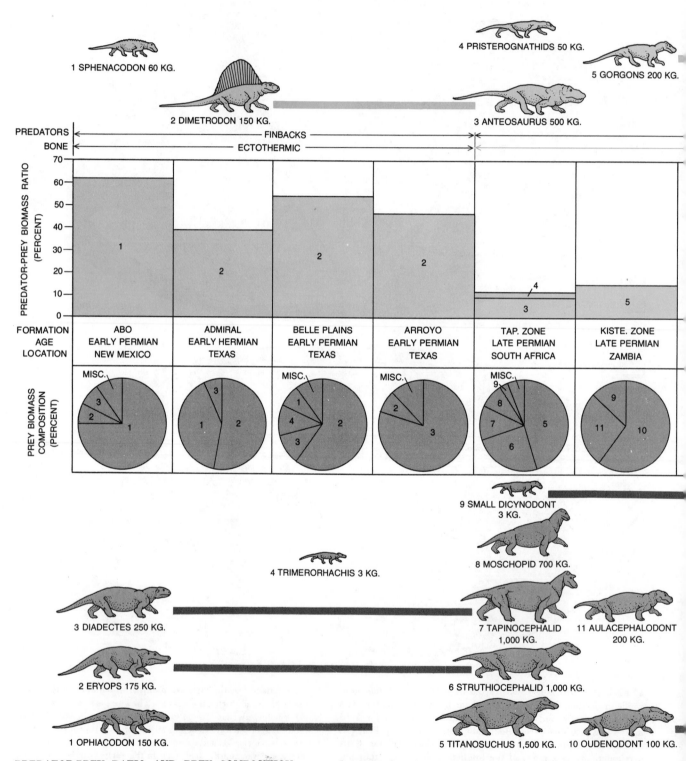

PREDATOR-PREY RATIO AND PREY COMPOSITION are shown on these two pages and the next two pages for a number of fossil communities, each representing a particular time zone and depositional environment. The predator (top) and prey (bottom) animals involved at each site are illustrated. For each deposit the histogram (color) gives the predator biomass as a percent of the

mammals and birds, on the other hand, reach a maximum predator-prey biomass ratio of only from 1 to 3 percent—whether they are weasel and mouse or lion and zebra [*see bottom illustration on page 128*].

Some fossil deposits yield hundreds or thousands of individuals representing a single community; their live body weight can be calculated from the reconstruction of complete skeletons, and the total predator-prey biomass ratios are then easily worked out. Predator-prey ratios are powerful tools for paleophysiology because they are the direct result of predator metabolism.

The Age of Ectothermy

The paleobioenergetic methodology I have outlined can be tested by analyzing

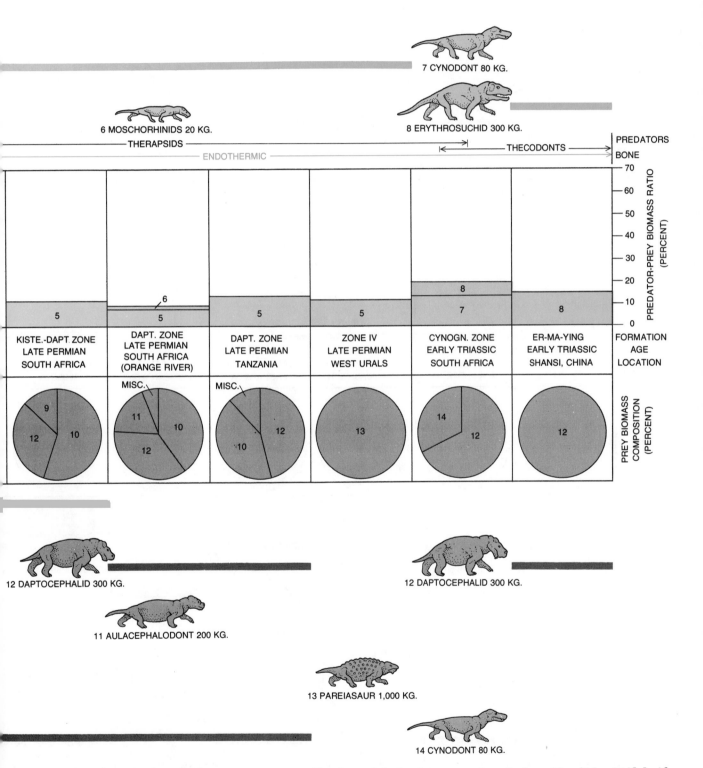

total prey biomass, in other words, the predator-prey ratio. The pie charts give the composition of the available prey. Note the sudden drop in the predator-prey ratios during the transition from the finback pelycosaurs to the early therapsids, which coincided with the first appearance of endothermic bone and also with the invasion of cold southern Gondwanaland by early therapsids of all sizes.

the first land-vertebrate predator-prey system, the early Permian communities of primitive reptiles and amphibians. The first predators capable of killing relatively large prey were the finback pelycosaurs of the family Sphenacodontidae, typified by *Dimetrodon*, whose tall-spined fin makes it popular with car-

toonists. Although this family included the direct ancestors of mammal-like reptiles and hence of mammals, the sphenacodonts themselves had a very primitive level of organization, with a limb anatomy less advanced than that of living lizards. Finback bone histology was emphatically ectothermic, with a low den-

sity of blood vessels, few Haversian canals and the distinct growth rings that are common in specimens from seasonally arid climates.

One might suspect that finbacks and their prey would be confined to warm, equable climates, and early Permian paleogeography offers an excellent op-

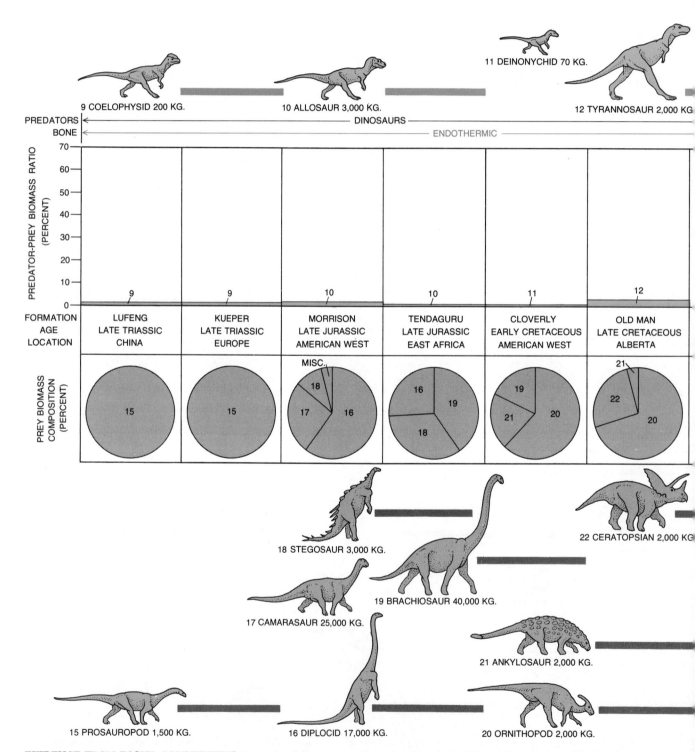

EVIDENCE FROM FOSSIL COMMUNITIES is continued from the preceding two pages. The animals are of course not all drawn to the same scale; their adult weights are given. The drawings are presented with the same limb-stride positions in each to emphasize

portunity to test this prediction. During the early part of the period ice caps covered the southern tips of the continental land masses, all of which were part of the single southern supercontinent Gondwanaland, and glacial sediment is reported at the extreme northerly tip of the Permian land mass in Siberia by Russian geologists [*see illustration on page 129*]. The Permian Equator crossed what are now the American Southwest, the Maritime Provinces of Canada and western Europe. Here are found sediments produced in very hot climates: thick-bedded evaporite salts and fully oxidized, red-stained mudstones. The latitudinal temperature gradient in the Permian must have been at least as steep as it is at present. Three Permian floral zones reflect the strong poleward temperature gradient. The Angaran flora of Siberia displays wood with growth rings from a wet environment, implying a moist climate with cold winters. The

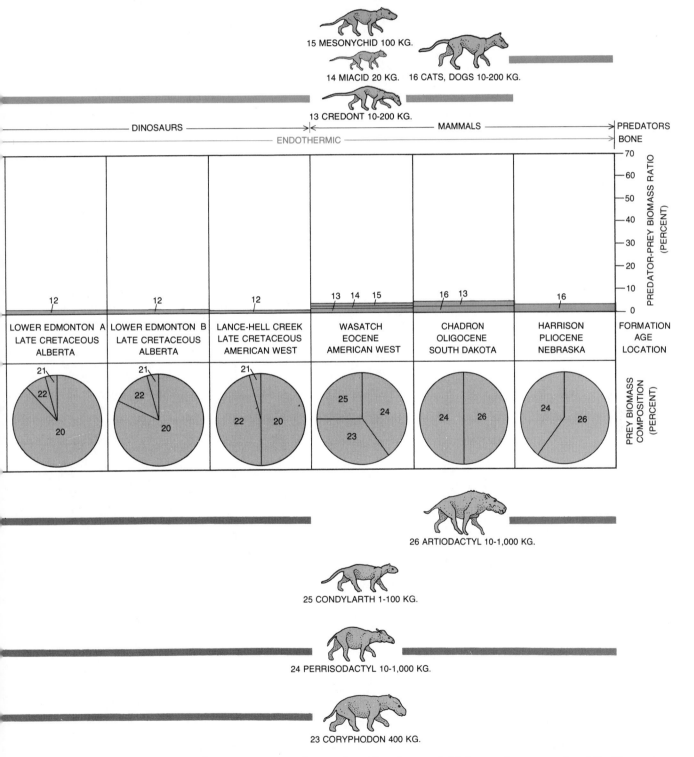

relative limb length. Long-limbed, fast-sprinting vertebrates of large size appeared only with the dinosaurs, in the middle Triassic. Note the remarkably low predator-prey ratios of the dinosaurs, as low as or lower than those of the Cenozoic (and modern) mammals.

LONGISQUAMA, a small animal whose fossil was discovered in middle Triassic lake beds in Turkestan by the Russian paleontologist A. Sharov, was a thecodont. Its body was covered by long overlapping scales that were keeled, suggesting that they constituted a structural stage in the evolution of feathers. The long devices along the back were V-shaped in cross section; they may have served as parachutes and also as threat devices, as shown here.

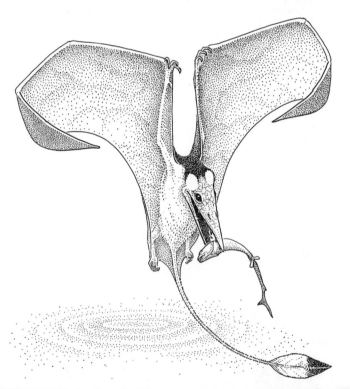

SORDUS PILOSUS, also found by Sharov, was a pterosaur: a flying reptile of the Jurassic period that was a descendant of thecodonts or of very early dinosaurs. Superbly preserved fossils show that the animal was insulated with a dense growth of hair or hairlike feathers; hence the name, which means "hairy devil." Insulation strongly suggests endothermy.

Euramerian flora of the equatorial region had two plant associations: wet swamp communities with no growth rings in the wood, implying a continuous warm-moist growing season, and semiarid, red-bed-evaporite communities with some growth rings, reflecting a tropical dry season. In glaciated Gondwanaland the peculiar *Glossopteris* flora dominated, with wood from wet environments showing sharp growth rings.

The ectothermy of the finbacks is confirmed by their geographic zonation. Finback communities are known only from near the Permian Equator; no large early Permian land vertebrates of any kind are found in glaciated Gondwanaland. (One peculiar little fish-eating reptile, *Mesosaurus*, is known from southern Gondwanaland, and its bone has sharp growth rings. The animal must have fed and reproduced during the Gondwanaland summer and then burrowed into the mud of lagoon bottoms to hibernate, much as large snapping turtles do today in New England.)

Excellent samples of finback communities are available for predator-prey studies, thanks largely to the lifework of the late Alfred Sherwood Romer of Harvard University. In order to derive a predator-prey ratio from a fossil community one simply calculates the number of individuals, and thus the total live weight, represented by all the predator and prey specimens that are found together in a sediment representing one particular environment. In working with scattered and disarticulated skeletons it is best to count only bones that have about the same robustness, and hence the same preservability, in both predator and prey. The humerus and the femur are good choices for finback communities: they are about the same size with respect to the body in the prey and the predator and should give a ratio that faithfully represents the ratio of the animals in life.

In the earlier early Permian zones the most important finback prey were semiaquatic fish-eating amphibians and reptiles, particularly the big-headed amphibian *Eryops* and the long-snouted pelycosaur *Ophiacodon*. As the climate became more arid in Europe and America these water-linked forms decreased in numbers, and the fully terrestrial herbivore *Diadectes* became the chief prey genus. In all zones from all environments the calculated biomass ratio of predator to prey in finback communities is very high: from 35 to 60 percent, the same range seen in living ectothermic spiders and lizards.

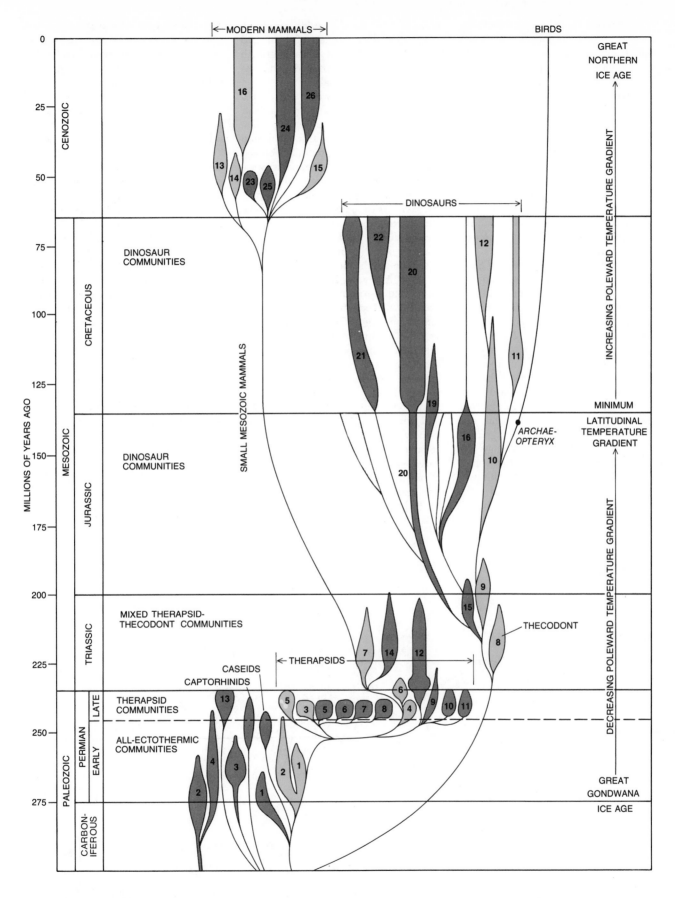

PREDATOR-PREY SYSTEMS of land vertebrates and the paths of descent of successive groups are diagrammed with the predators (*color*) and the prey animals (*gray*) numbered to refer to the groups named and pictured in the illustrations on pages 130 through 133. The relative importance of the live biomass represented by fossils is indicated by the width of the gray and colored pathways.

All three of the paleobioenergetic indicators agree: the finback pelycosaurs and their contemporaries were ectotherms with low heat production and a lizardlike physiology that confined their distribution to the Tropics.

Therapsid Communities

The mammal-like reptiles (order Therapsida), descendants of the finbacks, made their debut at the transition from the early to the late Permian and immediately became the dominant large land vertebrates all over the world. The three metabolism-measuring techniques show that they were endotherms.

The earliest therapsids retained many finback characteristics but had acquired limb adaptations that made possible a trotting gait and much higher running speeds. From early late Permian to the middle Triassic one line of therapsids became increasingly like primitive mammals in all details of the skull, the teeth and the limbs, so that some of the very advanced mammal-like therapsids (cynodonts) are difficult to separate from the first true mammals. The change in physiology, however, was not so gradual. Detailed studies of bone histology conducted by Armand Riqles of the University of Paris indicate that the bioenergetic transition was sudden and early: all the finbacks had fully ectothermic bone; all the early therapsids—and there is an extraordinary variety of them—had fully endothermic bone, with no growth rings and with closely packed blood vessels and Haversian canals.

The late Permian world still had a severe latitudinal temperature gradient; some glaciation continued in Tasmania, and the southern end of Gondwanaland retained its cold-adapted *Glossopteris* flora. If the earliest therapsids were equipped with endothermy, they would presumably have been able to invade southern Africa, South America and the other parts of the southern cold-temperature realm. They did exactly that. A rich diversity of early therapsid families has been found in the southern Cape District of South Africa, in Rhodesia, in Brazil and in India—regions reaching to 65 degrees south Permian latitude [*see illustration on page 129*]. Early therapsids as large as rhinoceroses were common there, and many species grew to an adult weight greater than 10 kilograms, too large for true hibernation. These early therapsids must have had physiological adaptations that enabled them to feed in and move through the snows of the cold Gondwanaland winters. There were also some ectothermic holdovers from the early Permian that survived into the late Permian, notably the immense herbivorous caseid pelycosaurs and the big-headed, seed-eating captorhinids. As one might predict, large species of these two ectothermic families were confined to areas near the late Permian Equator; big caseids and captorhinids are not found with the therapsids in cold Gondwanaland. In the late Permian, then, there was a "modern" faunal zonation of large vertebrates, with endothermic therapsids and some big ectotherms in the Tropics giving way to an all-endothermic therapsid fauna in the cold south.

In the earliest therapsid communities of southern Africa, superbly represented in collections built up by Lieuwe Boonstra of the South African Museum and by James Kitching of the University of the Witwatersrand, the predator-prey ratios are between 9 and 16 percent. That is much lower than in early Permian finback communities. Equally low ratios are found for tropical therapsids from the U.S.S.R. even though the prey species there were totally different from those of Africa. The sudden decrease in predator-prey ratios from finbacks to early therapsids coincides exactly with the sudden change in bone histology from ectothermic to endothermic reported by Riqles and also with the sudden invasion of the southern cold-temperate zone by a rich therapsid fauna. The conclusion is unavoidable that even early therapsids were endotherms with high heat production.

It seems certain, moreover, that in the

ARCHAEOPTERYX, generally considered the first bird, is known from late Jurassic fossils that show its feather covering clearly. In spite of its very birdlike appearance, *Archaeopteryx* was closely related to certain small dinosaurs (*see illustration on next page*) and could not fly. The presence of insulation in the thecodont *Longisquama* and in *Sordus* and *Archaeopteryx*, which were descendants of thecodonts, indicates that insulation and endothermy were acquired very early, probably in early Triassic thecodonts.

cold Gondwanaland winters the therapsids would have required surface insulation. Hair is usually thought of as a late development that first appeared in the advanced therapsids, but it must have been present in the southern African endotherms of the early late Permian. How did hair originate? Possibly the ancestors of therapsids had touch-sensitive hairs scattered over the body as adaptations for night foraging; natural selection could then have favored increased density of hair as the animals' heat production increased and they moved into colder climates.

The therapsid predator-prey ratios, although much lower than those of ectotherms, are still about three times higher than those of advanced mammals today. Such ratios indicate that the therapsids achieved endothermy with a moderately high heat production, far higher than in typical reptiles but still lower than in most modern mammals. Predator-prey ratios of early Cenozoic communities seem to be lower than those of therapsids, and so one might conclude that a further increase in metabolism occurred somewhere between the advanced therapsids of the Triassic and the mammals of the post-Cretaceous era. Therapsids may have operated at a lower body temperature than most living mammals do, and thus they may have saved energy with a lower thermostat setting. This

suggestion is reinforced by the low body temperature of the most primitive living mammals: monotremes (such as the spiny anteater) and the insectivorous tenrecs of Madagascar; they maintain a temperature of about 30 degrees Celsius instead of the 36 to 39 degrees of most modern mammals.

Thecodont Transition

The vigorous and successful therapsid dynasty ruled until the middle of the Triassic. Then their fortunes waned and a new group, which was later to include the dinosaurs, began to take over the roles of large predators and herbivores. These were the Archosauria, and the first wave of archosaurs were the thecodonts. The earliest thecodonts, small and medium-sized animals found in therapsid communities during the Permian-Triassic transition, had an ectothermic bone histology. In modern ecosystems the role played by large freshwater predators seems to be one in which ectothermy is competitively superior to endothermy; the low metabolic rate of ectotherms may be a key advantage because it allows much longer dives. Two groups of thecodonts became large freshwater fish-eaters: the phytosaurs, which were confined to the Triassic, and the crocodilians, which remain successful today. Both groups have ectothermic bone.

(The crocodilian endothermy was either inherited directly from the first thecodonts or derived secondarily from endothermic intermediate ancestors.) In most of the later, fully terrestrial advanced thecodonts, on the other hand, Riqles discovered a typical endothermic bone histology; the later thecodonts were apparently endothermic.

The predator-prey evidence for thecodonts is scanty. The ratios are hard to compute because big carnivorous cynodonts and even early dinosaurs usually shared the predatory role with thecodonts. One sample from China that has only one large predator genus, a big-headed erythrosuchid thecodont, does give a ratio of about 10 percent, which is in the endothermic range. The zonal evidence is clearer. World climate was moderating in the Triassic (the glaciers were gone), but a distinctive flora and some wood growth rings suggest that southern Gondwanaland was not yet warm all year. What is significant in this regard is the distribution of phytosaurs, the big ectothermic fish-eating thecodonts. Their fossils are common in North America and Europe (in the Triassic Tropics) and in India, which was warmed by the equatorial Tethys Ocean, but they have not been found in southern Gondwanaland, in southern Africa or in Argentina, even though a rich endothermic thecodont fauna did exist

DINOSAURIAN ANCESTRY of *Archaeopteryx* (*left*), and thus of birds, is indicated by its close anatomical relation to such small dinosaurs as *Microvenator* (*right*) and *Deinonychus*; John H. Ostrom of Yale University demonstrated that they were virtually identical in all details of joint anatomy. The long forelimbs of *Archaeopteryx* were probably used for capturing prey, not for flight.

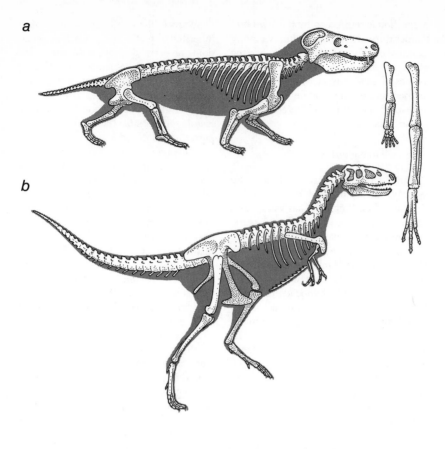

LIMB LENGTHS of dinosaurs are compared with those of two ecologically equivalent therapsids. The limbs were relatively longer in the dinosaurs and the appended muscles were larger, indicating that the dinosaurs had a larger capacity for high levels of exercise metabolism. The two top drawings represent the animals as if they were the same weight; the adult carnivorous therapsid *Cynognathus* (*a*) actually weighed 100 kilograms and the juvenile dinosaur *Albertosaurus* (*b*) 600 kilograms. Two herbivores, therapsid *Struthiocephalus* (*c*) and horned dinosaur *Centrosaurus* (*d*), weighed about 1,500 kilograms.

there.

Did some of the thecodonts have thermal insulation? Direct evidence comes from the discoveries of A. Sharov of the Academy of Sciences of the U.S.S.R. Sharov found a partial skeleton of a small thecodont and named it *Longisquama* for its long scales: strange parachutelike devices along the back that may have served to break the animal's fall when it leaped from trees. More important is the covering of long, overlapping, keeled scales that trapped an insulating layer of air next to its body [*see top illustration on page 134*]. These scales lacked the complex anatomy of real feathers, but they are a perfect ancestral stage for the insulation of birds. Feathers are usually assumed to have appeared only late in the Jurassic with the first bird, *Archaeopteryx*. The likelihood that some thecodonts had insulation is supported, however, by another of Sharov's discoveries: a pterosaur, or flying reptile, whose fossils in Jurassic lake beds still show the epidermal covering. This beast (appropriately named *Sordus pilosus*, the "hairy devil") had a dense growth of hair or hairlike feathers all over its body and limbs. Pterosaurs are descendants of Triassic thecodonts or perhaps of very primitive dinosaurs. The insulation in both *Sordus* and *Longisquama*, and the presence of big erythrosuchid thecodonts at the southern limits of Gondwanaland, strongly suggest that some endothermic thecodonts had acquired insulation by the early Triassic.

The Dinosaurs

Dinosaurs, descendants of early thecodonts, appeared first in the middle Triassic and by the end of the period had replaced thecodonts and the remaining therapsids as the dominant terrestrial vertebrates. Zonal evidence for endothermy in dinosaurs is somewhat equivocal. The Jurassic was a time of climatic optimum, when the poleward temperature gradient was the gentlest that has prevailed from the Permian until the present day. In the succeeding Cretaceous period latitudinal zoning of oceanic plankton and land plants seems, however, to have been a bit sharper. Rhinoceros-sized Cretaceous dinosaurs and big marine lizards are found in the rocks of the Canadian far north, within the Cretaceous Arctic Circle. Dale A. Russell of the National Museums of Canada points out that at these latitudes the sun would have been below the horizon for months at a time. The environment of the dinosaurs would have been far severer than the environment of the marine reptiles because of the lack of a wind-chill factor in the water and because of the ocean's temperature-buffering effect. Moreover, locomotion costs far less energy per kilometer in water than on land, so that the marine reptiles could have migrated away from the arctic winter. These considerations suggest, but do not prove, that arctic dinosaurs must have been able to cope with cold stress.

Dinosaur bone histology is less equivocal. All dinosaur species that have been investigated show fully endothermic bone, some with a blood-vessel density higher than that in living mammals. Since bone histology separates endotherms from ectotherms in the Permian and the Triassic, this evidence alone should be a strong argument for the endothermy of dinosaurs. Yet the predator-prey ratios are even more compelling. Dinosaur carnivore fossils are exceedingly rare. The predator-prey ratios for dinosaur communities in the Triassic, Jurassic and Cretaceous are usually from 1 to 3 percent, far lower even than those of therapsids and fully as low as those in large samples of fossils from advanced mammal communities in the Cenozoic. I am persuaded that all the available quantitative evidence is in favor of high heat production and a large annual energy budget in dinosaurs.

Were dinosaurs insulated? Explicit evidence comes from a surprising source: *Archaeopteryx*. As an undergraduate a decade ago I was a member of a paleontological field party led by John H. Ostrom of Yale University. Near Bridger, Mont., Ostrom found a remarkably preserved little dinosaurian carnivore, *Deinonychus*, that shed a great deal of light on carnivorous dinosaurs in general. A few years later, while looking for pterosaur fossils in European museums, Ostrom came on a specimen of *Archaeopteryx* that had been mislabeled for years as a flying reptile, and he noticed extraordinary points of resemblance between *Archaeopteryx* and carnivorous dinosaurs. After a detailed anatomical analysis Ostrom has now established beyond any reasonable doubt that the immediate ancestor of *Archaeopteryx* must have been a small dinosaur, perhaps one related to *Deinonychus*. Previously it had been thought that the ancestor of *Archaeopteryx*, and thus of birds, was a thecodont rather far removed from dinosaurs themselves.

Archaeopteryx was quite thoroughly feathered, and yet it probably could not fly: the shoulder joints were identical with those of carnivorous dinosaurs and were adapted for grasping prey, not for

the peculiar arc of movement needed for wing-flapping. The feathers were probably adaptations not for powered flight or gliding but primarily for insulation. *Archaeopteryx* is so nearly identical in all known features with small carnivorous dinosaurs that it is hard to believe feathers were not present in such dinosaurs. Birds inherited their high metabolic rate and most probably their feathered insulation from dinosaurs; powered flight probably did not evolve until the first birds with flight-adapted shoulder joints appeared during the Cretaceous, long after *Archaeopteryx*.

It has been suggested a number of times that dinosaurs could have achieved a fairly constant body temperature in a warm environment by sheer bulk alone; large alligators approach this condition in the swamps of the U.S. Gulf states. This proposed thermal mechanism would not give rise to endothermic bone histology or low predator-prey ratios, however, nor would it explain arctic di-

nosaurs or the success of many small dinosaur species with an adult weight of between five and 50 kilograms.

Dinosaur Brains and Limbs

Large brain size and endothermy seem to be linked; most birds and mammals have a ratio of brain size to body size much larger than that of living reptiles and amphibians. The acquisition of endothermy is probably a prerequisite for the enlargement of the brain because the proper functioning of a complex central nervous system calls for the guarantee of a constant body temperature. It is not surprising that endothermy appeared before brain enlargement in the evolutionary line leading to mammals. Therapsids had small brains with reptilian organization; not until the Cenozoic did mammals attain the large brain size characteristic of most modern species. A large brain is certainly not necessary for endothermy, since the physio-

	BONE HISTOLOGY	PRESENT IN TEMPERATE ZONE	PREDATOR–PREY RATIO	LIMB LENGTH
FINBACKS, OTHER EARLY PERMIAN LAND VERTEBRATES		NO	50 PERCENT	SHORT
LATE PERMIAN CASEIDS AND BIG CAPTORHINIDS		NO	NOT APPLICABLE	SHORT
LATE PERMIAN– EARLY TRIASSIC THERAPSIDS		YES	10 PERCENT	SHORT
EARLIEST THECODONTS		?	?	SHORT
MOST LAND THECODONTS		YES UP TO 600 KILOGRAMS	10 PERCENT	SHORT
FRESHWATER THECODONTS		NO UP TO 500 KILOGRAMS	?	SHORT
DINOSAURS		YES	1-3 PERCENT	LONG
CENOZOIC MAMMALS		YES	1-5 PERCENT	LONG

PALEOBIOENERGETIC EVIDENCE is summed up here. The appropriate blocks are shaded to show whether the available data constitute evidence for ectothermy (*gray*) or endothermy (*color*) according to criteria discussed in the text of this article. Caseids and captorhinids are herbivores, so that there is no predator-prey ratio. There are early thecodonts in temperate-zone deposits, but they are small and so the evidence is not significant.

logical feedback mechanisms responsible for thermoregulation are deep within the "old" region of the brain, not in the higher learning centers. Most large dinosaurs did have relatively small brains. Russell has shown, however, that some small and medium-sized carnivorous dinosaurs had brains as large as or larger than modern birds of the same body size.

Up to this point I have concentrated on thermoregulatory heat production. Metabolism during exercise can also be read from fossils. Short bursts of intense exercise are powered by anaerobic metabolism within muscles, and the oxygen debt incurred is paid back afterward by the heart-lung system. Most modern birds and mammals have much higher

levels of maximum aerobic metabolism than living reptiles and can repay an oxygen debt much faster. Apparently this difference does not keep small ectothermic animals from moving fast: the top running speeds of small lizards equal or exceed those of small mammals. The difficulty of repaying oxygen debt increases with increasing body size, however, and the living large reptiles (crocodilians, giant lizards and turtles) have noticeably shorter limbs, less limb musculature and lower top speeds than many large mammals, such as the big cats and the hoofed herbivores.

The early Permian ectothermic dynasty was also strikingly short-limbed; evidently the physiological capacity for

high sprinting speeds in large animals had not yet evolved. Even the late therapsids, including the most advanced cynodonts, had very short limbs compared with the modern-looking running mammals that appeared early in the Cenozoic. Large dinosaurs, on the other hand, resembled modern running mammals, not therapsids, in locomotor anatomy and limb proportions. Modern, fast-running mammals utilize an anatomical trick that adds an extra limb segment to the forelimb stroke. The scapula, or shoulder blade, which is relatively immobile in most primitive vertebrates, is free to swing backward and forward and thus increase the stride length. Jane A. Peterson of Harvard has shown that

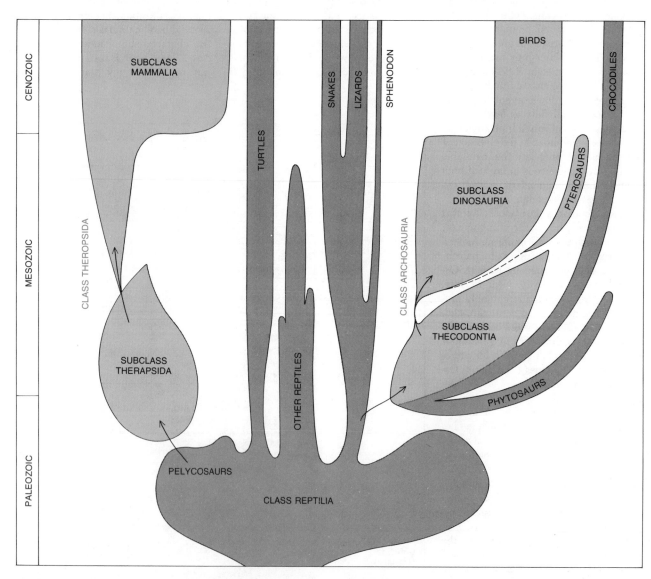

RECLASSIFICATION of land vertebrates (excluding the Amphibia) is suggested by the author on the basis of bioenergetic and anatomical evidence. The critical break comes with the development of endothermy (*color*), which is competitively superior to ectothermy (*gray*) for large land vertebrates. Therapsids were endothermic, closer in physiology to mammals than to today's reptiles. Birds almost certainly inherited their bioenergetics (as well as their joint anatomy) from dinosaurs. The new classes presented here, the Theropsida and the Archosauria, reflect energetic evolution more faithfully than the traditional groupings (*see illustration on page 126*). The width of the pathways representing the various groups is proportional to the biomass represented by their fossils.

living chameleonid lizards have also evolved scapular swinging, although its details are different from those in mammals. Quadrupedal dinosaurs evolved a chameleon-type scapula, and they must have had long strides and running speeds comparable to those of big savanna mammals today.

When the dinosaurs fell at the end of the Cretaceous, they were not a senile, moribund group that had played out its evolutionary options. Rather they were vigorous, still diversifying into new orders and producing a variety of big-brained carnivores with the highest grade of intelligence yet present on land. What caused their fall? It was not competition, because mammals did not begin to diversify until after all the dinosaur groups (except birds!) had disappeared. Some geochemical and microfossil evidence suggests a moderate drop in ocean temperature at the transition from the Cretaceous to the Cenozoic, and so cold has been suggested as the reason. But the very groups that would have been most sensitive to cold, the large crocodilians, are found as far north as Saskatchewan and as far south as Argentina before and immediately after the end of the Cretaceous. A more likely reason is the draining of shallow seas on the continents and a lull in mountain-building activity in most parts of the world, which would have produced vast stretches of monotonous topography. Such geological events decrease the variety of habitats that are available to land animals, and thus increase competition. They can also cause the collapse of intricate, high-ly evolved ecosystems; the larger animals seem to be the more affected. At the end of the Permian similar changes had been accompanied by catastrophic extinctions among therapsids and other land groups. Now, at the end of the Cretaceous, it was the dinosaurs that suffered a catastrophe; the mammals and birds, perhaps because they were so much smaller, found places for themselves in the changing landscape and survived.

The success of the dinosaurs, an enigma as long as they were considered "cold-blooded," can now be seen as the predictable result of the superiority of their high heat production, high aerobic exercise metabolism and insulation. They were endotherms. Yet the concept of dinosaurs as ectotherms is deeply entrenched in a century of paleontological literature. Being a reptile connotes being an ectotherm, and the dinosaurs have always been classified in the subclass Archosauria of the class Reptilia; the other land-vertebrate classes were the Mammalia and the Aves. Perhaps, then, it is time to reclassify.

Taxonomic Conclusion

What better dividing line than the invention of endothermy? There has been no more far-reaching adaptive breakthrough, and so the transition from ectothermy to endothermy can serve to separate the land vertebrates into higher taxonomic categories. For some time it has been suggested that the therapsids should be removed from the Reptilia and joined with the Mammalia; in the light of the sudden increase in heat production and the probable presence of hair in early therapsids, I fully agree. The term Theropsida has been applied to mammals and their therapsid ancestors. Let us establish a new class Theropsida, with therapsids and true mammals as two subclasses [see illustration on preceding page].

How about the class Aves? All the quantitative data from bone histology and predator-prey ratios, as well as the dinosaurian nature of Archaeopteryx, show that all the essentials of avian biology—very high heat production, very high aerobic exercise metabolism and feathery insulation—were present in the dinosaur ancestors of birds. I do not believe birds deserve to be put in a taxonomic class separate from dinosaurs. Peter Galton of the University of Bridgeport and I have suggested a more reasonable classification: putting the birds into the Dinosauria. Since bone histology suggests that most thecodonts were endothermic, the thecodonts could then be joined with the Dinosauria in a great endothermic class Archosauria, comparable to the Theropsida. The classification may seem radical at first, but it is actually a good deal neater bioenergetically than the traditional Reptilia, Aves and Mammalia. And for those of us who are fond of dinosaurs the new classification has a particularly happy implication: The dinosaurs are not extinct. The colorful and successful diversity of the living birds is a continuing expression of basic dinosaur biology.

14

The Ancestors of Mammals

by Edwin H. Colbert
March 1949

In the Permian and Triassic Periods lived the therapsids and the ictidosaurs, a curious group of reptiles with many mammalian characteristics

I T IS difficult to see much in common between a modern reptile, such as a crocodile, and a modern mammal, such as a dog. Anatomically and physiologically they seem to be about as far apart as possible for four-footed animals. But if we go back in geologic time we find a close connection between some of the early mammals and certain reptiles. Improbable as it may seem, the fossil record shows that the earliest mammals were descended from reptilian ancestors.

It was somewhat more than a century ago that the bones of mammal-like reptiles were first discovered in South Africa by Andrew Geddes Bain, a well-known fossil collector of the period. Bain's specimens were noted and described by the great English anatomist and paleontologist, Sir Richard Owen. The significance of Bain's findings was overlooked for decades.

Charles Darwin's disciple Thomas Huxley suggested that the mammals had arisen from amphibians, a conclusion to which he was led by his studies in comparative anatomy. But in the years between about 1870 and 1884 Owen and the brilliant U. S. paleontologist Edward Drinker Cope, after studying certain fossil reptiles from South Africa, independently reached another conclusion which has been strengthened with every passing year—namely that the ancestry of the mammals is to be found in these fossil reptiles of the Permian and Triassic Periods, some 150 to 230 million years ago.

T HE physiological, reproductive and anatomical differences between living reptiles and mammals are readily apparent. First, the reptiles are "cold-blooded" animals in which the internal body temperature varies more or less directly with the temperature of their environment; mammals are "warm-blooded," with a fairly constant body heat and an outer covering of hair to insulate them. The reptiles, as a result of their lack of temperature control, are for the most part sluggish, by comparison with the active mammals. Most of the reptiles lay eggs from which the young are hatched, though some retain the eggs within the body of the female

so that the young are born alive. In most mammals, the embryo develops and is nourished within the uterus of the mother. And the mammals of course are distinguished by the property from which they derive their name: they suckle their young.

Many of the anatomical differences between modern reptiles and mammals are reflections of physiological or reproductive differences. The reptiles are typified by a relatively small and simple brain, whereas mammals have a large brain. Modern reptiles have a single bony joint at the base of the skull, the occipital condyle, to articulate the head with the backbone; mammals have two condyles. The lower jaw in reptiles consists of several elements, one of which, the articular bone, works against the quadrate bone of the skull to form the articulation between skull and jaw. In mammals there is a single jawbone, the dentary, which articulates directly with the squamosal bone of the skull. Reptiles have a single bone in the middle ear; mammals have a chain of three. The teeth of a reptile are generally pretty much alike, and they are renewed by many "generations," so that a replacement is on hand for any tooth that may drop out. The teeth of mammals are differentiated into incisors, canines, premolars and molars, and a mammal is limited to two sets—the "milk" teeth and the later permanent teeth.

There are also important differences in the rest of the skeleton. The vertebrae of a reptile are fairly uniform, whereas in a mammal they are strongly differentiated in the neck, thorax and lumbar regions. A reptile's long bones usually can continue to grow throughout the life of the animal; a mammal, on the other hand, has separate "epiphyses" at the ends of the bones which become fused with the bones as the animal reaches maturity and prevent further growth. In reptiles the number of bones in the fingers and toes varies, while in mammals they are limited to two bones in the thumb and big toe and three bones in each of the other fingers and toes.

Because of these many differences, the early anatomists failed to see the relationship between mammals and reptiles.

But with the discovery of many bones of mammal-like reptiles in South Africa, notably by the physician-paleontologist Robert Broom, a multitude of likenesses appeared. Although by far the greatest number and variety of these reptiles has been excavated from South Africa, by now they have also been found in many parts of the world, including North and South America, Russia, England and western China. It is therefore apparent that during the last stages of the Paleozoic Era and the first stages of the Mesozoic they were spread over almost all the earth.

The reptiles from which the mammals are believed to have come belong to two orders, known as the therapsids and the ictidosaurs. Of particular interest among the therapsids is the suborder called the theriodonts, so named from their "beast"-like or mammal-like teeth. An especially important genus of theriodont is *Cynognathus* (meaning dog-jaw).

T O ANYONE familiar with modern reptiles, *Cynognathus* seems most unreptilian. In life this animal must have been very different in appearance and in actions from the reptiles of today. Even the fossilized bones show this, for in *Cynognathus* many anatomical characters bridge the structural gap between reptile and mammal.

Cynognathus was a rather large animal with a long, doglike skull, as big as the skull of a wolf. Evidently it was carnivorous, for its skull is armed with sharp, strong teeth well adapted to seizing and tearing its prey. These teeth, quite unlike those of the carnivorous crocodile, are not evenly spaced and uniform but rather are separated into differentiated groups like the teeth of many mammals. They must surely have functioned like mammals' teeth. In the front of the jaws are small, conical incisor teeth for nipping or biting. Behind the incisors there is a gap, followed by a single, large, daggerlike canine in each of the upper and lower jaws. Like the canine in present-day wolves or foxes, it must have been the great, slashing knife that formed the principal weapon of the animal. Behind it are the cheek teeth, which in mammals are known as the pre-

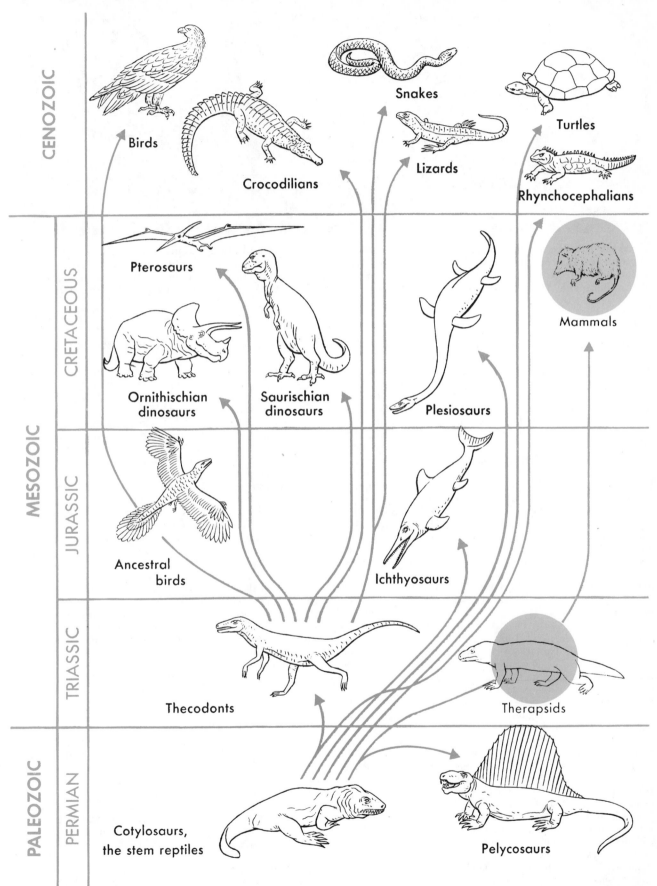

CENOZOIC

MESOZOIC

PALEOZOIC

CRETACEOUS

JURASSIC

TRIASSIC

PERMIAN

Birds

Crocodilians

Snakes

Lizards

Turtles

Rhynchocephalians

Pterosaurs

Ornithischian
dinosaurs

Saurischian
dinosaurs

Plesiosaurs

Mammals

Ancestral
birds

Ichthyosaurs

Thecodonts

Therapsids

Cotylosaurs,
the stem reptiles

Pelycosaurs

MAMMAL-LIKE REPTILES occupy a relatively minor position in the entire reptilian line of descent. At lower right are the therapsids, with the ictidosaurs one of the reptilian orders with mammalian characteristics. The early mammal which is shown in the Cretaceous Period (*near top*) is rather similar to the present-day opossum.

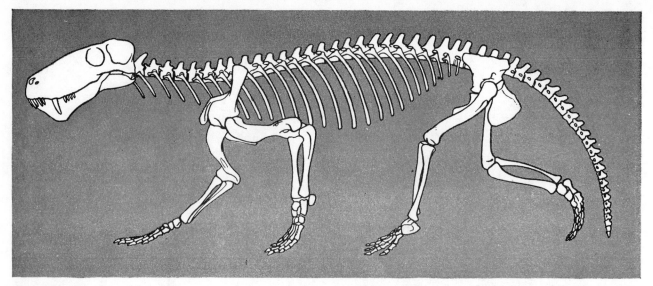

LYCAENOPS is a typical species among the mammal-like reptiles. It is one of the theriodonts, meaning that it had beastlike teeth. The theriodonts are a suborder of the therapsids. One of the distinguishing features of *Lycaenops*, and other mammal-like reptiles, is a pair of long teeth used as slashing instruments. This restoration was made by the author and artist John C. Germann at New York's American Museum of Natural History.

molars and molars. In *Cynognathus*, as in the mammals, these teeth are complex, with several cusps forming the crown of each tooth. Evidently they were useful for cutting food into small pieces, so that it could be quickly assimilated by the digestive system. This is indeed a contrast to the living reptiles, which bolt their food and then digest it slowly.

There are various other mammal-like characters in the skull of *Cynognathus*. For instance, as a corollary to its perfected dentition, this animal, like the mammals, had a secondary bony palate separating the respiratory tract from the alimentary passage. This obviously added to the efficiency and speed of eating, which were important to a relatively active animal. Again, in *Cynognathus* as in the mammals, there are two condyles joining the skull to the first vertebra of the backbone.

Various mammal-like characters appear in the skeleton of *Cynognathus* behind the skull. Thus there is a considerable degree of specialization of the vertebrae: those in the neck are different from the rib-bearing trunk vertebrae, and one can see the beginnings of a ribless lumbar region, as in the mammals. The shoulder blade has a strong spine along its front edge—something quite new for the reptiles, and a forerunner of the spine that is so distinctive of the mammalian shoulder blade. In the pelvic girdle the ilium is elongated, and much of this elongation is forward, so that the bone begins to develop a shape prophetic of the mammalian ilium. The limbs and the feet are in certain respects rather mammal-like. Evidently *Cynognathus* had a somewhat mammal-like posture, with the body car-

ried well above the ground and the feet pulled in toward the midline to give strong support and increase the efficiency of walking.

Yet in spite of all these advances, *Cynognathus* is distinctly a reptile, and it retains many of its ancient reptilian features. Thus it has a full complement of reptile bones in the skull, and to a large degree in the skeleton as well. There is little of the loss and coalescence of bones so typical of mammalian structure. The lower jaw is formed of several bones instead of a single one, and the skull is hinged to the lower jaw in the typical reptilian arrangement. The finger and toe bones are entirely reptilian.

For all that, *Cynognathus* represents a great forward step in evolutionary development. Moreover, certain other theriodont reptiles related to *Cynognathus* show advances in characters where *Cynognathus* was conservative. For instance, a theriodont known as *Bauria*, which in many respects is less mammal-like than *Cynognathus*, has the same number of bones in its toes as a mammal has. Thus while no one theriodont reptile approaches completely the mammalian type of structure, the theriodonts as a group clearly exhibit a trend in that direction.

THE approach toward the mammals is carried even further in the ictidosaurs, of which we unfortunately know all too little. The ictidosaurs possess to an even more advanced degree the various mammal-like characters that are distinctive of the theriodonts; in addition, they show trends toward the mammals in certain characters that are still thoroughly reptilian in the theriodonts. For instance,

the ictidosaurs have lost some of the skull bones characteristic of the theriodonts and have a very much more mammalian skull pattern. They show a further advance in the bony secondary palate. The ictidosaurs also have a greatly enlarged dentary bone in the lower jaw—a step toward the single jawbone of the mammals. But they still have other bones in the lower jaw and a typically reptilian hinge between skull and jaw.

Discoveries in two localities on opposite sides of the earth in recent years have extended greatly our knowledge of the ictidosaurs. In China, Dr. Chung Chien Young has described ictidosaur skulls and bones that he unearthed in Yünnan; in England Walter Kuhne, by the most painstaking methods, has also obtained skulls and parts of skeletons of these very important predecessors of the mammals. These ictidosaurs bear many structural resemblances to the platypus of Australia, the most primitive of living mammals.

At what stage in their evolution did these mammal-like reptiles cross the threshold into full mammalian status? We can assume that the reptile ceased to be a reptile and became a mammal when it had established a constant body temperature and an insulating coat of hair, and when reproduction had reached an advanced stage of development, especially when the female had begun to lactate and suckle its young. Unfortunately we can obtain no direct evidence on these changes, for temperature controls, hair, mammary glands and other soft parts of the anatomy are not preserved as fossils.

Considering only the bones, we can say that the evolving animal had reached a mammalian stage when it achieved the combination of structural features already indicated as characteristic of mammals, that is, a full differentiation of the vertebrae, a fused pelvis, perfected feet, epiphyses on the long bones, a double occipital condyle at the base of the skull, a perfected secondary palate, a chain of three small bones in the middle ear, and a single lower jawbone articulating with the skull. Of these features, the last is perhaps the most significant.

In the mammal-like reptiles, there was a progressive reduction of the quadrate and articular bones which formed the reptilian articulation between skull and jaw; in the ictidosaurs these elements were very small indeed. Now from a synthesis of embryological and paleontological evidence we know that in the final transition from reptile to mammal, there was a remarkable transformation in these bones. As the reptile became more nearly a mammal, the two bones finally abandoned their functions as articulating elements between the skull and the jaw, and entered the middle ear. The quadrate became the mammalian incus bone, the articular became the mammalian malleus, and together with the stapes the three bones formed the chain of ear ossicles that is characteristic of the mammals. At the same time a new joint was formed between the squamosal of the skull and the dentary, the only bone left in the lower jaw.

None of the ictidosaurs had quite attained this stage of evolution, and so by definition they can still be regarded as reptiles. Yet at that point the distinction between reptile and mammal had become so narrow that the line of demarcation between them must be based, perhaps arbitrarily, on this difference between a single ear ossicle and three ear ossicles. Though seemingly insignificant, the difference is so constant that it serves admirably as a reference point at which to draw the very fine line between reptiles and mammals.

THE first mammals probably looked very much like the mammal-like reptiles that were their immediate ancestors. It was in the Triassic Period of earth history that the change from reptile to mammal was made, and from the succeeding Jurassic Period on, we can see mammals sharing the earth with reptiles. The Triassic Period was still early in the Mesozoic Era, the great age of dinosaurs. For a long time, geologically speaking, the early mammals were destined to live in a world full of reptiles. It was a lush, tropical world, in which the giant reptiles were supreme and the mammals relatively insignificant. But as the conditions on the land changed, the earth became more suitable for mammals than for reptiles. By the time the mammals became dominant on the earth, their reptilian ancestors had long been extinct.

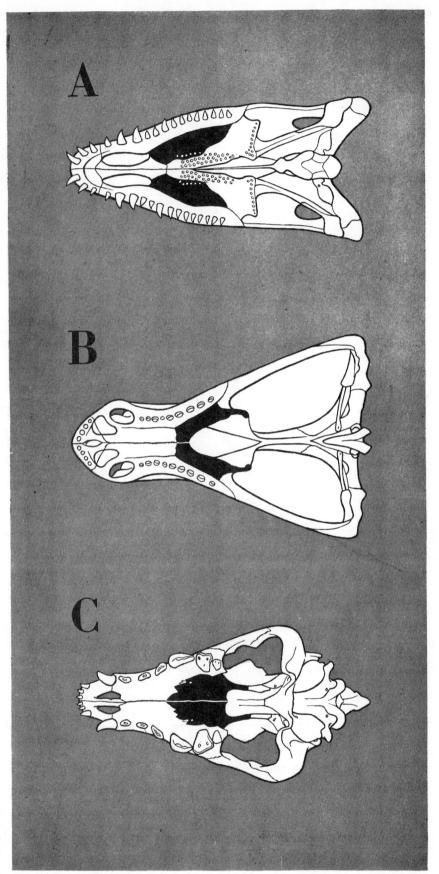

DEVELOPMENT of secondary palate (*black*) illustrates evolutionary position of the mammal-like reptiles. Early mammal-like reptile *Dimetrodon* (A) had tiny palate, with nostrils opening into mouth. Later mammal-like reptile *Cynognathus* (B) had closed nasal passage. Dog has full palate.

15   The Cave Bear

by Björn Kurtén
March 1972

This large-headed species lived from the Pyrenees to the Caspian Sea during the Ice Age. One cave alone has yielded the remains of 30,000 such bears. What caused the species to become extinct?

Many species of bears spend part of their lives in caves, but only one has ever been known as the cave bear. That species is *Ursus spelaeus*, which is now extinct. It was given this name by its 18th-century discoverers because they based their description on a skull found in a cave. The name became widely established during the next century as European fossil collectors turned up thousands of the animals' bones in caves all the way from the Spanish Pyrenees in the west nearly to the coast of the Caspian Sea in the east. In most of the caves where the bones of cave bears were found they comprised 90 or even 99.9 percent of all the fossils present. The largest single accumulation was discovered in the Austrian province of Steiermark, in a cave near Mixnitz known as the Dragon's Lair. The deposit contained the fossil remains of no fewer than 30,000 cave bears!

By paleontological standards *Ursus spelaeus* is a species with a very brief history. The first cave bears probably evolved at a time near the end of the second great ice advance of the Pleistocene epoch. That was the Mindel glaciation, which began some 700,000 years ago. The bones of an immediately ancestral bear species have been found in older fossil strata. Deposits laid down in the subsequent interglacial period, the Mindel-Riss, contain a few fossil remains of a true cave bear. A bear skull preserved in Mindel-Riss sediments at Swanscombe in England shows the domed forehead that is characteristic of the species, and other cave bear fossils of similar age have been found in a cave in Württemberg in Germany.

It is from sites of later Pleistocene times, however, that the cave bear fossils have come in the greatest abundance. For example, the very numerous remains found in the Dragon's Lair were evi-

dently deposited there during the final Pleistocene ice advance, the 60,000-year Würm glaciation that ended some 12,000 years ago. This fossil accumulation and others of equivalent antiquity make it plain that the species flourished during the late Pleistocene. Yet by the end of the Würm glaciation, or at most a few hundreds of years later, *Ursus spelaeus* was entirely extinct. What can the fossil record tell us about the life of the cave bear? What accounts for its ultimate disappearance?

Kurt Ehrenberg of the Vienna Natural History Museum has made a detailed study of the thousands of fossils from the Dragon's Lair. His findings present a vivid picture of cave bear life. The most numerous specimens from the site are bear teeth. Ehrenberg found that the unworn milk teeth of very young bears—newborn animals and possibly even some fetal ones—were relatively abundant. The presence of this earliest age class at the site indicates that the cave bears came to the Dragon's Lair to hibernate during the winter months. It is during this winter interval that living bears of the Temperate Zone drop their cubs; there is no reason to believe the same was not true of the cave bears. The thousands of milk teeth at the site are the remains of bear cubs whose lives ended before they had ever seen the world outside the cave. Because the bones of newborn animals are extremely fragile almost no trace of the young cubs except teeth has been pre-

served in the Dragon's Lair. Ehrenberg was successful, however, in recovering an almost complete skeleton of a seven-month-old cave bear cub the size of a St. Bernard puppy—scarcely two feet long and a foot high—from another cave deposit in Steiermark [*see illustration on page 148*].

One-year-old bears, animals that had summered for one season outside the cave and then returned for their first hibernation, made up the next age class at the Dragon's Lair. A few of them still had some of their milk teeth but all had begun to cut their permanent teeth. The bones of the one-year-old bears are as rare as those of the newborn ones, so that almost no evidence of pathology is available to explain why bears of this age failed to survive their second winter.

Ehrenberg has identified two additional age classes—two-year-olds and three-year-olds—but thereafter he has difficulty distinguishing between sexually mature cave bears of different ages. No detailed census was made at the Dragon's Lair. Studies at other caves, however, indicate that roughly 70 percent of the bear fossils were from cubs that had died before reaching sexual maturity at age four or five. Nearly 30 percent of the remaining fossils were from bears that had died when they were quite old. Only a few of the fossils represented bears in the prime of life that had either been maimed in some fashion or were suffering from disease. Evidently once a cub reached maturity it might

PALEOLITHIC PORTRAITS of bears, by artists of the period who left their works in the caves of western Europe, include the engraved image of a cave bear (*rendered in the top illustration on the opposite page*), a species characterized by its domed forehead. The 20-inch-long likeness is one of the many Ice Age animal images at La Combarelle, near Les Eyzies in France. A brown bear (*bottom illustration*), with its receding forehead, is an Ice Age animal that still flourishes today. Its 12-inch-long image, outlined in black pigment, appears on the stalagmite-coated wall of a cave at Santimamiñe, a site near Santander in Spain.

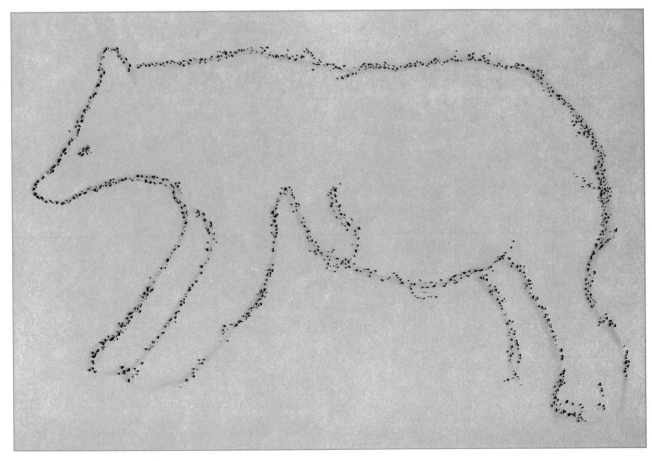

reasonably be expected to survive for another 15 years or so. Taking the cave bear population as a whole, the gross annual mortality appears to have been about 20 percent.

Cave bears were free-ranging during most of the year; they took shelter in caves only during their winter hibernation. Each spring they dispersed, the males moving off alone and the females traveling with their surviving cubs, and began a season of feeding and fat accumulation. Their diet seems to have been primarily vegetarian, to judge by the animals' dentition and the evidence of tooth wear. In flesh-eaters such as the polar bear the cheek teeth are reduced in size and the cusps of four teeth farther forward are very sharp; these sharp teeth enable the animal to rip and shred its prey. In the cave bears quite the opposite is true. The cheek teeth are greatly developed and the cusps of the teeth that are sharp in flesh-eaters are blunt. The cheek teeth of adult cave bears are also excessively worn. The big tooth crowns are ground

down; in some of the older animals they are completely worn away and even the roots of the teeth show signs of wear.

Although female cave bears were consistently smaller than the males, both were formidable animals. From nose to tail they were about the same size as a grizzly bear, but their body was much heavier, with a deep, barrel-like chest. Their paws were short but broad, and the toed-in stance was more pronounced than it is among other bears. The cave bear's most striking characteristic, seen in all but a few fossil skulls, was the doming of the forehead region. This bulge did not come from any enlargement of the braincase; it was the result of oversize nasal sinus cavities. These increased the height of the skull; probably providing better leverage for the temporal muscles connected to the lower jaw. Because the hindmost cheek teeth did most of the work in chewing, the muscles would have been oriented more vertically than in other bears, making a higher skull advantageous.

At the end of the feeding season the cave bear would select a winter den

and hibernate. The caves show no evidence of strict segregation by sex. In most natural populations of mammals males and females are born in approximately equal numbers; the cave bear apparently was no exception. As a result the fossil remains in many caves show a 50–50 ratio of male to female. This, however, was not true everywhere. In upper strata of the Dragon's Lair males

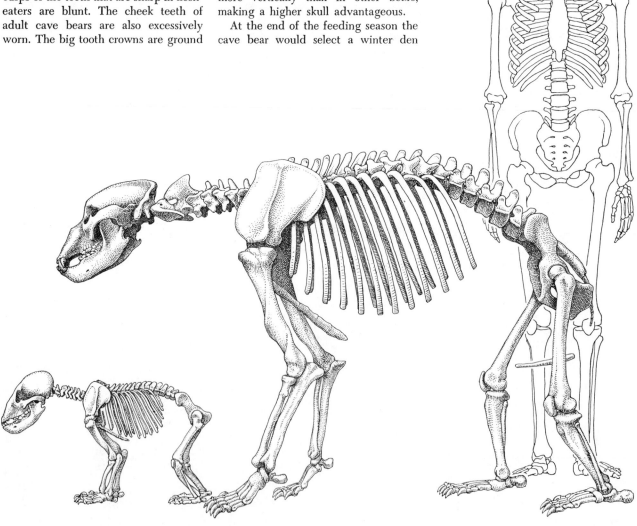

CAVE BEAR SKELETONS are those of an adult male, some five feet long from nose to tail and four feet high, in the collection of the Smithsonian Institution, and of a seven-month-old cub, only two feet long and one foot high, in the collection of the Vienna Natural History Museum. For purposes of comparison a human skeleton five feet 10 inches in height is included. Between the hind legs of the adult bear is the animal's penis bone. Many of these fragile bones had reknit after fracturing during the bear's lifetime.

outnumbered females three to one; for the fossil collection as a whole the ratio is 60 percent male to 40 percent female. At Cueva del Toll in Spain 52 percent of the fossils are male. At Cotencher in Switzerland the male percentage is 48; at Gondenans, Montolivot and Saint-Bras in France the percentages are respectively 44, 33 and 28. When the dimensions of these caves are compared, it appears that sexual preference may explain the different ratios. Females predominated in the small caves and males in the large ones. Perhaps females with cubs to guard tended to select smaller, more easily defended caves where danger was not likely to lurk around the corner.

In some instances the sex ratios are quite bizarre, reflecting a human bias rather than any natural phenomenon. For example, a cave bear site at Hohlestein in Germany was excavated in the 19th century by a party from the Stuttgart Natural History Museum in cooperation with some amateur fossil-hunters. The fossils were then divided among the museum and the private collectors, with the museum having first pick. As a result the Hohlestein fossils at the museum are almost all large, showy specimens, which is to say males. Most of the private collections have since been lost, so that the true ratio of male to female cave bears at Hohlestein will never be known. Another joint venture undertaken by the Stuttgart museum, at a cave bear site named the Sibyls' Cave, gave first pick of specimens to the amateur diggers; the leftover fossils that the museum received were almost exclusively those of female cave bears. One unwary paleontologist, analyzing the collection without taking this fact into account, came to the conclusion that the bears of the Sibyls' Cave were a dwarf race.

The cave bears' way of life, as revealed by the fossil record, thus seems to have been a simple enough annual cycle of wandering and feeding from spring to fall and hibernating and whelping in shelter during the winter months. Once past the perils of immaturity, little lay ahead for the adult except an uneventful repetition of the cycle for a decade or two. If the same bear made its winter den in the same cave two years in succession, this would perhaps have been no more than coincidence. Nonetheless, the large number of remains in many of the caves suggests that few of them were unoccupied for long.

Why is it that the very young and the very old bears were the ones most likely to die in their winter den? It must be that the bears of these age classes were the ones most likely to have experienced an unsuccessful summer season, and hence to have failed to build a store of fat sufficient to last them through the long and severe Ice Age winter. It is easy to see how this fate could befall the very old; among bears of advancing years tooth wear was increasingly severe. The completely worn-down teeth characteristic of old bears must have greatly hampered their summer feeding and left them ill-prepared for winter's rigors.

As for the high mortality rate among immature bears, some or perhaps most of the deaths can be attributed to simple inexperience and inadequate summer feeding. Yet accident, conflict and disease—which are apparent among the relatively rare remains of mature bears—may also have played a part. The roster of known cave bear diseases owes much to the careful fossil analyses of the Austrian pathologist Richard Breuer. Among his findings are instances of gout and related changes (which sometimes led to the fusion of limb bones and vertebrae), of rickets and of tooth decay and damage accompanied by inflammation of the jaw. For example, one well-preserved skull from a Swiss cave, which I have before me as I write, belonged to a powerful young male probably no more than five or six years old. One of its canine teeth is broken and there are traces of serious inflammation around the stump. If a sore jaw meant a bad summer season, the broken canine may have been the cause of the young male's death. If death from any such disability should

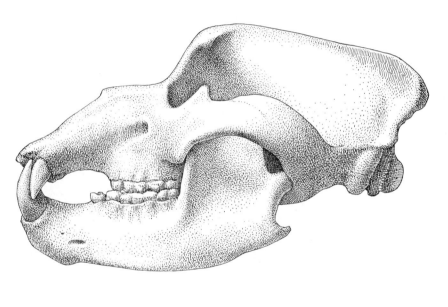

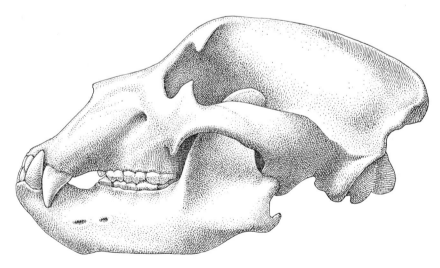

MALE BEAR SKULLS are shown here at the same scale for comparison; the cave bear skull (*top*) is actually some 20 inches long and the brown bear skull (*bottom*) is only 18 inches long. The more rounded forehead of the cave bear is due to an enlargement of the nasal sinus cavities. Its cheek teeth are larger and fewer in number than the brown bear's.

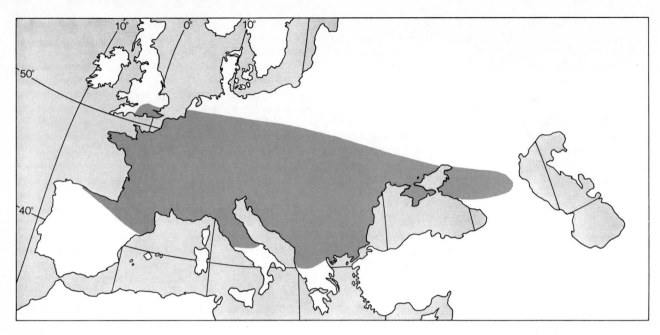

RESTRICTED TERRITORY of the cave bear (*color*) contrasts unfavorably with the far wider distribution of relatives such as the cosmopolitan brown bear. Even within its limited European territory the species was divided into small localized populations.

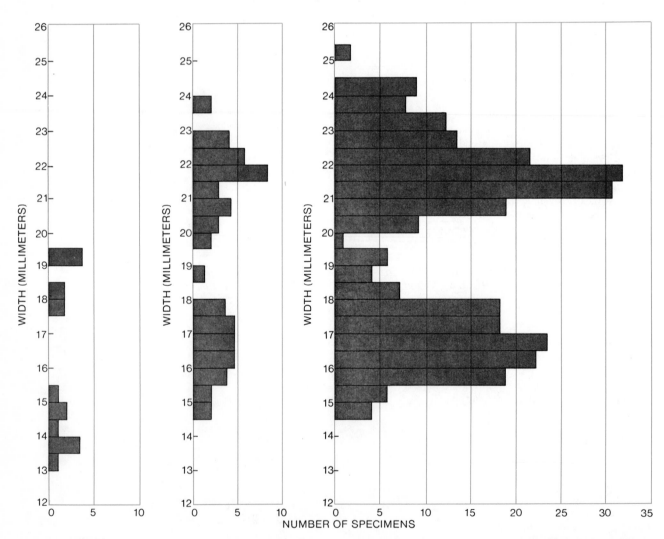

MALE CAVE BEARS were consistently larger than the females in the same population. The graph shows the average width of cave bear lower canine teeth in a Ukrainian fossil collection (*center*), in the collection from the "Dragon's Lair" near Mixnitz in Austria (*right*) and in another Austrian collection from Dachstein (*left*). Each collection shows two clusters of measurements. The larger sizes comprise the male teeth, the smaller the female. The dimorphism is apparent even among the dwarf bears from Dachstein.

have happened to strike a nursing female, her cubs too would have been doomed.

Accidental deaths must have been far less common, but there is evidence of fatalities caused by rocks falling from the ceiling of caves. A famous cave at Gailenreuth, near Würzburg in Germany, has yielded evidence of another kind of mishap: many cave bears apparently fell down an open shaft there from time to time. Either the animals were killed by the fall or, unable to escape, they died of starvation.

Cave bears would rarely have been killed by other animals; a mature bear in good health was much too powerful an animal to have fallen prey to most other Ice Age carnivores. In terms of size the only species in late Pleistocene times that might have been antagonists of the cave bear were the cave lion (*Felis leo spelaea*) and the leopard (*Felis pardus*). Conflicts between bears and lions or leopards are virtually unknown among living species, and there is no evidence to suggest that circumstances were different during the Ice Age.

There is evidence of fights between rival male cave bears during the mating season. Like many living carnivorous mammals, the cave bear had a penis bone. When these rather fragile bones are found in a fossil collection, some of them show signs of having been fractured and then having reknit, apparently as a result of the mating-fight behavior of the animals. If the other injuries in these encounters were severe enough to interfere with the animals' summer fattening, death might have followed during the winter. It does not seem improbable that a certain number of cubs would have died each year because of chance contact with a hostile adult; such incidents are known among living bears.

The great number of cave bear remains might easily suggest that in late Pleistocene times there was a cave bear "population explosion." Any such conclusion would be wrong. As Wolfgang Soergel of the University of Freiburg has demonstrated, no such hypothesis is needed. For example, consider the largest accumulation of all, the fossils from the Dragon's Lair. The death of no more than one cave bear at that site every other winter during the 60,000 years of the Würm glaciation would have been enough to produce the fossils found there.

Calculating on the basis of an estimated cave bear mortality rate of 20 percent per year, how large a standing population was needed in order to sup-

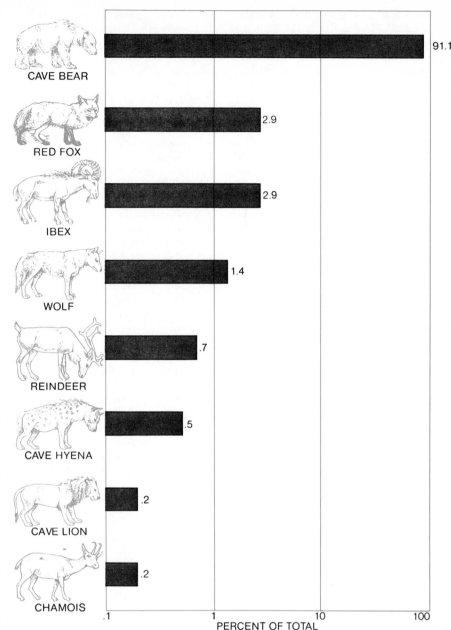

PREPONDERANCE OF CAVE BEARS among the fossil animals unearthed in the caves of Europe is typified by this inventory from Tischofer Cave, near Kurstein in Austria. Eight species of mammals, represented by some 420 individuals, were collected from one level there. Bears outnumbered even the next most abundant species more than 30 to one.

ply the necessary carcasses? The year-to-year census at the Dragon's Lair, it turns out, might never have exceeded two or three individuals—say a mother with one or two cubs one year and some solitary males the next. At other sites, where there are fewer cave bear fossils, the accumulations were evidently produced by even smaller numbers of bears taking up residence in a far more haphazard manner. The correct conclusion to draw, therefore, is that, far from growing explosively, the cave bear population of Ice Age Europe was probably always rather small.

At this point it is worth emphasizing a common item of paleontological knowledge. This is that the bones likeliest to be preserved as fossils are the bones of animals whose carcasses are protected against destructive processes soonest after death. Such protection can be the result of, among other causes, accidental drowning in a swamp, burial in flood sediment or tar or volcanic ash or, as is pertinent here, concealment in a cave. Now, it seems quite clear that most cave bear fossils are the remains of bears that died during winter hibernation. It appears probable, moreover, that relatively few mature bears died before they reached old age. Admittedly the bones of any cave bears that died outside their winter shelters had only a small chance of preservation, so that our ability to estimate the number of such deaths is extremely limited. Nonetheless, the rugged physique of the mature cave bear suggests that the mortality rate was low during this part of the animals' life cycle. Finally, the fossil record of the caves includes a significant number of old bears, senile animals that met their end nearly toothless and almost certainly hungry. When one combines the large number of juvenile cave bears that died in the winter before they reached sexual maturity with the smaller but still substantial number that died of old age, it seems likely that the final figure represents a large percentage of all the cave bears that ever lived. If this interpretation is correct, and the caves of Europe have indeed sheltered the bones of most Ice Age cave bears, that would go a long way toward explaining the mass occurrences of these fossils. It would also be an instance of selective species preservation unique in the fossil record.

What are the major factors that brought about the extinction of the cave bear? One is that as a species the cave bear inhabited a surprisingly restricted geographic range compared with the ranges of other bear species. For example, in the British Isles the remains of cave bears are found only in a narrow strip of southern England. Their penetration into Spain was similarly limited. They are unknown to the south of Monte Cassino in Italy or south of Macedonia in Greece or much north of 51 degrees north latitude in the rest of Europe. The species' most easterly extension was a narrow corridor running from east of the Sea of Azov nearly to the Caspian Sea. The geographic range of the cave bear's evolutionary predecessors, *Ursus minimus* of the late Pliocene and *Ursus etruscus* of the early Pleistocene, was also essentially European. The other bear species that rose from the same stock, the Asiatic and American black bears and the cosmopolitan brown bear, have a much wider range.

Not only was the cave bear restricted in range as a species but also individual cave bears, unlike most large carnivores, had a rather small home territory. The evidence is the development of several local races of cave bears, most of them distinguished by subnormal size. Unlike the Stuttgart museum's hand-picked assortment of females from the Sibyls' Cave, these animals were genuine dwarfs. The males are about the size of normal female cave bears and the dwarf females are smaller still. The skull from the Mindel-Riss interglacial deposits at Swanscombe is that of a dwarf male. The female of the same dwarf English race may be represented by a skull from the lowest level in Kent's Cavern at Torquay; it is the smallest adult cave bear skull I have ever seen.

Austrian cave deposits of late Pleistocene age are notable both for their dwarf races of cave bears and for a puzzle the dwarfs present. If one selects some suitable index of the animal's size, such as the length of the crown of the last upper molar, and plots this measurement for each of the Austrian dwarf races according to the altitude at which it is found, an unmistakable negative correlation appears. The measurements range from a mean length of 45.5 millimeters at 1,000 meters above sea level to a mean length of 40 millimeters at 2,200 meters. Evidently the higher the site, the smaller the bear.

How is one to interpret this finding? It is scarcely credible not only that the bears sorted themselves out according to size and elevation but also that no valley bear ever went uphill, and no dwarf bear went downhill, to breed. A more logical assumption is that the larger and smaller races were not contemporaries; indeed, it is likely that the dwarf forms flourished only during warm intervals when the higher alpine sites where their remains are found were more inhabitable.

The result of isolation was not invariably dwarfism. The upper strata at Kent's Cavern and at Wookey Hole, another late Pleistocene cave deposit in England, contain the fossils of cave bears that may fairly be described as a giant race. Whether such races are large or small, however, their development bespeaks small home territories, stationary habits, breeding isolation and a minimum of contacts among the many widely scattered and not particularly abundant stocks of cave bears.

A species with a limited range that is further subdivided into a series of isolated races is ill-prepared to cope with the shock of drastic changes in climate and environment. The final millenniums of the Ice Age, which saw a European landscape that had consisted mainly of tundra, subarctic taiga and steppe transformed into all-enveloping Temperate Zone forest, brought with them just such a shock. Simultaneously the cave bears began to disappear from many regions. Some isolated groups may have vanished even before the changes became severe; when the number of animals in a local population falls below a critical minimum, the remaining animals are liable to accidental extinction.

There is some suggestive evidence to support the proposal that the cave bears' extinction was caused by environmental change. Rudolf Musil of the Moravian Museum in Brno has made an interesting discovery concerning the fossil remains in a bear cave in Czechoslovakia: the uppermost strata show an increase in the mortality rate for juvenile cave bears. In view of the fact that the normal juvenile mortality rate was already about 70 percent of births, an increase in the rate, presumably reflecting new environmental pressure, may in itself

CAVE BEAR IN THE FLESH probably looked very much as it is shown here. This is a sketch of the life-size reconstruction of a male at the Natural History Museum of Basel.

account for the cave bears' extinction.

Was Paleolithic man instrumental in the disappearance of the cave Bear? Most probably his major influence was indirect. Certain evidence found in Upper Paleolithic paintings, engravings and sculptures indicates that the early Europeans who produced these works were familiar with the appearance of both the cave bear and the brown bear [see illustrations on page 147]. Paleolithic hunters may have attacked cave bears; some bear skulls show lesions that could have been produced by stone-tipped projectiles. Such evidence of hunting is rare, however, and the possibility that Paleolithic hunters exterminated the cave bears is remote. There is one site in Germany (Taubach, near Weimar) where an accumulation of bear bones seems to reflect the hunting activities of Paleolithic man, but almost all the bones at the site are those of brown bears, not cave bears.

Much has been said and written about a supposed "bear cult" among the peoples of Upper Paleolithic Europe. Most of the evidence that is presented in support of the cult hypothesis, however, can just as easily be explained on other grounds. For example, cave bear skulls have been found in natural crannies and niches in the walls of caves, as if men had placed them there. Taking into consideration the year-to-year sequence of natural events in a bear cave, the Swiss speleologist F. E. Koby has concluded that it is unnecessary to invoke a human agency to account for such findings. When the onset of cool weather caused the bears to start looking for a winter den, they must often have entered caves containing bears that had died the pre-

ceding winter. By the end of summer, thanks to the scavenging of smaller animals, any such carcass would have been picked clean; the skeleton might already have been partly disarticulated, with some of its bones crushed by hyenas and wolves. Now the skeleton would be further trampled, broken up and swept aside as the new settlers prepared their den. Koby calls this disarticulation process "dry transport." He believes it can account for not only the discovery of cave bear skulls in cave crannies but also a second kind of evidence cited by bear-cult advocates: the occasional finding of a bear skull with limb bones pushed into the eye sockets, the nasal opening or the opening for the spinal cord.

Another piece of evidence that is advanced by supporters of the bear-cult hypothesis is a life-size clay sculpture of a headless bear found in the cave at Montespan in the French Pyrenees. Whether or not this example of Upper Paleolithic sculpture, which is certainly ambitious in scale, should be given the status of a cult object seems open to question. Other animal species are even more frequently represented in the art of the period.

There remains a discovery made in the 1920's by the Swiss prehistorian Emil Bächler in the Drachenloch cave near Vättis in the Tamina Valley of Switzerland. What Bächler found in the cave were several coffin-like enclosures made with slabs of stone. A diagram that he published in 1923 represents one of the enclosures as containing the skulls of two cave bears, visible in profile. Here too Koby has challenged the bear cultists. His criticism is based on a later diagram of the same stone enclosure published by Bächler in 1940. In

the enclosure there are now six cave bear skulls rather than two, and they are seen full face rather than in profile. One must add in fairness that this criticism does not in itself dismiss Bächler's original observation.

What if, in spite of such doubts, one accepts the hypothesis that an active and widespread bear cult flourished at the end of the Ice Age? Do the various bear skulls and limb bones that served as cult objects represent cave bears that were first hunted down and slain by the cultists and then preserved as trophies? In view of the number of bones already at the cultists' disposal in the bear caves, such a pattern of ritual trophy-hunting seems scarcely probable. Thus the bear cult, even if it did exist, could have flourished without producing significant pressure on the cave bear population.

Where Upper Paleolithic man may have helped to tilt the scale in favor of the cave bears' extinction was in competition for natural shelter. Unlike other bear species, which can and generally do hibernate successfully in an open-air den, the cave bears seem to have been entirely dependent on caves for winter refuge. Of course, not every such natural shelter was occupied in the winter by the people of the Upper Paleolithic. Nonetheless, human interference that affected only a percentage of the available bear caves at irregular intervals could have helped to reduce the numbers in local populations of cave bears below the minimum survival level. Ultimately this indirect human influence may have acted, in combination with the pressures of climatic and environmental change that marked the end of the Ice Age, to bring about the cave bears' extinction.

16   Homo Erectus

by William W. Howells
November 1966

This species, until recently known by a multiplicity of other names, was probably the immediate predecessor of modern man. It now seems possible that the transition took place some 500,000 years ago

In 1891 Eugène Dubois, a young Dutch anatomist bent on discovering early man, was examining a fossil-rich layer of gravels beside the Solo River in Java. He found what he was after: an ancient human skull. The next year he discovered in the same formation a human thighbone. These two fossils, now known to be more than 700,000 years old, were the first remains to be found of the prehistoric human species known today as *Homo erectus*. It is appropriate on the 75th anniversary of Dubois's discovery to review how our understanding of this early man has been broadened and clarified by more recent discoveries of fossil men of similar antiquity and the same general characteristics, so that *Homo erectus* is now viewed as representing a major stage in the evolution of man. Also of interest, although of less consequence, is the way in which the name *Homo erectus,* now accepted by many scholars, has been chosen after a long period during which "scientific" names for human fossils were bestowed rather capriciously.

Man first received his formal name in 1758, when Carolus Linnaeus called him *Homo sapiens*. Linnaeus was trying simply to bring order to the world of living things by distinguishing each species of plant and animal from every other and by arranging them all in a hierarchical system. Considering living men, he recognized them quite correctly as one species in the system. The two centuries that followed Linnaeus saw first the establishment of evolutionary theory and then the realization of its genetic foundations; as a result ideas on the relations of species as units of plant and animal life have become considerably more complex. For example, a species can form two or more new species, which Linnaeus originally thought was impossible. By today's definition a spe-

cies typically consists of a series of local or regional populations that may exhibit minor differences of form or color but that otherwise share a common genetic structure and pool of genes and are thus able to interbreed across population lines. Only when two such populations have gradually undergone so many different changes in their genetic makeup that the likelihood of their interbreeding falls below a critical point are they genetically cut off from each other and do they become separate species. Alternatively, over a great many generations an equivalent amount of change will take place in the same population, so that its later form will be recognized as a species different from the earlier. This kind of difference, of course, cannot be put to the test of interbreeding and can only be judged by the physical form of the fossils involved.

In the case of living man there is no reason to revise Linnaeus' assignment: *Homo sapiens* is a good, typical species. Evolution, however, was not in Linnaeus' ken. He never saw a human fossil, much less conceived of men different from living men. Between his time and ours the use of the Linnaean system of classification as applied to man and his relatives past and present became almost a game. On grasping the concept of evolution, scholars saw that modern man must have had ancestors. They were prepared to anticipate the actual discovery of these ancestral forms, and perhaps the greatest anticipator was the German biologist Ernst Haeckel. Working on the basis of fragmentary information in 1889, when the only well-known fossil human remains were the comparatively recent bones discovered 25 years earlier in the Neander Valley of Germany, Haeckel drew up a theoretical ancestral line for man. The line began among some postu-

lated extinct apes of the Miocene epoch and reached *Homo sapiens* by way of an imagined group of "ape-men" (Pithecanthropi) and a group of more advanced but still speechless early men (Alali) whom he visualized as the worldwide stock from which modern men had evolved [*see illustration on page 156*]. A creature combining these various presapient attributes took form in the pooled imagination of Haeckel and his compatriots August Schleicher and Gabriel Max. Max produced a family portrait, and the still-to-be-discovered ancestor was given the respectable Linnaean name *Pithecanthropus alalus*.

Were he living today Haeckel would never do such a thing. It is now the requirement of the International Code of Zoological Nomenclature that the naming of any new genus or species be supported by publication of the specimen's particulars together with a description showing it to be recognizably different from any genus or species previously known. Haeckel was rescued from retroactive embarrassment, however, by Dubois, who gave Haeckel's genus name to Java man. The skull was too large to be an ape's and apparently too small to be a man's; the name *Pithecanthropus* seemed perfectly appropriate. On the other hand, the thighbone from the same formation was essentially modern; its possessor had evidently walked upright. Dubois therefore gave his discovery the species name *erectus*. Since Dubois's time the legitimacy of his finds has been confirmed by the discovery in Java (by G. H. R. von Koenigswald between 1936 and 1939 and by Indonesian workers within the past three years) of equally old and older fossils of the same population.

In the 50 years between Dubois's discovery and the beginning of World

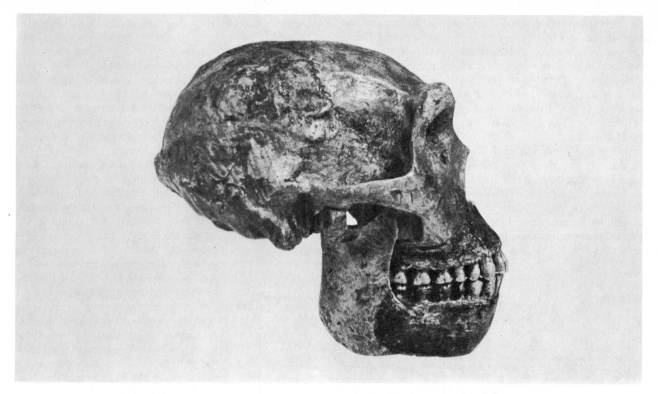

JAVA MAN, whose 700,000-year-old remains were unearthed in 1891 by Eugène Dubois, is representative of the earliest *Homo erectus* population so far discovered. This reconstruction was made recently by G. H. R. von Koenigswald and combines the features of the more primitive members of this species of man that he found in the lowest (Djetis) fossil strata at Sangiran in central Java during the 1930's. The characteristics that are typical of *Homo erectus* include the smallness and flatness of the cranium, the heavy browridge and both the sharp bend and the ridge for muscle attachment at the rear of the skull. The robustness of the jaws adds to the species' primitive appearance. In most respects except size, however, the teeth of *Homo erectus* resemble those of modern man.

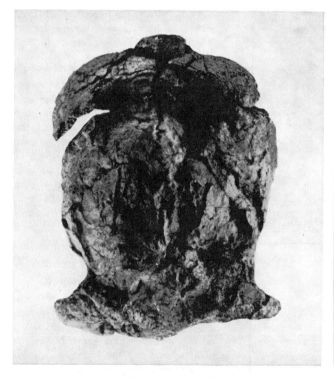

LANTIAN MAN is the most recently found *Homo erectus* fossil. The discovery consists of a jawbone and this skullcap (*top view, browridge at bottom*) from which the occipital bone (*top*) is partially detached. Woo Ju-kang of the Chinese Academy of Sciences in Peking provided the photograph; this fossil man from Shensi may be as old as the earliest specimens of *Homo erectus* from Java.

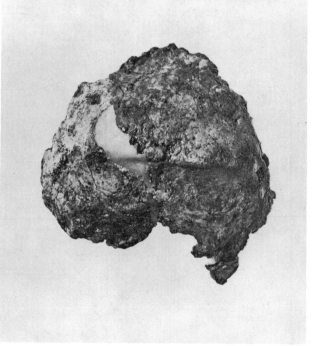

OCCIPITAL BONE found at Vértesszöllös in Hungary in 1965 is 500,000 or more years old. The only older human fossil in Europe is the Heidelberg jaw. The bone forms the rear of a skull; the ridge for muscle attachment (*horizontal line*) is readily apparent. In spite of this primitive feature and its great age, the skull fragment from Vértesszöllös has been assigned to the species *Homo sapiens*.

War II various other important new kinds of human fossil came into view. For our purposes the principal ones (with some of the Linnaean names thrust on them) were (1) the lower jaw found at Mauer in Germany in 1907 (*Homo heidelbergensis* or *Palaeanthropus*), (2) the nearly complete skull found at Broken Hill in Rhodesia in 1921 (*Homo rhodesiensis* or *Cyphanthropus*), (3) various remains uncovered near Peking in China, beginning with one tooth in 1923 and finally comprising a collection representing more than 40 men, women and children by the end of 1937 (*Sinanthropus pekinensis*), and (4) several skulls found in 1931 and 1932 near

Ngandong on the Solo River not far from where Dubois had worked (*Homo soloensis* or *Javanthropus*). This is a fair number of fossils, but they were threatened with being outnumbered by the names assigned to them. The British student of early man Bernard G. Campbell has recorded the following variants in the case of the Mauer jawbone alone: *Palaeanthropus heidelbergensis, Pseudhomo heidelbergensis, Protanthropus heidelbergensis, Praehomo heidelbergensis, Praehomo europaeus, Anthropus heidelbergensis, Maueranthropus heidelbergensis, Europanthropus heidelbergensis* and *Euranthropus*.

Often the men responsible for these

redundant christenings were guilty merely of innocent grandiloquence. They were not formally declaring their conviction that each fossil hominid belonged to a separate genus, distinct from *Homo*, which would imply an enormous diversity in the human stock. Nonetheless, the multiplicity of names has interfered with an understanding of the evolutionary significance of the fossils that bore them. Moreover, the human family trees drawn during this period showed a fundamental resemblance to Haeckel's original venture; the rather isolated specimens of early man were stuck on here and there like Christmas-tree ornaments. Although the arrangements evinced a vague consciousness of evolution, no scheme was presented that intelligibly interpreted the fossil record.

At last two questions came to the fore. First, to what degree did the fossils really differ? Second, what was the difference among them over a period of time? The fossil men of the most recent period—those who had lived between roughly 100,000 and 30,000 years ago—were Neanderthal man, Rhodesian man and Solo man. They have been known traditionally as *Homo neanderthalensis, Homo rhodesiensis* and *Homo soloensis*, names that declare each of the three to be a separate species, distinct from one another and from *Homo sapiens*. This in turn suggests that if Neanderthal and Rhodesian populations had come in contact, they would probably not have interbred. Such a conclusion is difficult to establish on the basis of fossils, particularly when they are few and tell very little about the geographical range of the species. Today's general view is a contrary one. These comparatively recent fossil men, it is now believed, did not constitute separate species. They were at most incipient species, that is, subspecies or variant populations that had developed in widely separated parts of the world but were still probably able to breed with one another or with *Homo sapiens*.

It was also soon recognized that the older Java and Peking fossils were not very different from one another. The suggestion followed that both populations be placed in a single genus (*Pithecanthropus*) and that the junior name (*Sinanthropus*) be dropped. Even this, however, was one genus too many for Ernst Mayr of Harvard University. Mayr, whose specialty is the evolutionary basis of biological classification, declared that ordinary zoological standards would not permit Java and Peking man to occupy

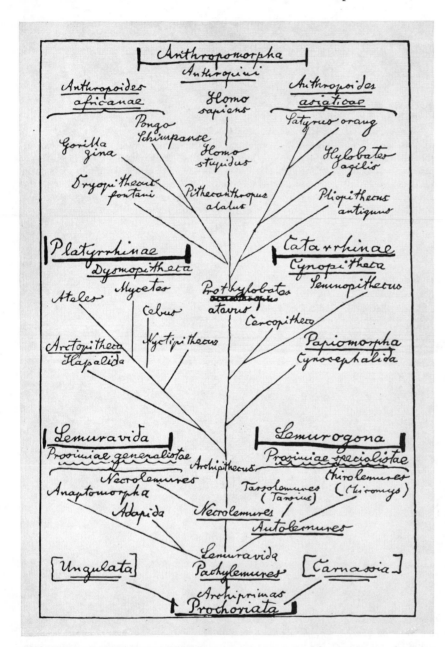

THE NAME "PITHECANTHROPUS," or ape-man, was coined by the German biologist Ernst Haeckel in 1889 for a postulated precursor of *Homo sapiens*. Haeckel placed the ape-man genus two steps below modern man on his "tree" of primate evolution, adding the species name *alalus*, or "speechless," because he deemed speech an exclusively human trait.

GRADE	EUROPE	NORTH AFRICA	EAST AFRICA	SOUTH AFRICA	EAST ASIA	SOUTHEAST ASIA
(5)	*HOMO SAPIENS* (VERTESSZÖLLÖS)					
(4)						*(HOMO ERECTUS SOLOENSIS)*
3	*HOMO ERECTUS HEIDELBERGENSIS*	*HOMO ERECTUS MAURITANICUS*	*HOMO ERECTUS LEAKEYI*		*HOMO ERECTUS PEKINENSIS*	
2						*HOMO ERECTUS ERECTUS*
1			*HOMO ERECTUS HABILIS*	*HOMO ERECTUS CAPENSIS*	*(HOMO ERECTUS LANTIANENSIS)*	*HOMO ERECTUS MODJOKERTENSIS*

EIGHT SUBSPECIES of *Homo erectus* that are generally accepted today have been given appropriate names and ranked in order of evolutionary progress by the British scholar Bernard G. Campbell. The author has added Lantian man to Campbell's lowest *Homo erectus* grade and provided a fourth grade to accommodate Solo man, a late but primitive survival. The author has also added a fifth grade for the *Homo sapiens* fossil from Vértesszöllös (*color*). Colored area suggests that Heidelberg man is its possible forebear.

a genus separate from modern man. In his opinion the amount of evolutionary progress that separates *Pithecanthropus* from ourselves is a step that allows the recognition only of a different species. After all, Java and Peking man apparently had bodies just like our own; that is to say, they were attacking the problem of survival with exactly the same adaptations, although with smaller brains. On this view Java man is placed in the genus *Homo* but according to the rules retains his original species name and so becomes *Homo erectus*. Under the circumstances Peking man can be distinguished from him only as a subspecies: *Homo erectus pekinensis*.

The simplification is something more than sweeping out a clutter of old names to please the International Commission on Zoological Nomenclature. The reduction of fossil hominids to not more than two species and the recognition of *Homo erectus* has become increasingly useful as a way of looking at a stage of human evolution. This has been increasingly evident in recent years, as human fossils have continued to come to light and as new and improved methods of dating them have been developed. It is now possible to place both the old discoveries and the new ones much more precisely in time, and that is basic to establishing the entire pattern of human evolution in the past few million years.

To consider dating first, the period during which *Homo erectus* flourished occupies the early middle part of the Pleistocene epoch. The evidence that now enables us to subdivide the Pleistocene with some degree of confidence is of several kinds. For example, the fossil animals found in association with fossil men often indicate whether the climate of the time was cold or warm. The comparison of animal communities is also helpful in correlating intervals of time on one continent with intervals on another. The ability to determine absolute dates, which makes possible the correlation of the relative dates derived from sequences of strata in widely separated localities, is another significant development. Foremost among the methods of absolute dating at the moment is one based on the rate of decay of radioactive potassium into argon. A second method showing much promise is the analysis of deep-sea sediments; changes in the forms of planktonic life embedded in samples of the bottom reflect worldwide temperature changes. When the absolute ages of key points in sediment sequences are determined by physical or chemical methods, it ought to be possible to assign dates to all the major events of the Pleistocene. Such methods have already suggested that the epoch began more than three million years ago and that its first major cold phase (corresponding to the Günz

glaciation of the Alps) may date back to as much as 1.5 million years ago. The period of time occupied by *Homo erectus* now appears to extend from about a million years ago to 500,000 years ago in terms of absolute dates, or from some time during the first interglacial period in the Northern Hemisphere to about the end of the second major cold phase (corresponding to the Mindel glaciation of the Alps).

On the basis of the fossils found before World War II, with the exception of the isolated and somewhat peculiar Heidelberg jaw, *Homo erectus* would have appeared to be a human population of the Far East. The Java skulls, particularly those that come from the lowest fossil strata (known as the Djetis beds), are unsurpassed within the entire group in primitiveness. Even the skulls from the strata above them (the Trinil beds), in which Dubois made his original discovery, have very thick walls and room for only a small brain. Their cranial capacity probably averages less than 900 cubic centimeters, compared with an average of 500 c.c. for gorillas and about 1,400 c.c. for modern man. The later representatives of Java man must be more than 710,000 years old, because potassium-argon analysis has shown that tektites (glassy stones formed by or from meteorites) in higher strata of the same formation are of that age.

The Peking fossils are younger, prob-

158

ably dating to the middle of the second Pleistocene cold phase, and are physically somewhat less crude than the Java ones. The braincase is higher, the face shorter and the cranial capacity approaches 1,100 c.c., but the general construction of skull and jaw is similar. The teeth of both Java man and Peking man are somewhat larger than modern man's and are distinguished by traces of an enamel collar, called a cingulum, around some of the crowns. The latter is an ancient and primitive trait in man and apes.

Discoveries of human fossils after World War II have added significantly to the picture of man's distribution at this period. The pertinent finds are the following:

1949: Swartkrans, South Africa. Jaw and facial fragments, originally given the name *Telanthropus capensis*. These were found among the copious remains at this site of the primitive subhumans known as australopithecines. The fossils were recognized at once by the late Robert Broom and his colleague John T. Robinson as more advanced than the australopithecines both in size and in traits of jaw and teeth. Robinson has now assigned *Telanthropus* to *Homo erectus*, since that is where he evidently belongs.

1955: Ternifine, Algeria. Three jaws and a parietal bone, given the name *Atlanthropus mauritanicus*, were found under a deep covering of sand on the clay floor of an ancient pond by Camille Arambourg. The teeth and jaws show a strong likeness to the Peking remains.

1961: Olduvai Gorge, Tanzania. A skullcap, not formally named but identified as the Bed II Hominid, was discovered by L. S. B. Leakey. Found in a context with a provisional potassium-argon date of 500,000 years ago, the skull's estimated cranial capacity is 1,000 c.c. Although differing somewhat in detail, it has the general characteristics of the two Far Eastern subspecies of *Homo erectus*. At lower levels in this same important site were found the remains of a number of individuals with small skulls, now collectively referred to as "Homo habilis."

1963–1964: Lantian district, Shensi, China. A lower jaw and a skullcap were found by Chinese workers at two separate localities in the district and given the name *Sinanthropus lantianensis*. Animal fossils indicate that the Lantian sites are older than the one that yielded Peking man and roughly as old as the lowest formation in Java. The form of the skull and jaw accords well with this

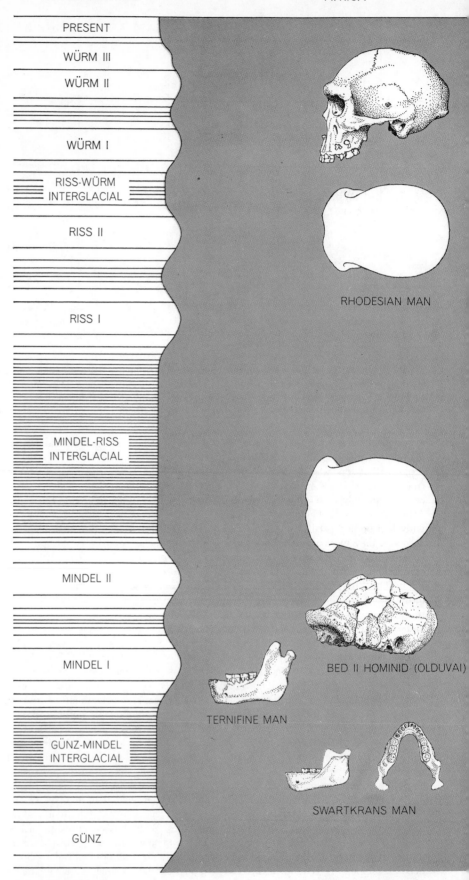

PRESENT

WÜRM III

WÜRM II

WÜRM I

RISS-WÜRM INTERGLACIAL

RISS II

RISS I

MINDEL-RISS INTERGLACIAL

MINDEL II

MINDEL I

GÜNZ-MINDEL INTERGLACIAL

GÜNZ

RHODESIAN MAN

BED II HOMINID (OLDUVAI)

TERNIFINE MAN

SWARTKRANS MAN

FOSSIL EVIDENCE for the existence of a single species of early man instead of several species and genera of forerunners of *Homo sapiens* is presented in this array of individual remains whose age places them in the interval of approximately 500,000 years that separates the first Pleistocene interglacial period from the end of the second glacial period (*see scale at left*). The earliest *Homo erectus* fossils known, from Java and China, belong to the first interglacial period; the earliest *Homo erectus* remains from South Africa may be equally

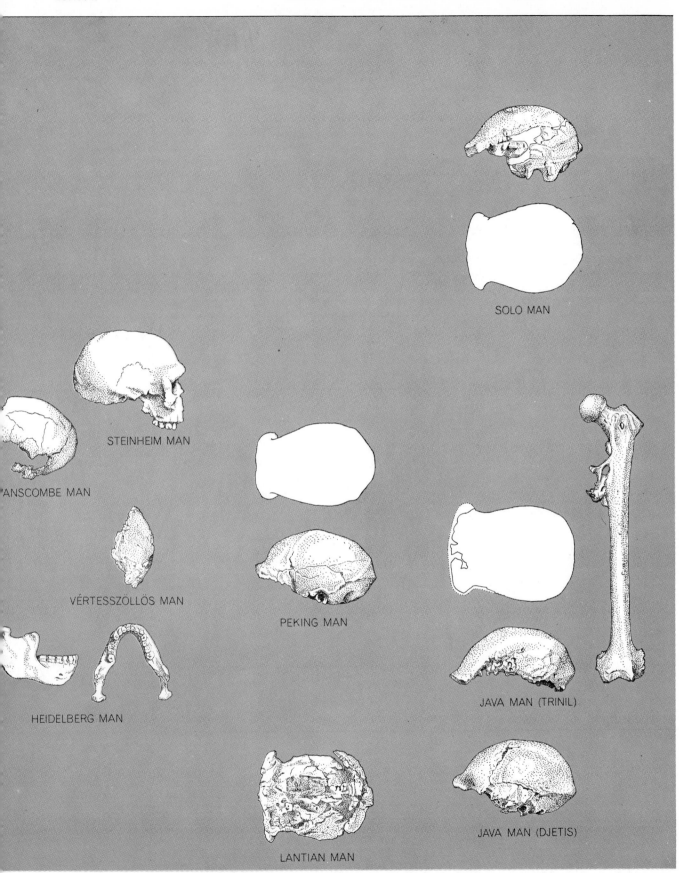

SOLO MAN

STEINHEIM MAN

ANSCOMBE MAN

VÉRTESSZÖLLÖS MAN

PEKING MAN

HEIDELBERG MAN

JAVA MAN (TRINIL)

LANTIAN MAN

JAVA MAN (DJETIS)

old. Half a million years later *Homo erectus* continued to be represented in China by the remains of Peking man and in Africa by the skull from Olduvai Gorge. In the intervening period this small-brained precursor of modern man was not the only human species inhabiting the earth, nor did *Homo erectus* become extinct when the 500,000-year period ended. One kind of man who had apparent-

ly reached the grade of *Homo sapiens* in Europe by the middle or later part of the second Pleistocene glacial period was unearthed recently at Vértesszöllös in Hungary. In the following interglacial period *Homo sapiens* is represented by the Steinheim and Swanscombe females. Solo man's remains indicate that *Homo erectus* survived for several hundred thousand years after that.

dating; both are distinctly more primitive than the Peking fossils. Both differ somewhat in detail from the Java subspecies of *Homo erectus*, but the estimated capacity of this otherwise large skull (780 c.c.) is small and close to that of the earliest fossil cranium unearthed in Java.

1965: Vértesszöllös, Hungary. An isolated occipital bone (in the back of the skull) was found by L. Vértes. This skull fragment is the first human fossil from the early middle Pleistocene to be unearthed in Europe since the Heidelberg jaw. It evidently dates to the middle or later part of the Mindel glaciation and thus falls clearly within the *Homo erectus* time zone as defined here. The bone is moderately thick and shows a well-defined ridge for the attachment of neck muscles such as is seen in all the *erectus* skulls. It is unlike *erectus* occipital bones, however, in that it is both large and definitely less angled; these features indicate a more advanced skull.

In addition to these five discoveries, something else of considerable importance happened during this period. The Piltdown fraud, perpetrated sometime before 1912, was finally exposed in 1953. The detective work of J. S. Weiner, Sir Wilfrid Le Gros Clark and Kenneth Oakley removed from the fossil record a supposed hominid with a fully apelike jaw and manlike skull that could scarcely be fitted into any sensible evolutionary scheme.

From this accumulation of finds, many of them made so recently, there emerges a picture of men with skeletons like ours but with brains much smaller, skulls much thicker and flatter and furnished with protruding brows in front and a marked angle in the rear, and with teeth somewhat larger and exhibiting a few slightly more primitive traits. This picture suggests an evolutionary level, or grade, occupying half a million years of human 'history and now seen

to prevail all over the inhabited Old World. This is the meaning of *Homo erectus*. It gives us a new foundation for ideas as to the pace and the pattern of human evolution over a critical span of time.

Quite possibly this summary is too tidy; before the 100th anniversary of the resurrection of *Homo erectus* is celebrated complications may appear that we cannot perceive at present. Even today there are a number of fringe problems we cannot neglect. Here are some of them.

What was the amount of evolution taking place within the *erectus* grade? There is probably a good deal of accident of discovery involved in defining *Homo erectus*. Chance, in other words, may have isolated a segment of a continuum, since finds from the time immediately following this 500,000-year period are almost lacking. It seems likely, in fact practically certain, that real evolutionary progress was taking place,

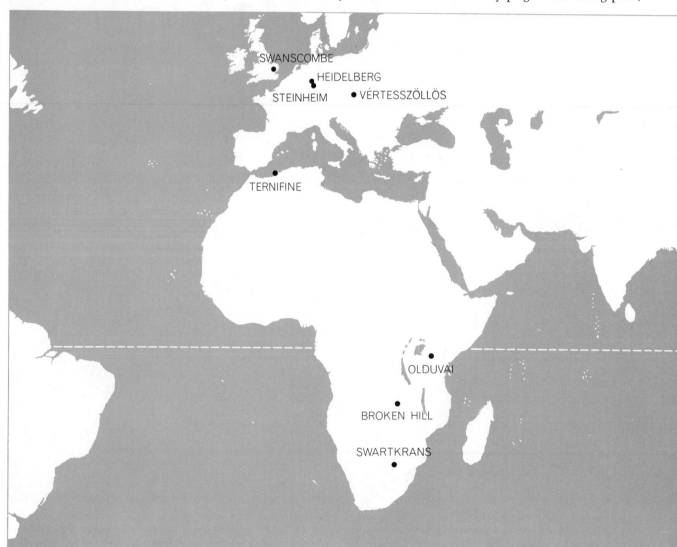

DISTRIBUTION of *Homo erectus* seemed to be confined mainly to the Far East and Southeast Asia on the basis of fossils unearthed before World War II; the sole exception was the Heidelberg jaw. Postwar findings in South, East and North Africa, as well as dis-

but the tools made by man during this period reveal little of it. As for the fossils themselves, the oldest skulls—from Java and Lantian—are the crudest and have the smallest brains. In Java, one region with some discernible stratigraphy, the later skulls show signs of evolutionary advance compared with the earlier ones. The Peking skulls, which are almost certainly later still, are even more progressive. Bernard Campbell, who has recently suggested that all the known forms of *Homo erectus* be formally recognized as named subspecies, has arranged the names in the order of their relative progressiveness. I have added some names to Campbell's list; they appear in parentheses in the illustration on page 157. As the illustration indicates, the advances in grade seem indeed to correspond fairly well with the passage of time.

What are the relations of *Homo erectus* to Rhodesian and Solo man? This is a point of particular importance, be-

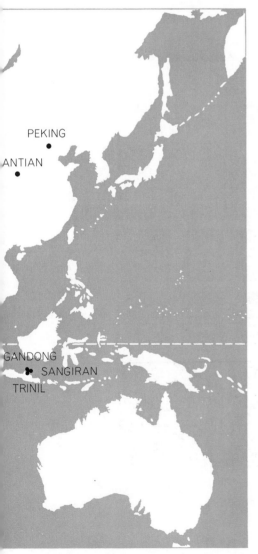

covery of a new *Homo erectus* site in northern China, have extended the species' range.

cause both the African and the Javanese fossils are much younger than the date we have set as the general upward boundary for *Homo erectus*. Rhodesian man may have been alive as recently as 30,000 years ago and may have actually overlapped with modern man. Solo man probably existed during the last Pleistocene cold phase; this is still very recent compared with the time zone of the other *erectus* fossils described here. Carleton S. Coon of the University of Pennsylvania deems both late fossil men to be *Homo erectus* on the basis of tooth size and skull flatness. His placing of Rhodesian man is arguable, but Solo man is so primitive, so like Java man in many aspects of his skull form and so close to Peking man in brain size that his classification as *Homo erectus* seems almost inevitable. The meaning of his survival hundreds of thousands of years after the period I have suggested, and his relation to the modern men who succeeded him in Southeast Asia in recent times, are unanswered questions of considerable importance.

Where did *Homo erectus* come from? The Swartkrans discovery makes it clear that he arose before the last representatives of the australopithecines had died out at that site. The best present evidence of his origin is also from Africa; it consists of the series of fossils unearthed at Olduvai Gorge by Leakey and his wife and called Homo habilis. These remains seem to reflect a transition from an australopithecine level to an *erectus* level about a million years ago. This date seems almost too late, however, when one considers the age of *Homo erectus* finds elsewhere in the world, particularly in Java.

Where did *Homo erectus* go? The paths are simply untraced, both those that presumably lead to the Swanscombe and Steinheim people of Europe during the Pleistocene's second interglacial period and those leading to the much later Rhodesian and Neanderthal men. This is a period lacking useful evidence. Above all, the nature of the line leading to living man—*Homo sapiens* in the Linnaean sense—remains a matter of pure theory.

We may, however, have a clue. Here it is necessary to face a final problem. What was the real variation in physical type during the time period of *Homo erectus*? On the whole, considering the time and space involved, it does not appear to be very large; the similarity of the North African jaws to those of Peking man, for example, is striking in spite of the thousands of

miles that separated the two populations. The Heidelberg jaw, however, has always seemed to be somewhat different from all the others and a little closer to modern man in the nature of its teeth. The only other European fossil approaching the Heidelberg jaw in antiquity is the occipital bone recently found at Vértesszöllös. This piece of skull likewise appears to be progressive in form and may have belonged to the same general kind of man as the Heidelberg jaw, although it is somewhat more recent in date.

Andor Thoma of Hungary's Kossuth University at Debrecen in Hungary, who has kindly given me information concerning the Vértesszöllös fossil, will publish a formal description soon in the French journal *L'Anthropologie*. He estimates that the cranial capacity was about 1,400 c.c., close to the average for modern man and well above that of the known specimens of *Homo erectus*. Although the occipital bone is thick, it is larger and less sharply angled than the matching skull area of Rhodesian man. It is certainly more modern-looking than the Solo skulls. I see no reason at this point to dispute Thoma's estimate of brain volume. He concludes that Vértesszöllös man had in fact reached the *sapiens* grade in skull form and brain size and accordingly has named him a subspecies of *Homo sapiens*.

Thoma's finding therefore places a population of more progressive, *sapiens* humanity contemporary with the populations of *Homo erectus* 500,000 years ago or more. From the succeeding interglacial period in Europe have come the Swanscombe and Steinheim skulls, generally recognized as *sapiens* in grade. They are less heavy than the Hungarian fossil, more curved in occipital profile and smaller in size; they are also apparently both female, which would account for part of these differences.

The trail of evidence is of course faint, but there are no present signs of contradiction; what we may be seeing is a line that follows *Homo sapiens* back from Swanscombe and Steinheim to Vértesszöllös and finally to Heidelberg man at the root. This is something like the Solo case in reverse, a *Homo sapiens* population surprisingly early in time, in contrast to a possible *Homo erectus* population surprisingly late. In fact, we are seeing only the outlines of what we must still discover. It is easy to perceive how badly we need more fossils; for example, we cannot relate Heidelberg man to any later Europeans until we find some skull parts to add to his solitary jaw.

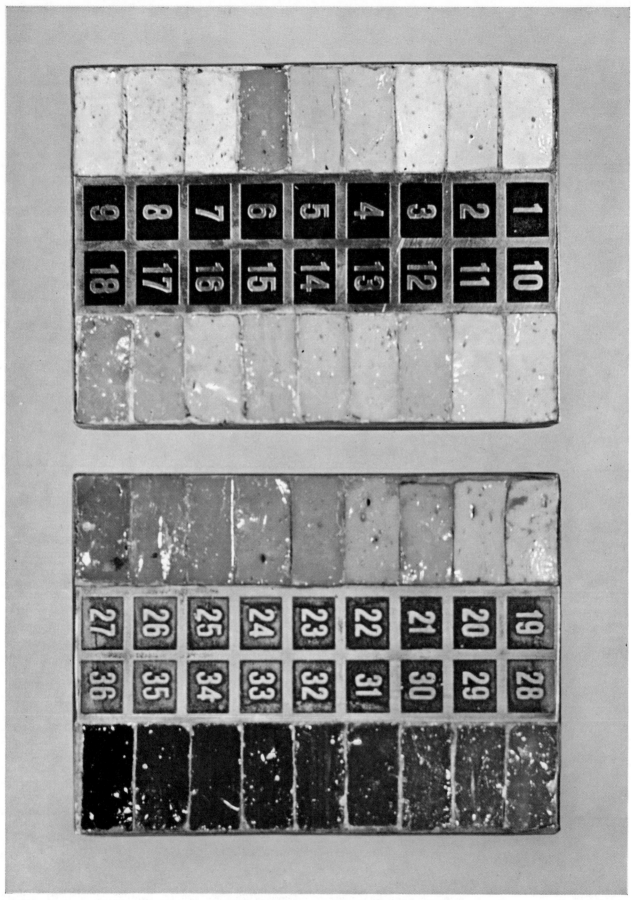

COLOR OF HUMAN SKIN is sometimes measured by physical anthropologists on the von Luschan scale. Reproduced here somewhat larger than natural size, the scale consists of numbered ceramic tiles which are compared visually to color of the underside of subject's forearm. Both sides of the scale are shown; colors range from almost pure white (*top right*) to black (*bottom left*).

The Distribution of Man

by William W. Howells
September 1960

Homo sapiens arose in the Old World, but has since become the most widely distributed of all animal species. In the process he has differentiated into three principal strains

Men with chins, relatively small brow ridges and small facial skeletons, and with high, flat-sided skulls, probably appeared on earth in the period between the last two great continental glaciers, say from 150,000 to 50,000 years ago. If the time of their origin is blurred, the place is no less so. The new species doubtless emerged from a number of related populations distributed over a considerable part of the Old World. Thus *Homo sapiens* evolved as a species and began to differentiate into races at the same time.

In any case, our direct ancestor, like his older relatives, was at once product and master of the crude pebble tools that primitive human forms had learned to use hundreds of thousands of years earlier. His inheritance also included a social organization and some level of verbal communication.

Between these hazy beginnings and the agricultural revolution of about 10,-000 years ago *Homo sapiens* radiated over most of the earth, and differentiated into clearly distinguishable races. The processes were intimately related. Like the forces that had created man, they reflected both the workings of man's environment and of his own invention. So much can be said with reasonable confidence. The details are another matter. The when, where and how of the origin of races puzzle us not much less than they puzzled Charles Darwin.

A little over a century ago a pleasingly simple explanation of races enjoyed some popularity. The races were separate species, created by God as they are today. The Biblical account of Adam and Eve was meant to apply only to Caucasians. Heretical as the idea might be, it was argued that the Negroes appearing in Egyptian monuments, and the skulls of the ancient Indian mound-builders of Ohio, differed in no way from their living descendants, and so there could have been no important change in the only slightly longer time since the Creation itself, set by Archbishop Ussher at 4004 B.C.

With his *Origin of Species*, Darwin undid all this careful "science" at a stroke. Natural selection and the immense stretch of time provided by the geological time-scale made gradual evolution seem the obvious explanation of racial or species differences. But in his later book, *The Descent of Man*, Darwin turned his back on his own central notion of natural selection as the cause of races. He there preferred sexual selection, or the accentuation of racial features through long-established ideals of beauty in different segments of mankind. This proposition failed to impress anthropologists, and so Darwin's demolishing of the old views left something of a void that has never been satisfactorily filled.

Not for want of trying. Some students continued, until recent years, to insist that races are indeed separate species, or even separate genera, with Whites descended from chimpanzees, Negroes from gorillas and Mongoloids from orangutans. Darwin himself had already argued against such a possibility when a contemporary proposed that these same apes had in turn descended from three different monkey species. Darwin pointed out that so great a degree of convergence in evolution, producing thoroughgoing identities in detail (as opposed to, say, the superficial resemblance of whales and fishes) simply could not be expected. The same objection applies to a milder hypothesis, formulated by the late Franz Weidenreich during the 1940's. Races, he held, descended separately, not from such extremely divergent parents as the several great apes, but from the less-separated lines of fossil men. For example, Peking man led to the Mongoloids, and Rhodesian man to the "Africans." But again there are more marked distinctions between those fossil men than between living races.

Actually the most reasonable—I should say the only reasonable—pattern suggested by animal evolution in general is that of racial divergence within a stock already possessing distinctive features of *Homo sapiens*. As I have indicated, such a stock had appeared at the latest by the beginning of the last glacial advance and almost certainly much earlier, perhaps by the end of the preceding glaciation, which is dated at some 150,000 years ago.

Even if fossil remains were more plentiful than they are, they might not in themselves decide the questions of time and place much more accurately. By the time *Homo sapiens* was common enough to provide a chance of our finding some of his fossil remains, he was probably already sufficiently widespread as to give only a general idea of his "place of origin." Moreover, bones and artifacts may concentrate in misleading places. (Consider the parallel case of the australopithecine "man-apes" known so well from the Lower Pleistocene of South Africa. This area is thought of as their home. In fact the region actually was a geographical *cul-de-sac*, and merely a good fossil trap at that time. It is now clear that such prehumans were widespread not only in Africa but also in Asia. We have no real idea of their first center of dispersion, and we should assume that our earliest knowledge of them is not from the actual dawn of their existence.)

In attempting to fix the emergence

of modern races of man somewhat more precisely we can apply something like the chronological reasoning of the pre-Darwinians. The Upper Paleolithic invaders of Europe (*e.g.*, the Cro-Magnons) mark the definite entrance of *Homo sapiens,* and these men were already stamped with a "White" racial nature at about 35,000 B.C. But a recently discovered skull from Liukiang in China, probably of the same order of age, is definitely not Caucasian, whatever else it may be. And the earliest American fossil men, perhaps 20,000 years old, are recognizable as Indians. No other remains are certainly so old; we cannot now say anything about the first Negroes. Thus racial differences are definitely older than 35,000 years. And yet—this is sheer guess—the more successful *Homo sapiens* would probably have overcome the other human types, such as Neanderthal and Rhodesian men, much earlier if he had reached his full development long before. But these types survived well into the last 50,000 years. So we might assume that *Homo sapiens,* and his earliest racial distinctions, is a product of the period between the last two glaciations, coming into his own early during the last glaciation.

When we try to envisage the causes of racial development, we think today of four factors: natural selection, genetic drift, mutation and mixture (interbreeding). With regard to basic divergence at the level of races, the first two are undoubtedly the chief determinants. If forces of any kind favor individuals of one genetic complexion over others, in the sense that they live and reproduce more successfully, the favored individuals will necessarily increase their bequest of genes to the next generation relative to the rest of the population. That is selection; a force with direction.

Genetic drift is a force without direction, an accidental change in the gene proportions of a population. Other things being equal, some parents just have more offspring than others. If such variations can build up, an originally homogeneous population may split into two different ones by chance. It is somewhat as though there were a sack containing 50 red and 50 white billiard balls, each periodically reproducing itself, say by doubling. Suppose you start a new population, drawing out 50 balls without looking. The most likely single result would be 25 of each color, but it is more likely that you would end up with some other combination, perhaps as extreme as 20 reds and 30 whites. After this population divides, you make a new drawing, and so on. Of course at each

subsequent step the departure from the then-prevailing proportion is as likely to favor red as white. Nevertheless, once the first drawing has been made with the above result, red has the better chance of vanishing. So it is with genes for hereditary traits.

Both drift and selection should have stronger effects the smaller and more isolated the population. It is easy to imagine them in action among bands of ancient men, living close to nature. (It would be a great mistake, however, to imagine that selection is not also effective in modern populations.) Hence we can look upon racial beginnings as part accident, part design, design meaning any pattern of minor change obedient to natural selection.

Darwin was probably right the first time, then, and natural selection is more important in racial adaptation than he himself later came to think. Curiously, however, it is extremely difficult to find demonstrable, or even logically appealing, adaptive advantages in racial features. The two leading examples of adaptation in human physique are not usually considered racial at all. One is the tendency among warm-blooded animals of the same species to be larger in colder parts of their territory. As an animal of a given shape gets larger, its inner bulk increases faster than its outer surface,

DISTRIBUTION OF MAN and his races in three epochs is depicted in the maps on these and the following two pages. Key to the races appears in legend below. Solid blue areas in map at top represent glaciers. According to available evidence, it is believed that by 8000 B.C. (*map at top*) early Mongoloids had already spread from the Old World to the New World, while late Mongoloids inhabited a large part of northern Asia. Distribution in A.D. 1000 (*map at bottom*) has late Mongoloids dominating Asia, northern Canada and southern Greenland, and early Mongoloids dominating the Americas. The Pygmies and Bushmen of Africa began a decline that has continued up to the present (*see map on next two pages*).

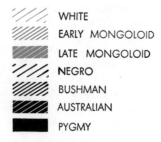

WHITE
EARLY MONGOLOID
LATE MONGOLOID
NEGRO
BUSHMAN
AUSTRALIAN
PYGMY

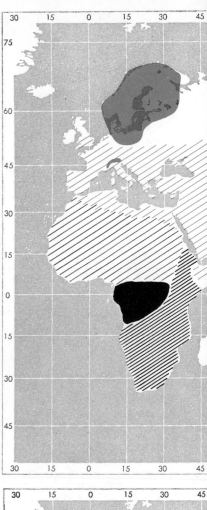

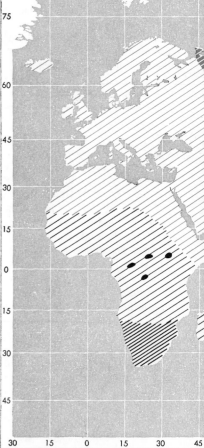

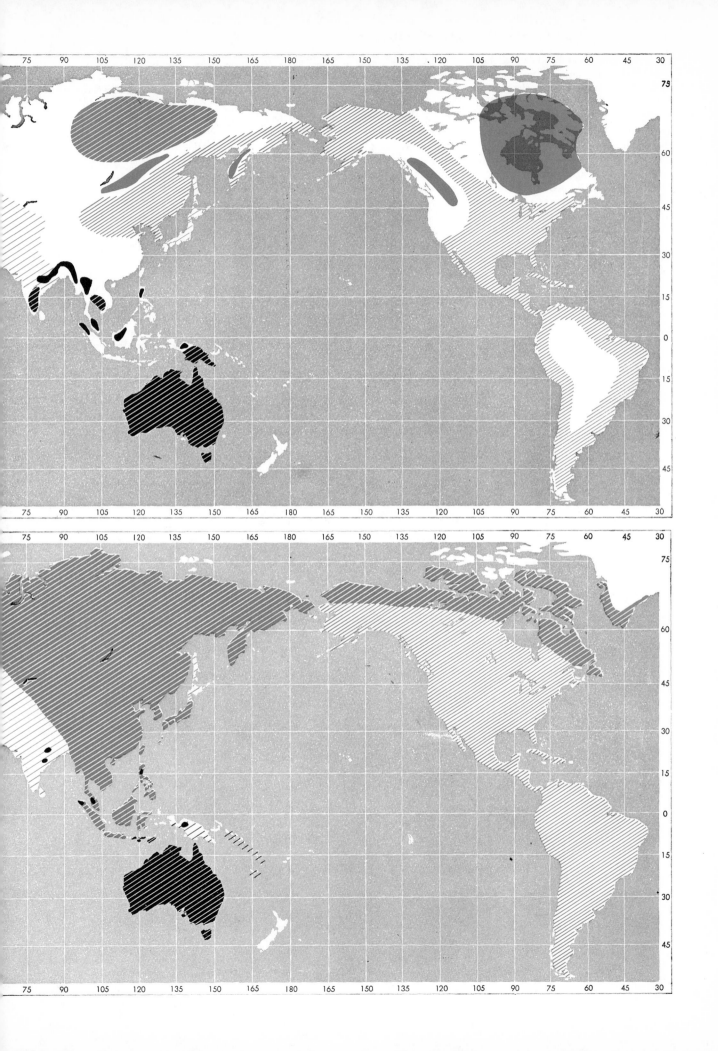

so the ratio of heat produced to heat dissipated is higher in larger individuals. It has, indeed, been shown that the average body weight of man goes up as annual mean temperature goes down, speaking very broadly, and considering those populations that have remained where they are a long time. The second example concerns the size of extremities (limbs, ears, muzzles). They are smaller in colder parts of the range and larger in warmer, for the same basic reason—heat conservation and dissipation. Man obeys this rule also, producing lanky, long-limbed populations in hot deserts and dumpy, short-limbed peoples in the Arctic.

This does not carry us far with the major, historic races as we know them. Perhaps the most striking of all racial features is the dark skin of Negroes. The color of Negro skin is due to a concentration of melanin, the universal human pigment that diffuses sunlight and screens out its damaging ultraviolet component. Does it not seem obvious that in the long course of time the Negroes, living astride the Equator in Africa and in the western Pacific, developed their dark skins as a direct response to a strong sun? It makes sense. It would be folly to deny that such an adaptation is present. But a great deal of the present Negro habitat is shade forest and not bright sun, which is in fact strongest in the deserts some distance north of the Equator. The Pygmies are decidedly forest dwellers, not only in Africa but in their several habitats in southeastern Asia as well.

At any rate there is enough doubt to have called forth other suggestions. One is that forest hunters needed protective coloration, both for stalking and for their protection from predators; dark skin would have lowest visibility in the patchy light and shade beneath the trees. Another is that densely pigmented skins may have other qualities—e.g., resistance to infection—of which we are unaware.

A more straightforward way out of the dilemma is to suppose that the Negroes are actually new to the Congo forest, and that they served their racial apprenticeship hunting and fishing in the sunny grasslands of the southern Sahara. If so, their Pygmy relatives might represent the first accommodation of the race to the forest, before agriculture but after dark skin had been acquired. Smaller size certainly makes a chase after game through the undergrowth less exhausting and faster. As for woolly hair, it is easy to see it (still without proof) as an excellent, nonmatting insulation against solar heat. Thick Negro lips? Every suggestion yet made has a zany sound. They may only be a side effect of some properties of heavily pigmented

WHITE
EARLY MONGOLOID
LATE MONGOLOID
NEGRO
BUSHMAN
AUSTRALIAN
PYGMY

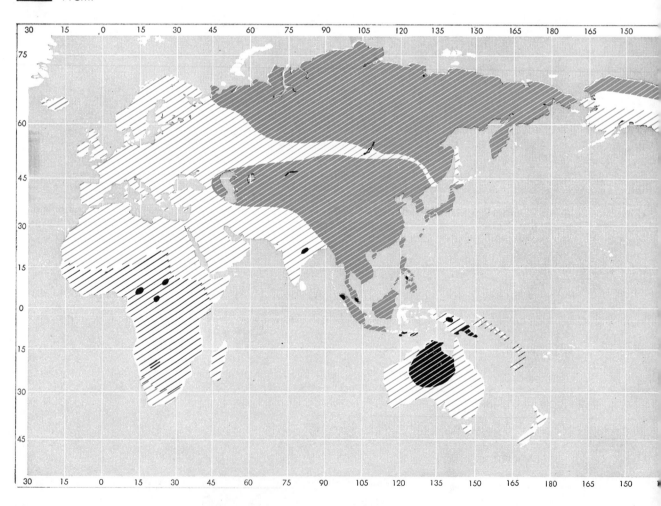

PRESENT DISTRIBUTION OF RACES OF MAN reflects dominance of White, late Mongoloid and Negro races. Diffusion of Whites has been attended by decline of early Mongoloids in America, Bushmen in Africa and indigenous population in Australia.

skin (ability to produce thick scar tissue, for example), even as blond hair is doubtless a side effect of the general depigmentation of men that has occurred in northern Europe.

At some remove racially from Negroes and Pygmies are the Bushmen and Hottentots of southern Africa. They are small, or at least lightly built, with distinctive wide, small, flat faces; they are rather infantile looking, and have a five-cornered skull outline that seems to be an ancient inheritance. Their skin is yellowish-brown, not dark. None of this has been clearly interpreted, although the small size is thought to be an accommodation to water and food economy in the arid environment. The light skin, in an open sunny country, contradicts the sun-pigment theory, and has in fact been used in favor of the protective-coloration hypothesis. Bushmen and background blend beautifully for color, at least as human beings see color.

Bushmen, and especially Hottentots, have another dramatic characteristic:

Narrow band of Whites in Asia represents Russian colonization of southern Siberia.

steatopygia. If they are well nourished, the adult women accumulate a surprising quantity of fat on their buttocks. This seems to be a simple storehouse mechanism reminiscent of the camel's hump; a storehouse that is not distributed like a blanket over the torso generally, where it would be disadvantageous in a hot climate. The characteristic nicely demonstrates adaptive selection working in a human racial population.

The Caucasians make the best argument for skin color as an ultraviolet screen. They extend from cloudy northern Europe, where the ultraviolet in the little available sunlight is not only acceptable but desirable, down to the fiercely sun-baked Sahara and peninsular India. All the way, the correspondence with skin color is good: blond around the Baltic, swarthy on the Mediterranean, brunet in Africa and Arabia, dark brown in India. Thus, given a long enough time of occupation, and doubtless some mixture to provide dark-skinned genes in the south, natural selection could well be held responsible.

On the other hand, the Caucasians' straight faces and often prominent noses lack any evident adaptive significance. It is the reverse with the Mongoloids, whose countenances form a coherent pattern that seems consistent with their racial history. From the standpoint of evolution it is Western man, not the Oriental, who is inscrutable. The "almond" eyes of the Mongoloid are deeply set in protective fat-lined lids, the nose and forehead are flattish and the cheeks are broad and fat-padded. In every way, it has been pointed out, this is an ideal mask to protect eyes, nose and sinuses against bitterly cold weather. Such a face is the pole toward which the peoples of eastern Asia point, and it reaches its most marked and uniform expression in the cold northeastern part of the continent, from Korea north.

Theoretically the Mongoloid face developed under intense natural selection some time during the last glacial advance among peoples trapped north of a ring of mountain glaciers and subjected to fierce cold, which would have weeded out the less adapted, in the most classic Darwinian fashion, through pneumonia and sinus infections. If the picture is accurate, this face type is the latest major human adaptation. It could not be very old. For one thing, the population would have had to reach a stage of advanced skill in hunting and living to survive at all in such cold, a stage probably not attained before the Upper Paleolithic (beginning about 35,000 B.C.). For an-

other, the adaptation must have occurred after the American Indians, who are Mongoloid but without the transformed face, migrated across the Bering Strait. (Only the Eskimos reflect the extension of full-fledged, recent Mongoloids into America.) All this suggests a process taking a relatively small number of generations (about 600) between 25,000 and 10,000 B. C.

The discussion so far has treated human beings as though they were any mammal under the influence of natural selection and the other forces of evolution. It says very little about why man invaded the various environments that have shaped him and how he got himself distributed in the way we find him now. For an understanding of these processes we must take into account man's own peculiar abilities. He has created culture, a milieu for action and development that must be added to the simplicities of sun, snow, forest or plain.

Let us go back to the beginning. Man started as an apelike creature, certainly vegetarian, certainly connected with wooded zones, limited like all other primates to tropical or near-tropical regions. In becoming a walker he had begun to extend his range. Tools, social rules and intelligence all progressed together; he learned to form efficient groups, armed with weapons not provided by nature. He started to eat meat, and later to cook it; the more concentrated diet widened his possibilities for using his time; the hunting of animals beckoned him still farther in various directions.

All this was probably accomplished during the small-brained australopithecine stage. It put man on a new plane, with the potential to reach all parts of the earth, and not only those in which he could find food ready to his hand, or be comfortable in his bare skin. He did not actually reach his limits until the end of the last glaciation, and in fact left large tracts empty for most of the period. By then he had become *Homo sapiens*, with a large brain. He had tools keen enough to give him clothes of animal skin. He had invented projectiles to widen the perimeter of his striking power: bolas, javelins with spear throwers, arrows with bows. He was using dogs to widen the perimeter of his senses in tracking. He had found what could be eaten from the sea and its shores. He could move only slowly, and was probably by no means adventurous. But hunting territory was precious, and the surplus of an expanding population had

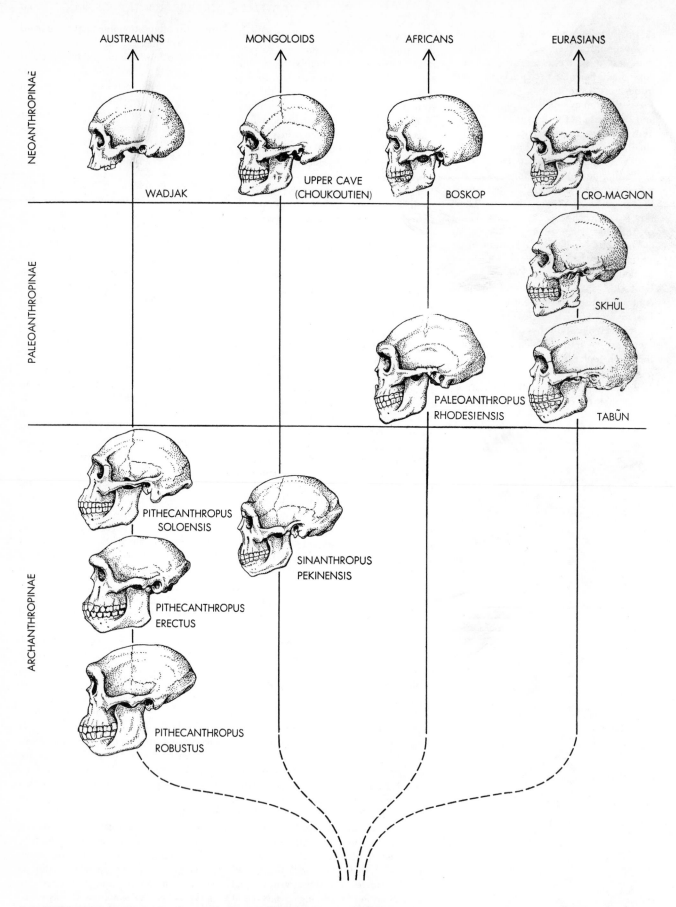

AUSTRALIANS   MONGOLOIDS   AFRICANS   EURASIANS

NEOANTHROPINAE

WADJAK   UPPER CAVE (CHOUKOUTIEN)   BOSKOP   CRO-MAGNON

PALEOANTHROPINAE

SKHŪL

PALEOANTHROPUS RHODESIENSIS   TABŪN

ARCHANTHROPINAE

PITHECANTHROPUS SOLOENSIS

SINANTHROPUS PEKINENSIS

PITHECANTHROPUS ERECTUS

PITHECANTHROPUS ROBUSTUS

POLYPHYLETIC SCHOOL of anthropology, chiefly identified with Franz Weidenreich, conceives modern races of man descending from four ancestral lines. According to this school, ancestors of Australians (*left*) include *Pithecanthropus soloensis* (Solo man) and *Pithecanthropus erectus* (Java man). Original ancestor of Mongoloids is *Sinanthropus pekinensis* (Peking man); of Africans, *Paleoanthropus rhodesiensis* (Rhodesian man). Four skulls at top are early *Homo sapiens*. Alternative theory is shown on next page.

to stake out new preserves wherever there was freedom ahead. So this pressure, and man's command of nature, primitive though it still was, sent the hunters of the end of the Ice Age throughout the Old World, out into Australia, up into the far north, over the Bering Strait and down the whole length of the Americas to Tierra del Fuego. At the beginning of this dispersion we have brutes barely able to shape a stone tool; at the end, the wily, self-reliant Eskimo, with his complicated traps, weapons and sledges and his clever hunting tricks.

The great racial radiation carried out by migratory hunters culminated in the world as it was about 10,000 years ago. The Whites occupied Europe, northern and eastern Africa and the Near East, and extended far to the east in Central Asia toward the Pacific shore. Negroes occupied the Sahara, better watered then, and Pygmies the African equatorial forest; south, in the open country, were Bushmen only. Other Pygmies, the Negritos, lived in the forests of much of India and southeastern Asia; while in the open country of these areas and in Australia were men like the present Australian aborigines: brown, beetle-browed and wavy-haired. Most of the Pacific was empty. People such as the American Indians stretched from China and Mongolia over Alaska to the Straits of Magellan; the more strongly Mongoloid peoples had not yet attained their domination of the Far East.

During the whole period the human population had depended on the supply of wild game for food, and the accent had been on relative isolation of peoples and groups. Still close to nature (as we think of nature), man was in a good position for rapid small-scale evolution, both through natural selection and through the operation of chance in causing differences among widely separated tribes even if selection was not strong.

Then opened the Neolithic period, the beginning of a great change. Agriculture was invented, at first inefficient and feeble, but in our day able to feed phenomenally large populations while freeing them from looking for food. The limit on local numbers of people was gradually removed, and with it the necessity for the isolation and spacing of groups and the careful observation of boundaries. Now, as there began to be surpluses available for trading, connections between communities became more useful. Later came a spreading of bonds from higher centers of trade and of authority. Isolation gave way to contact,

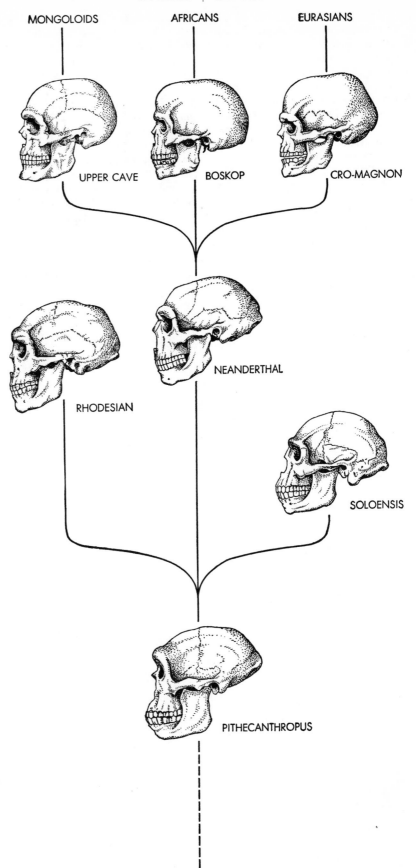

UNILINEAR OR "HAT-RACK" SCHOOL predicates three races descending from single ancestral line, as opposed to polyphyletic theory depicted at left. Rhodesian, Neanderthal and Solo man all descend from *Pithecanthropus*. Neanderthal is ancestor of early *Homo sapiens* (Upper Cave, Boskop and Cro-Magnon) from which modern races descended.

even when contact meant war.

The change was not speedy by our standards, though in comparison with the pace of the Stone Age it seems like a headlong rush. The new economy planted people much more solidly, of course. Farmers have been uprooting and displacing hunters from the time of the first planters to our own day, when Bushman survivors are still losing reservation land to agriculturalists in southwestern Africa. These Bushmen, a scattering of Australian aborigines, the Eskimos and a few other groups are the only representatives of their age still in place. On the other hand, primitive representatives of the Neolithic level of farming still live in many places after the thousands of years since they first became established there.

Nevertheless mobility increased and has increased ever since. Early woodland farmers were partly nomadic, moving every generation following exhaustion of the soil, however solidly fixed they may have been during each sojourn. The Danubians of 6,000 years ago can be traced archeologically as they made the same kind of periodic removes as central Africans, Iroquois Indians and pioneer Yankee farmers. Another side of farming—animal husbandry—gave rise to pastoral nomadism. Herders were much lighter of foot, and historically have tended to be warlike and domineering. With irrigation, villages could settle forever and evolve into the urban centers of high civilizations. Far from immobilizing man, however, these centers served

as fixed bases from which contact (and conflict) worked outward.

The rest of the story is written more clearly. New crops or new agricultural methods opened new territories, such as equatorial Africa, and the great plains of the U. S., never successfully farmed by the Indians. New materials such as copper and tin made places once hopeless for habitation desirable as sources of raw material or as way stations for trade. Thus an island like Crete rose from nothing to dominate the eastern Mediterranean for centuries. Well before the earliest historians had made records, big population shifts were taking place. Our mental picture of the aboriginal world is actually a recent one. The Bantu Negroes moved into central and

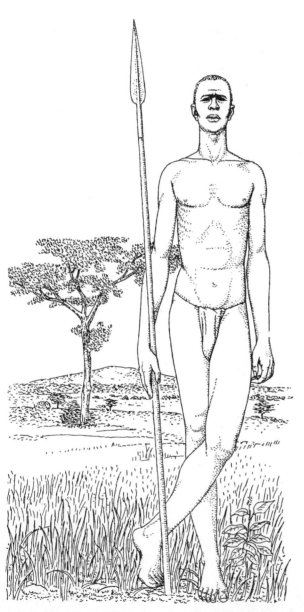 

HUMAN ADAPTATION TO CLIMATE is typified by Nilotic Negro of the Sudan (*left*) and arctic Eskimo (*right*). Greater body surface of Negro facilitates dissipation of unneeded body heat; proportionately greater bulk of the Eskimo conserves body heat.

southern Africa, peoples of Mongoloid type went south through China and into Japan, and ancient folk of Negrito and Australoid racial nature were submerged by Caucasians in India. Various interesting but inconsequential trickles also ran hither and yon; for example, the migration of the Polynesians into the far Pacific.

The greatest movement came with the advent of ocean sailing in Europe. (The Polynesians had sailed the high seas earlier, of course, but they had no high culture, nor did Providence interpose a continent across their route at a feasible distance, as it did for Columbus.) The Europeans poured out on the world. From the 15th to the 19th centuries they compelled other civilized peoples to accept contact, and subjected or erased the uncivilized. So today, once again, we have a quite different distribution of mankind from that of 1492.

It seems obvious that we stand at the beginning of still another phase. Contact is immediate, borders are slamming shut and competition is fierce. Biological fitness in races is now hard to trace, and even reproduction is heavily controlled by medicine and by social values. The racial picture of the future will be determined less by natural selection and disease resistances than by success in government and in the adjustment of numbers. The end of direct European dominance in Africa and Asia seems to mean the end of any possibility of the infiltration and expansion of the European variety of man there, on the New World model. History as we know it has been largely the expansion of the European horizon and of European peoples. But the end in China of mere absorption of Occidental invention, and the passionate self-assertion of the African tribes, make it likely that racial lines and territories will again be more sharply drawn than they have been for centuries. What man will make of himself next is a question that lies in the province of prophets, not anthropologists.

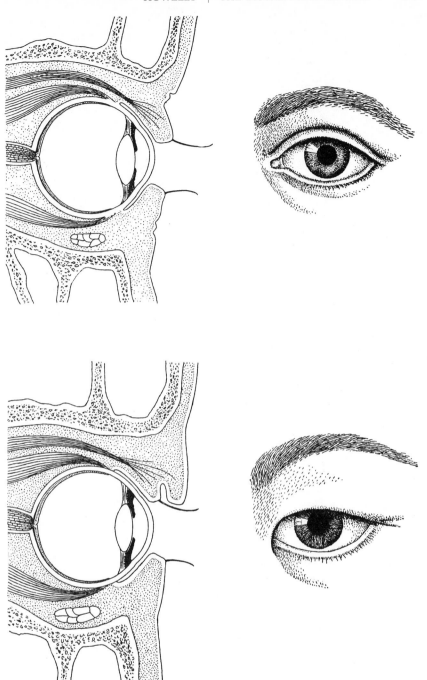

"ALMOND" EYE OF MONGOLOID RACES is among latest major human adaptations to environment. The Mongoloid fold, shown in lower drawings, protects the eye against the severe Asian winter. Drawings at top show the Caucasian eye with its single, fatty lid.

V

SOME MAJOR PATTERNS IN THE HISTORY OF LIFE

V SOME MAJOR PATTERNS IN THE HISTORY OF LIFE

INTRODUCTION

In this final part, we want to look at some of the broader aspects of the fossil record, particularly during the last 600 million years of the Phanerozoic. In Part I, we saw how organisms interact with their environment—or, as Dobzhansky put it, how each species occupies an adaptive niche in the economy of nature. In a static, unchanging world, we would expect life to "top out" in kinds and numbers of animals and plants. But the world is not static. In the animate world, variations generated by mutation and sexual reproduction continuously provide biological novelties better suited to coping with the environment. In the inanimate world, there is constant change owing to the ongoing processes of geological activity. Consequently, we can predict—at least in a general way—that the history of life will flow and ebb in concert with the dynamic evolution of the planet Earth.

What then are some of the overall tendencies of life's history since its inception billions of years ago? First, there has been an increase in its organizational complexity, from singlecelled creatures to multicellular; from simple tissues such as the outer layer, or ectoderm, of jellyfish, to complex organs such as the eyes of an insect; from the sluggish nerve nets of earthworms to the elaborate central nervous system of a dog. The apotheosis of this tendency is, of course, the human organism with its conscious and thinking mentality.

Another obvious trend in the history of life is the steady expansion into more and more different kinds of environments: from marine waters into fresh, from coastal wetlands into dry upland plateaus, from the earth's surface into the trees and sky above and into the soil and caves below, and, most recently, our own tentative ventures out into space. With this expansion of environments has come a corresponding increase in diversity of organisms, for, as more and more adaptive niches are sought out and occupied, there is an increase in species. The increase in diversity has also been self-stimulating, because as more and more kinds of animals and plants evolve, owing to new physical opportunities, still other organisms will appear, to capitalize on these new biological opportunities.

A third major tendency in the history of life is the more or less continual turnover of species, due to the constant changes experienced by the earth during its history. If we think of a species as a finely tuned adjustment of organisms to the local environment, we can expect that changes in the environment will often occur faster than the organisms can adapt to it and evolve with it. Thus, if environmental change oversteps the adaptive response of a species, that species will become extinct. But the unoccupied niche will not remain unfilled for long, because, given the inherent variation of organisms, natural selection will produce another species to fill it. If we look at higher—that is, more inclusive—biological categories, such as orders, classes, and phyla, we find that this turnover is dampened. Thus, although there are no bivalved clam *species* surviving from the Cambrian, this *class* of molluscs has, indeed, survived from then. Why? Because the basic general design of the bivalves has remained adaptive to an aquatic way of life. (Consider the wheel, which has always remained useful, even though many specific variations of the wheel have come and gone.)

Although extinction is the natural outcome for virtually all species—if not the higher categories of life to which they belong—there are some few instances where species, or at least genera, have persisted for very long times. These so-called living fossils represent broadly adapted forms that can hang on in the face of environmental change. This knowledge should temper the gloom of anticipating the end of our own species. For the quintessence of what makes humans a distinct species is our very ability to cope with our environment. And because this coping is as much cultural as genetic, we can respond rapidly to change. Moreover, given our intellect, we can also presumably foresee changes in advance and prepare for them. Whether, in fact, we humans realize this potential to preserve our species remains to be seen.

"Crises in the History of Life," by Newell, discusses both change and persistence of organisms as revealed by the fossil record. As Newell points out, wholesale disappearances of ancient floras and faunas were interpreted by earlier scientists as the result of periodic, sudden, violent catastrophes or revolutions. Although paleontologists subsequently eschewed using such possible revolutions or catastrophes as explanations for mass extinctions, the real causes eluded them, and still do today, even though we are closer to the answers.

Newell clearly states the most baffling aspect of mass extinction: The demise of large proportions of organisms that inhabit a wide range of habitats. The very variety of life becoming extinct at the end of the Permian and Cretaceous requires an equally large-scale cause. What we postulate as a factor effecting extinction for bottom-dwelling invertebrates must also affect terrestrial vertebrates. A corollary complication is that periods of major extinction for animals and plants rarely coincide.

Newell suggests that, whatever factor we invoke, it must be correlated to the restriction or destruction of habitat for the species that is becoming extinct. Loss of habitat decreases the geographic range and spatial variety for the occupying organisms. It also reduces not only the total numbers of individuals but, equally critical, the numbers of relatively isolated breeding groups. For Newell, the transgression and regression of shallow seas on and off the continents throughout geologic time would cause broad destruction of habitats, with accompanying mass extinctions. Although this article was written just before the theory of plate tectonics was formulated, Newell was alert to the role that tectonics played in shaping the ocean basins and in producing changes in sea level world wide. Finally, Newell emphasizes the role that key species play in a community, so that their extinction might, like falling dominoes, wipe out other interdependent species.

"Plate Tectonics and the History of Life in the Oceans," by Valentine and Moores, examines the environmental implications and biological responses to global changes wrought by continental fragmentation and reassembly. We would, of course, expect plate tectonics to have major impact on the habitats of the world. In this article, Valentine and Moores explain what those impacts are and how they influence the diversity of life. In general, they argue that when continents are large and few, climates are more seasonal or fluctuating than when the continents are small and dispersed. The shallow waters surrounding such continents would thus be highly variable environments or rather constant environments, respectively. And, when environments are constant or stable, there is more diversity of life, because organisms can afford to specialize, thereby producing more species. Variable environments, on the other hand, favor generalists that can tolerate wide ranges of physical and chemical conditions, thus reducing the total numbers of species. In addition, food resources vary by latitude, so that continents in the tropics have steady food production year round, whereas in high latitudes food production fluctuates seasonally. Consequently, equatorial marine faunas are much more diverse than boreal faunas. Finally, continents that are fragmented will have their shallow shelves more isolated from one another than if the continents are assembled into one supercontinent with a single shelf running around it. As we noted in Part I, geographic isolation of populations leads to differentiation of species, and so there will be more diversity of life when continents are dispersed than when they are welded together.

With this line of reasoning, Valentine and Moores show how the rise and fall in the diversity of shallow marine life correlates with continental drift and plate tectonics. The initial expansion of life in the late Precambrian and early Cambrian is due, they believe, to the breakup of a large supercontinent called Pangaea. Continental fragmentation led to stability of environments and food resources as well as isolated marine shelves, all of which, in turn, brought about a great increase in life's diversity. Continental reassembly in the late Paleozoic had opposite effects and culminated in the Permian crisis among marine invertebrates. No doubt, Newell's concept of the restriction of the shallow marine habitat was also important. (See Simberloff, 1974, for a quantitative corroboration of Newell's hypothesis.)

The Mesozoic breakup of the two great northern and southern continents, Laurasia and Gondwanaland, led to renewed diversity, and continued separation of the continents in the Cenozoic brought about still greater increases in diversity. As plausible as this idea is—namely, that the drifting apart and coming together of continents has caused significant changes in diversity of marine organisms—not all paleontologists agree. Some interpret the rise, decline, recovery, and expansion of marine fossils throughout the Phanerozoic as simply a reflection of the corresponding abundance of marine sedimentary rocks (see Raup, 1976).

Kurtén's article, "Continental Drift and Evolution," further presses home the importance of continental fragmentation in generating diversity among terrestrial vertebrates, such as reptiles and mammals. Here the argument runs as follows: Given a specific area of land surface, there will be some more or less equilibrium number of vertebrate types evolving there. Thus, in the late Mesozoic we can recognize about a dozen higher categories of reptiles, slightly more than half originating on the southern continent of Gondwanaland and less than half first appearing in the northern continent of Laurasia. The subsequent fragmentation and separation of these two supercontinents into South America, Africa, Australia, India, and Antarctica, on the one hand, and into North America and Eurasia, on the other, led to more than a doubling of the higher categories of mammals that succeeded the reptiles. Rather than attributing this increase in diversity to either something inherent in the mammalian condition or to more habitat variation of the continents, Kurtén ascribes it essentially to ecological duplication of basic adaptive types on continents isolated geographically from each other.

Strong support for Kurtén's hypothesis comes from the *loss* of diversity that resulted when North and South America were rejoined after tens of millions of years of separation during most of mammalian evolution history. In the late Cenozoic, the Isthmus of Panama provided a corridor for migrating mammals, both north and south. This faunal interchange of previously isolated similar adaptive types led to widespread extinctions among the less successful. Whereas there were some fifty different mammalian families in the late Pliocene, just before the reconnection, ten of these families have become extinct in the few million years since the continents were rejoined. (See Flessa, 1975, for additional support of Kurtén's hypothesis.)

We thus conclude these readings as we began them: with discussion and interpretation of the evolutionary significance of the geographic distributions of organisms. As we noted in Part I, it was these biogeographic observations that had such strong influence on Darwin's own thoughts. The concept of plate tectonics, dealing as it does with global processes of geological evolution, provides a new frame of reference within which paleontologists can view the flow and ebb of ancient life over the last half billion years. As that great American paleontologist, George Gaylord Simpson, put it a generation ago, "The history of organisms runs parallel with, is environmentally contained in, and continuously interacts with the physical history of the earth."

SUGGESTED FURTHER READING

Flessa, K., 1975, "Area, Continental Drift, and Mammalian Diversity," *Paleobiology*, vol. 1, pp. 189–194. Like Simberloff, 1974 (below), except the time is the Cenozoic and the organisms are terrestrial mammals. Both articles strongly reinforce the concept that geographic area and biological diversity are rather closely linked. If this is true, the shifting of landmasses and changing sea levels associated with plate tectonics must surely have had important effects on the history and evolution of life.

Raup, D. M. 1976. "Species Diversity in the Phanerozoic: Tabulation and Interpretation," *Paleobiology*, vol. 2, pp. 279–297. One of several articles in recent years by Raup that challenges the notion that life's diversity as recorded in the fossil record has any more significance than merely reflecting the vicissitudes of sediment accumulation and fossil preservation.

Simberloff, D. S. 1974. "Permo-Triassic Extinctions: Effects of Area on Biotic Equilibrium," *Journal of Geology*, vol. 82, pp. 267–274. Convincing argument that restriction of the Permian shallow seas correlates with widespread extinction among marine organisms at the end of the Paleozoic.

Valentine, J. W. 1973. *Evolutionary Paleoecology of the Marine Biosphere*, Englewood Cliffs, N.J.: Prentice-Hall. A brilliant integration of the functional and ecological interactions of organisms at various levels from the individual, population, and community levels up to the total marine biosphere. The crux of the book, however, lies in the hypothesis of how continental configurations influence stability in the marine environment and how that, in turn, controls the diversity of life therein. The last chapter "tells all" about the last three-quarter billion years.

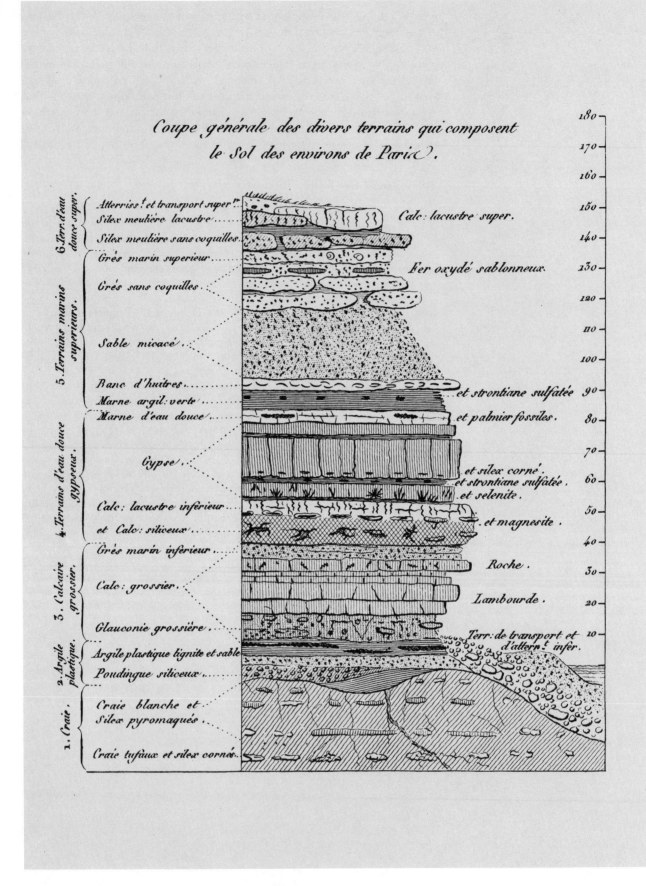

Coupe générale des divers terrains qui composent
le Sol des environs de Paris.

6. Terr. d'eau douce super.
Atterriss.t et transport super.r
Silex meulière lacustre ....
Calc: lacustre super.
Silex meulière sans coquilles.

5. Terrains marins superieurs.
Grès marin superieur ....
Grès sans coquilles ....
Fer oxydé sablonneux.
Sable micacé ....
Banc d'huitres ....
Marne argil. verte ....
et strontiane sulfatée
Marne d'eau douce ....
et palmier fossiles.

4. Terrains d'eau douce gypseux.
Gypse ....
et silex corné.
et strontiane sulfatée
et selenite.
Calc: lacustre inferieur ....
et Calc: siliceux ....
et magnesite.

3. Calcaire grossier.
Grès marin inferieur ....
Roche.
Calc: grossier ....
Lambourde.
Glauconie grossière ....
Terr: de transport et d'atterr.t infer.

2. Argile plastique.
Argile plastique lignite et sable
Poudingue siliceux ....

1. Craie.
Craie blanche et Silex pyromaqués ....
Craie tufaux et silex cornés.

"EVIDENCE" FOR CATASTROPHISM was adduced by the French naturalist Baron Georges Cuvier from his study of the Paris basin. He and Alexandre Brongniart published this diagram in 1822. Cuvier believed that the abrupt changes in the strata bespoke the occurrence of cataclysms. Although his observations were accurate, his conclusions are no longer generally accepted.

Crises in the History of Life

by Norman D. Newell
February 1963

*How is it that whole groups of animals have
simultaneously died out? Paleontologists are
returning to an earlier answer: natural catastrophe.
The catastrophes they visualize, however, are not
sudden but gradual*

The stream of life on earth has been continuous since it originated some three or four billion years ago. Yet the fossil record of past life is not a simple chronology of uniformly evolving organisms. The record is prevailingly one of erratic, often abrupt changes in environment, varying rates of evolution, extermination and repopulation. Dissimilar biotas replace one another in a kind of relay. Mass extinction, rapid migration and consequent disruption of biological equilibrium on both a local and a world-wide scale have accompanied continual environmental changes.

The main books and chapters of earth history—the eras, periods and epochs— were dominated for tens or even hundreds of millions of years by characteristic groups of animals and plants. Then, after ages of orderly evolution and biological success, many of the groups suddenly died out. The cause of these mass extinctions is still very much in doubt and constitutes a major problem of evolutionary history.

The striking episodes of disappearance and replacement of successive biotas in the layered fossil record were termed revolutions by Baron Georges Cuvier, the great French naturalist of the late 18th and early 19th centuries. Noting that these episodes generally correspond to unconformities, that is, gaps in the strata due to erosion, Cuvier attributed them to sudden and violent catastrophes. This view grew out of his study of the sequence of strata in the region of Paris. The historic diagram on the opposite page was drawn by Cuvier nearly 150 years ago. It represents a simple alternation of fossil-bearing rocks of marine and nonmarine origin, with many erosional breaks and marked interruptions in the sequence of fossils.

The objection to Cuvier's catastrophism is not merely that he ascribed events in earth history to cataclysms; many normal geological processes are at times cataclysmic. The objection is that he dismissed known processes and appealed to fantasy to explain natural phenomena. He believed that "the march of nature is changed and not one of her present agents could have sufficed to have effected her ancient works." This hypothesis, like so many others about extinction, is not amenable to scientific test and is hence of limited value. In fairness to Cuvier, however, one must recall that in his day it was widely believed that the earth was only a few thousand years old. Cuvier correctly perceived that normal geological processes could not have produced the earth as we know it in such a short time.

Now that we have learned that the earth is at least five or six billion years old, the necessity for invoking Cuverian catastrophes to explain geological history would seem to have disappeared. Nevertheless, a few writers such as Immanuel Velikovsky, the author of *Worlds in Collision,* and Charles H. Hapgood, the author of *The Earth's Shifting Crust,* continue to propose imaginary catastrophes on the basis of little or no historical evidence. Although it is well established that the earth's crust has shifted and that climates have changed, these changes almost certainly were more gradual than Hapgood suggests. Most geologists, following the "uniformitarian" point of view expounded in the 18th century by James Hutton and in the 19th by Charles Lyell, are satisfied that observable natural processes are quite adequate to explain the history of the earth. They agree, however, that these processes must have varied greatly in rate.

Charles Darwin, siding with Hutton and Lyell, also rejected catastrophism as an explanation for the abrupt changes in the fossil record. He attributed such changes to migrations of living organisms, to alterations of the local environment during the deposition of strata and to unconformities caused by erosion. Other important factors that are now given more attention than they were in Darwin's day are the mass extinction of organisms, acceleration of the rate of evolution and the thinning of strata due to extremely slow deposition.

The Record of Mass Extinctions

If we may judge from the fossil record, eventual extinction seems to be the lot of all organisms. Roughly 2,500 families of animals with an average longevity of somewhat less than 75 million years have left a fossil record. Of these, about a third are still living. Although a few families became extinct by evolving into new families, a majority dropped out of sight without descendants.

In spite of the high incidence of extinction, there has been a persistent gain in the diversity of living forms: new forms have appeared more rapidly than old forms have died out. Evidently organisms have discovered an increasing number of ecological niches to fill, and by modifying the environment they have produced ecological systems of great complexity, thereby making available still more niches. In fact, as I shall develop later, the interdependence of living organisms, involving complex chains of food supply, may provide an important key to the understanding of how relatively small changes in the environment could have triggered mass extinctions.

The fossil record of animals tells more about extinction than the fossil record of plants does. It has long been known

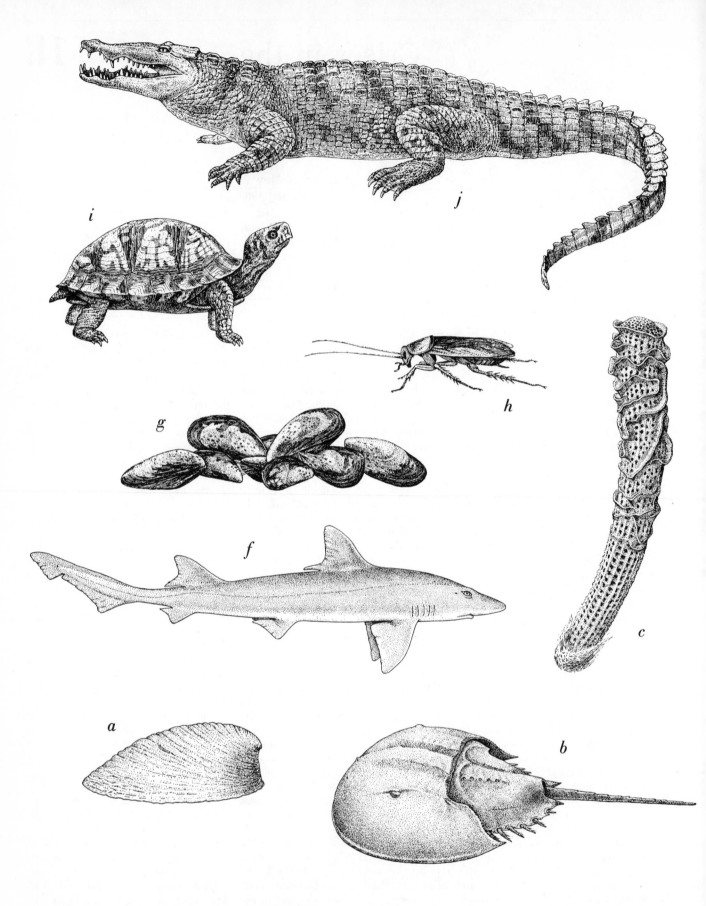

GALLERY OF HARDY ANIMALS contains living representatives of 11 groups that have weathered repeated crises in evolutionary history. Four of the groups can be traced back to the Cambrian period: the mollusk *Neopilina* (*a*), the horseshoe crab (*b*), the Venus's-flower-basket, *Euplectella* (*c*) and the brachiopod *Lingula* (*d*). One animal represents a group that goes back to the Ordovician period: the ostracode *Bairdia* (*e*). Two arose in the Devonian period: the shark (*f*) and the mussel (*g*). The cockroach

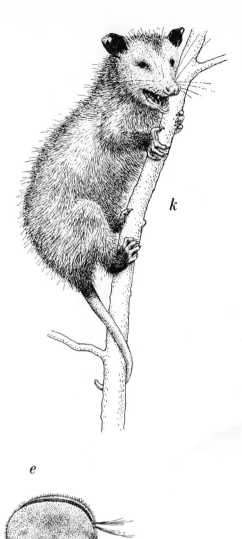

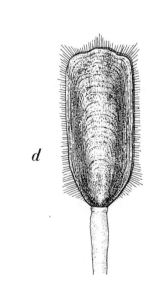

(h) goes back to the Pennsylvanian period. Two arose in the late Triassic: the turtle (i) and the crocodile (j). The opossum (k) appeared during the Cretaceous period.

that the major floral changes have not coincided with the major faunal ones. Each of the three successive principal land floras—the ferns and mosses, the gymnosperms and angiosperms—were ushered in by a short episode of rapid evolution followed by a long period of stability. The illustration on page 183 shows that once a major group of plants became established it continued for millions of years. Many groups of higher plants are seemingly immortal. Since green plants are the primary producers in the over-all ecosystem and animals are the consumers, it can hardly be doubted that the great developments in the plant kingdom affected animal evolution, but the history of this relation is not yet understood.

Successive episodes of mass extinction among animals—particularly the marine invertebrates, which are among the most abundant fossils—provide world-wide stratigraphic reference points that the paleontologist calls datums. Many of the datums have come to be adopted as boundaries of the main divisions of geologic time, but there remains some uncertainty whether the epochs of extinction constitute moments in geologic time or intervals of significant duration. In other words, did extinction occur over hundreds, thousands or millions of years? The question has been answered in many ways, but it still remains an outstanding problem.

A good example of mass extinction is provided by the abrupt disappearance of nearly two-thirds of the existing families of trilobites at the close of the Cambrian period. Before the mass extinction of these marine arthropods, which are distantly related to modern crustaceans, there were some 60 families of them. The abrupt disappearance of so many major groups of trilobites at one time has served as a convenient marker for defining the upper, or most recent, limit of the Cambrian period [see illustration on page 184].

Similar episodes of extinction characterize the history of every major group and most minor groups of animals that have left a good fossil record. It is striking that times of widespread extinction generally affected many quite unrelated groups in separate habitats. The parallelism of extinction between some of the aquatic and terrestrial groups is particularly remarkable [see illustration on page 186].

One cannot doubt that there were critical times in the history of animals. Widespread extinctions and consequent revolutionary changes in the course of

animal life occurred roughly at the end of the Cambrian; Ordovician, Devonian, Permian, Triassic and Cretaceous periods. Hundreds of minor episodes of extinction occurred on a more limited scale at the level of species and genera throughout geologic time, but here we shall restrict our attention to a few of the more outstanding mass extinctions.

At or near the close of the Permian period nearly half of the known families of animals throughout the world disappeared. The German paleontologist Otto Schindewolf notes that 24 orders and superfamilies also dropped out at this point. At no other time in history, save possibly the close of the Cambrian, has the animal world been so decimated. Recovery to something like the normal variety was not achieved until late in the Triassic period, 15 or 20 million years later.

Extinctions were taking place throughout Permian time and a number of major groups dropped out well before the end of the period, but many more survived to go out together, climaxing one of the greatest of all episodes of mass extinction affecting both land and marine animals. It was in the sea, however, that the decimation of animals was particularly dramatic. One great group of animals that disappeared at this time was the fusulinids, complex protozoans that ranged from microscopic sizes to two or three inches in length. They had populated the shallow seas of the world for 80 million years; their shells, piling up on the ocean floor, had formed vast deposits of limestone. The spiny productid brachiopods, likewise plentiful in the late Paleozoic seas, also vanished without descendants. These and many other groups dropped suddenly from a state of dominance to one of oblivion.

By the close of the Permian period 75 per cent of amphibian families and more than 80 per cent of the reptile families had also disappeared. The main suborders of these animals nonetheless survived the Permian to carry over into the Triassic.

The mass extinction on land and sea at the close of the Triassic period was almost equally significant. Primitive reptiles and amphibians that had dominated the land dropped out and were replaced by the early dinosaurs that had appeared and become widespread before the close of the period. It is tempting to conclude that competition with the more successful dinosaurs was an important factor in the disappearance of these early land animals, but what bearing could this have had on the equally impressive and

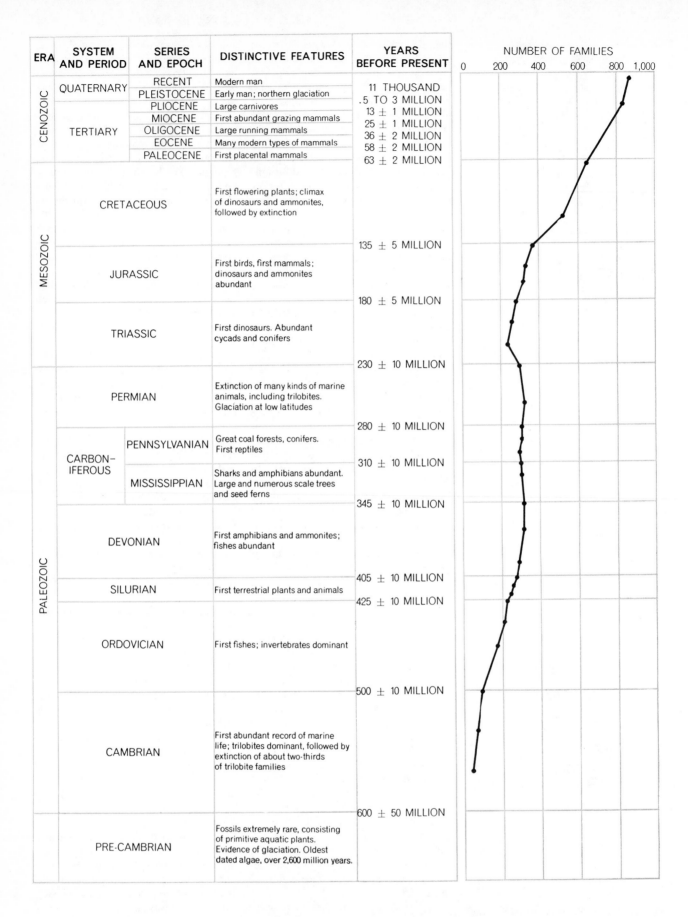

ERA	SYSTEM AND PERIOD	SERIES AND EPOCH	DISTINCTIVE FEATURES	YEARS BEFORE PRESENT
CENOZOIC	QUATERNARY	RECENT	Modern man	11 THOUSAND
		PLEISTOCENE	Early man; northern glaciation	.5 TO 3 MILLION
	TERTIARY	PLIOCENE	Large carnivores	13 ± 1 MILLION
		MIOCENE	First abundant grazing mammals	25 ± 1 MILLION
		OLIGOCENE	Large running mammals	36 ± 2 MILLION
		EOCENE	Many modern types of mammals	58 ± 2 MILLION
		PALEOCENE	First placental mammals	63 ± 2 MILLION
MESOZOIC	CRETACEOUS		First flowering plants; climax of dinosaurs and ammonites, followed by extinction	
	JURASSIC		First birds, first mammals; dinosaurs and ammonites abundant	135 ± 5 MILLION
	TRIASSIC		First dinosaurs. Abundant cycads and conifers	180 ± 5 MILLION
PALEOZOIC	PERMIAN		Extinction of many kinds of marine animals, including trilobites. Glaciation at low latitudes	230 ± 10 MILLION
	CARBON-IFEROUS	PENNSYLVANIAN	Great coal forests, conifers. First reptiles	280 ± 10 MILLION
		MISSISSIPPIAN	Sharks and amphibians abundant. Large and numerous scale trees and seed ferns	310 ± 10 MILLION
	DEVONIAN		First amphibians and ammonites; fishes abundant	345 ± 10 MILLION
	SILURIAN		First terrestrial plants and animals	405 ± 10 MILLION
	ORDOVICIAN		First fishes; invertebrates dominant	425 ± 10 MILLION
	CAMBRIAN		First abundant record of marine life; trilobites dominant, followed by extinction of about two-thirds of trilobite families	500 ± 10 MILLION
	PRE-CAMBRIAN		Fossils extremely rare, consisting of primitive aquatic plants. Evidence of glaciation. Oldest dated algae, over 2,600 million years.	600 ± 50 MILLION

GEOLOGICAL AGES can be dated by comparing relative amounts of radioactive elements remaining in samples of rock obtained from different stratigraphic levels. The expanding curve at the right indicates how the number of major families of fossil animals increased through geologic time. The sharp decline after the Permian reflects the most dramatic of several mass extinctions.

simultaneous decline in the sea of the ammonite mollusks? Late in the Triassic there were still 25 families of widely ranging ammonites. All but one became extinct at the end of the period and that one gave rise to the scores of families of Jurassic and Cretaceous time.

The late Cretaceous extinctions eliminated about a quarter of all the known families of animals, but as usual the plants were little affected. The beginning of a decline in several groups is discernible near the middle of the period, some 30 million years before the mass extinction at the close of the Cretaceous. The significant point is that many characteristic groups—dinosaurs, marine reptiles, flying reptiles, ammonites, bottom-dwelling aquatic mollusks and certain kinds of extinct marine plankton—were represented by several world-wide families until the close of the period. Schindewolf has cited 16 superfamilies and orders that now became extinct. Many world-wide genera of invertebrates and most of the known species of the youngest Cretaceous period drop out near or at the boundary between the Cretaceous and the overlying Paleocene rocks. On the other hand, many families of bottom-dwelling sea organisms, fishes and nautiloid cephalopods survived with only minor evolutionary modifications. This is also true of primitive mammals, turtles, crocodiles and most of the plants of the time.

In general the groups that survived each of the great episodes of mass extinction were conservative in their evolution. As a result they were probably able to withstand greater changes in environment than could those groups that disappeared, thus conforming to the well-known principle of "survival of the unspecialized," recognized by Darwin. But there were many exceptions and it does not follow that the groups that disappeared became extinct simply because they were highly specialized. Many were no more specialized than some groups that survived.

The Cretaceous period was remarkable for a uniform and world-wide distribution of many hundreds of distinctive groups of animals and plants, which was probably a direct result of low-lying lands, widespread seas, surprisingly uniform climate and an abundance of migration routes. Just at the top of the Cretaceous sequence the characteristic fauna is abruptly replaced by another, which is distinguished not so much by radically new kinds of animals as by the elimination of innumerable major groups that had characterized the late Cre-

taceous. The geological record is somewhat obscure at the close of the Cretaceous, but most investigators agree that there was a widespread break in sedimentation, indicating a brief but general withdrawal of shallow seas from the area of the continents.

Extinctions in the Human Epoch

At the close of the Tertiary period, which immediately preceded the Quaternary in which we live, new land connections were formed between North America and neighboring continents. The horse and camel, which had evolved in North America through Tertiary time, quickly crossed into Siberia and spread throughout Eurasia and Africa. Crossing the newly formed Isthmus of Panama at about the same time, many North American animals entered South America. From Asia the mammoth, bison, bear and large deer entered North America, while from the south came ground sloths and other mammals that had originated and evolved in South America. Widespread migration and concurrent episodes of mass extinction appear to mark the close of the Pliocene (some two or three million years ago) and the middle of the Pleistocene in both North America and

Eurasia. Another mass extinction, particularly notable in North America, occurred at the very close of the last extensive glaciation, but this time it apparently was not outstandingly marked by intercontinental migrations. Surprisingly, none of the extinctions coincided with glacial advances.

It is characteristic of the fossil record that immigrant faunas tend to replace the old native faunas. In some cases newly arrived or newly evolved families replaced old families quite rapidly, in less than a few million years. In other cases the replacement has been a protracted process, spreading over tens of millions or even hundreds of millions of years. We cannot, of course, know the exact nature of competition between bygone groups, but when they occupied the same habitat and were broadly overlapping in their ecological requirements, it can be assumed that they were in fact competitors for essential resources. The selective advantage of one competing stock over another may be so slight that a vast amount of time is required to decide the outcome.

At the time of the maximum extent of the continental glaciers some 11,000 years ago the ice-free land areas of the Northern Hemisphere supported a rich

HISTORY OF LAND PLANTS shows the spectacular rise of angiosperms in the last 135 million years. The bands are roughly proportional to the number of genera of plants in each group. Angiosperms are flowering plants, a group that includes all the common trees (except conifers), grasses and vegetables. Lower vascular plants include club mosses, quillworts and horsetails. The most familiar gymnosperms (naked-seed plants) are the conifers, or evergreens. The diagram is based on one prepared by Erling Dorf of Princeton University.

and varied fauna of large mammals comparable to that which now occupies Africa south of the Sahara. Many of the species of bears, horses, elks, beavers and elephants were larger than any of their relatives living today. As recently as 8,000 years ago the horse, elephant and camel families roamed all the continents but Australia and Antarctica. Since that time these and many other families have retreated into small regions confined to one or two continents.

In North America a few species dropped out at the height of the last glaciation, but the tempo of extinction stepped up rapidly between about 12,000 and 6,000 years ago, with a maximum rate around 8,000 years ago, when the climate had become milder and the glaciers were shrinking [see illustration on page 190]. A comparable, but possibly more gradual, loss of large mammals occurred at about the same time in Asia and Australia, but not in Africa. Many of the large herbivores and carnivores had been virtually world-wide through a great range in climate, only to become extinct within a few hundred years. Other organisms were generally unaffected by this episode of extinction.

On the basis of a limited series of radiocarbon dates Paul S. Martin of the University of Arizona has concluded that now extinct large mammals of North America began to disappear first in Alaska and Mexico, followed by those in the Great Plains. Somewhat questionable datings suggest that the last survivors may have lived in Florida only 2,000 to 4,000 years ago. Quite recently, therefore, roughly three-quarters of the North American herbivores disappeared, and most of the ecological niches that were vacated have not been filled by other species.

Glaciation evidently was not a significant agent in these extinctions. In the first place, they were concentrated during the final melting and retreat of the continental glaciers after the entire biota had successfully weathered a number of glacial and interglacial cycles. Second, the glacial climate certainly did not reach low latitudes, except in mountainous areas, and it is probable that the climate over large parts of the tropics was not very different from that of today.

Studies of fossil pollen and spores in many parts of the world show that the melting of the continental glaciers was

accompanied by a change from a rainy climate to a somewhat drier one with higher mean temperatures. As a result of these changes forests in many parts of the world retreated and were replaced by deserts and steppes. The changes, however, probably were not universal or severe enough to result in the elimination of any major habitat.

A number of investigators have proposed that the large mammals may have been hunted out of existence by prehistoric man, who may have used fire as a weapon. They point out that the mass extinctions coincided with the rapid growth of agriculture. Before this stage in human history a decrease in game supply would have been matched by a decrease in human populations, since man could not have destroyed a major food source without destroying himself.

In Africa and Eurasia, where man had lived in association with game animals throughout the Pleistocene, extinctions were not so conspicuously concentrated in the last part of the epoch. There was ample opportunity in the Old World for animals to become adapted to man through hundreds of thousands of years of coexistence. In the Americas and

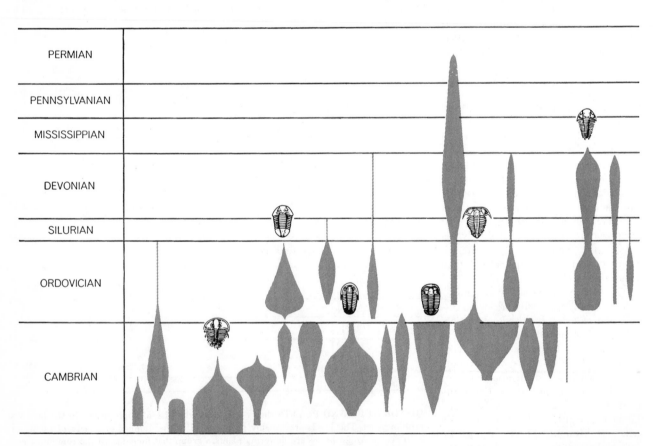

MASS EXTINCTION OF TRILOBITES, primitive arthropods, occurred at the close of the Cambrian period about 500 million years ago. During the Cambrian period hundreds of kinds of trilobites populated the shallow seas of the world. The chart depicts 15 superfamilies of Cambrian trilobites; the width of the shapes is roughly proportional to the number of members in each superfamily. Final extinction took place in the Permian. The chart is based on the work of H. B. Whittington of Harvard University.

Australia, where man was a comparative newcomer, the animals may have proved easy prey for the hunter.

We shall probably never know exactly what happened to the large mammals of the late Pleistocene, but their demise did coincide closely with the expansion of ancient man and with an abrupt change from a cool and moist to a warm and dry climate over much of the world. Possibly both of these factors contributed to this episode of mass extinction. We can only guess.

The Modern Crisis

Geological history cannot be observed but must be deduced from studies of stratigraphic sequences of rocks and fossils interpreted in the context of processes now operating on earth. It is helpful, therefore, to analyze some recent extinctions to find clues to the general causes of extinction.

We are now witnessing the disastrous effects on organic nature of the explosive spread of the human species and the concurrent development of an efficient technology of destruction. The human demand for space increases, hunting techniques are improved, new poisons are used and remote areas that had long served as havens for wildlife are now easily penetrated by hunter, fisherman, lumberman and farmer.

Studies of recent mammal extinctions show that man has been either directly or indirectly responsible for the disappearance, or near disappearance, of more than 450 species of animals. Without man's intervention there would have been few, if any, extinctions of birds or mammals within the past 2,000 years. The heaviest toll has been taken in the West Indies and the islands of the Pacific and Indian oceans, where about 70 species of birds have become extinct in the past few hundred years. On the continents the birds have fared somewhat better. In the same period five species of birds have disappeared from North America, three from Australia and one from Asia. Conservationists fear, however, that more North American birds will become extinct in the next 50 years than have in the past 5,000 years.

The savannas of Africa were remarkable until recently for a wealth of large mammals comparable only to the rich Tertiary and Pleistocene faunas of North America. In South Africa stock farming, road building, the fencing of grazing lands and indiscriminate hunting had wiped out the wild populations of large grazing mammals by the beginning of the 20th century. The depletion of ani-

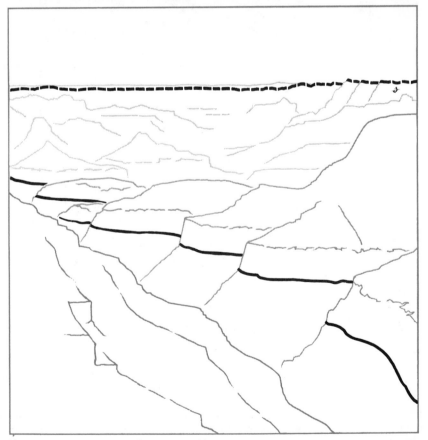

PALEONTOLOGICAL BOUNDARIES are clearly visible in this photograph of the Grand Canyon. The diagram below identifies the stratigraphic boundary between the Cambrian and Ordovician periods (*solid line*) and the top of the Permian rocks (*broken line*). These are world-wide paleontological division points, easily identified by marine fossils.

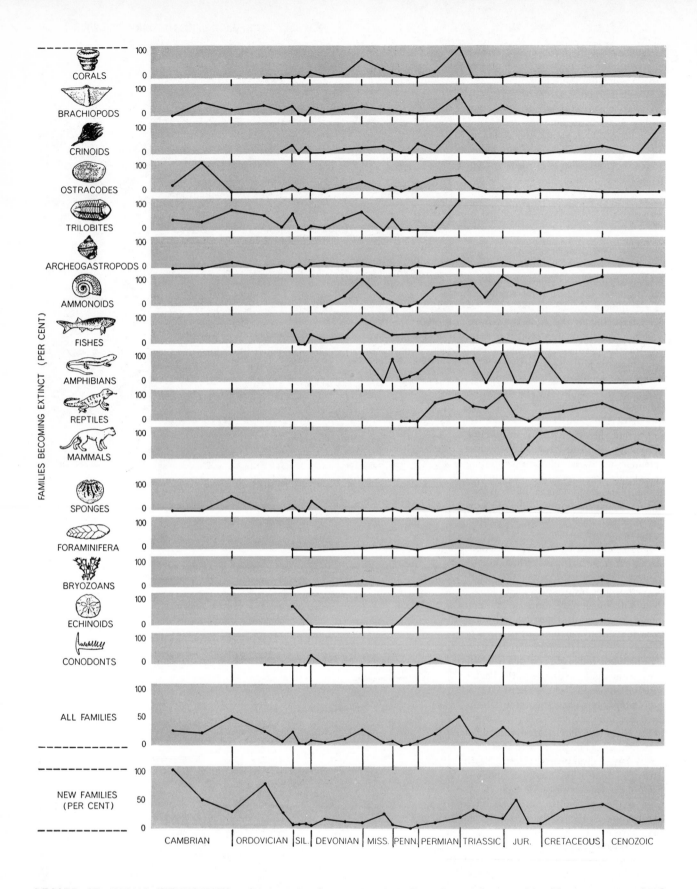

RECORD OF ANIMAL EXTINCTIONS makes it quite clear that the history of animals has been punctuated by repeated crises. The top panel of curves plots the ups and downs of 11 groups of animals from Cambrian times to the present. Massive extinctions took place at the close of the Ordovician, Devonian and Permian periods. The second panel shows the history of five other groups for which the evidence is less complete. (Curves are extrapolated between dots.) The next to bottom curve depicts the sum of extinctions for all the fossil groups plotted above (plus bivalves and caenogastropods). The bottom curve shows the per cent of new families in the main fossil groups. It indicates that periods of extinction were usually followed by an upsurge in evolutionary activity.

mals has now spread to Equatorial Africa as a result of poaching in and around the game reserves and the practice of eradicating game as a method of controlling human and animal epidemics. Within the past two decades it has become possible to travel for hundreds of miles across African grasslands without seeing any of the large mammals for which the continent is noted. To make matters worse, the great reserves that were set aside for the preservation of African wildlife are now threatened by political upheavals.

As a factor in extinction, man's predatory habits are supplemented by his destruction of habitats. Deforestation, cultivation, land drainage, water pollution, wholesale use of insecticides, the building of roads and fences—all are causing fragmentation and reduction in range of wild populations with resulting loss of environmental and genetic resources. These changes eventually are fatal to populations just able to maintain themselves under normal conditions. A few species have been able to take advantage of the new environments created by man, but for the most part the changes have been damaging.

Reduction of geographic range is prejudicial to a species in somewhat the same way as overpopulation. It places an increasing demand on diminishing environmental resources. Furthermore, the gene pool suffers loss of variability by reduction in the number of local breeding groups. These are deleterious changes, which can be disastrous to species that have narrow tolerances for one or more environmental factors. No organism is stronger than the weakest link in its ecological chain.

Man's direct attack on the organic world is reinforced by a host of competing and pathogenic organisms that he intentionally or unwittingly introduces to relatively defenseless native communities. Charles S. Elton of the University of Oxford has documented scores of examples of the catastrophic effects on established communities of man-sponsored invasions by pathogenic and other organisms. The scale of these ecological disturbances is world-wide; indeed, there are few unmodified faunas and floras now surviving.

The ill-advised introduction of predators such as foxes, cats, dogs, mongooses and rats into island communities has been particularly disastrous; many extinctions can be traced directly to this cause. Grazing and browsing domestic animals have destroyed or modified vegetation patterns. The introduction of

European mammals into Australia has been a primary factor in the rapid decimation of the native marsupials, which cannot compete successfully with placental mammals.

An illustration of invasion by a pathogenic organism is provided by an epidemic that in half a century has nearly wiped out the American sweet chestnut tree. The fungus infection responsible for this tragedy was accidentally introduced from China on nursery plants. The European chestnut, also susceptible to the fungus, is now suffering rapid decline, but the Chinese chestnut, which evolved in association with the blight, is comparatively immune.

Another example is provided by the marine eelgrass *Zostera,* which gives food and shelter to a host of invertebrates and fishes and forms a protective blanket over muddy bottoms. It is the

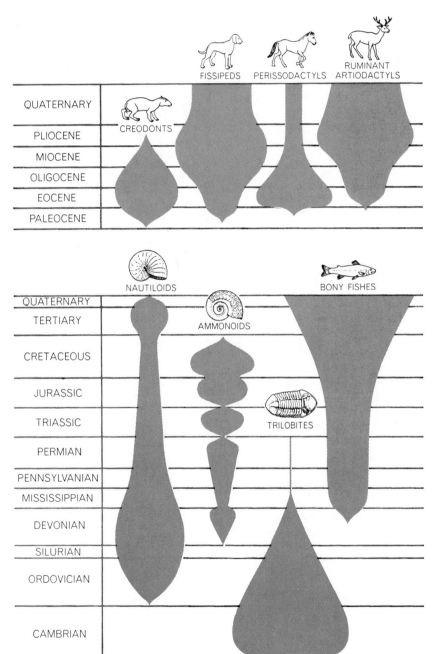

ECOLOGICAL REPLACEMENT appears to be a characteristic feature of evolution. The top diagram shows the breadth of family representation among four main groups of mammals over the last 60-odd million years. The bottom diagram shows a similar waxing and waning among four groups of marine swimmers, dating back to the earliest fossil records. The ammonoid group suffered near extinction twice before finally expiring. The diagrams are based on the work of George Gaylord Simpson of Harvard University and the author.

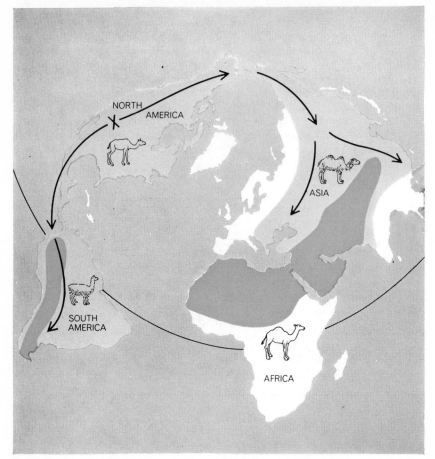

DISPERSAL OF CAMEL FAMILY from its origin (X) took place during Pleistocene times. Area in light color shows the maximum distribution of the family; dark color shows present distribution. This map is based on one in *Life: An Introduction to Biology*, by Simpson, C. S. Pittendrigh and L. H. Tiffany, published by Harcourt, Brace and Company.

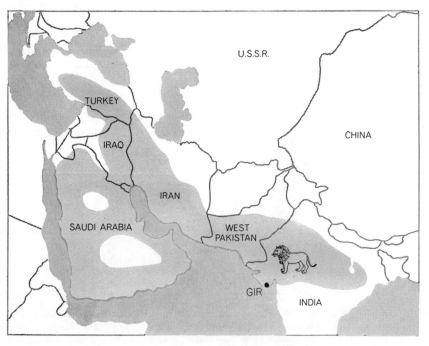

DISTRIBUTION OF ASIATIC LION has contracted dramatically just since 1800, when it roamed over large areas (*shown in color*) of the Middle East, Pakistan and India. Today the Asiatic lion is found wild only in Gir, a small game preserve in western India.

most characteristic member of a distinctive community that includes many plant and animal species. In the 1930's the eelgrass was attacked by a virus and was almost wiped out along the Atlantic shores of North America and Europe. Many animals and plants not directly attacked nevertheless disappeared for a time and the community was greatly altered. Resistant strains of *Zostera* fortunately escaped destruction and have slowly repopulated much of the former area. Eelgrass is a key member of a complex ecological community, and one can see that if it had not survived, many dependent organisms would have been placed in jeopardy and some might have been destroyed.

This cursory glance at recent extinctions indicates that excessive predation, destruction of habitat and invasion of established communities by man and his domestic animals have been primary causes of extinctions within historical time. The resulting disturbances of community equilibrium and shock waves of readjustment have produced ecological explosions with far-reaching effects.

The Causes of Mass Extinctions

It is now generally understood that organisms must be adapted to their environment in order to survive. As environmental changes gradually pass the limits of tolerance of a species, that species must evolve to cope with the new conditions or it will die. This is established by experiment and observation. Extinction, therefore, is not simply a result of environmental change but is also a consequence of failure of the evolutionary process to keep pace with changing conditions in the physical and biological environment. Extinction is an evolutionary as well as an ecological problem.

There has been much speculation about the causes of mass extinction; hypotheses have ranged from worldwide cataclysms to some kind of exhaustion of the germ plasm—a sort of evolutionary fatigue. Geology does not provide support for the postulated cataclysms and biology has failed to discover any compelling evidence that evolution is an effect of biological drive, or that extinction is a result of its failure. Hypotheses of extinction based on supposed racial old age or overspecialization, so popular among paleontologists a few generations ago and still echoed occasionally, have been generally abandoned for lack of evidence.

Of the many hypotheses advanced to explain mass extinctions, most are un-

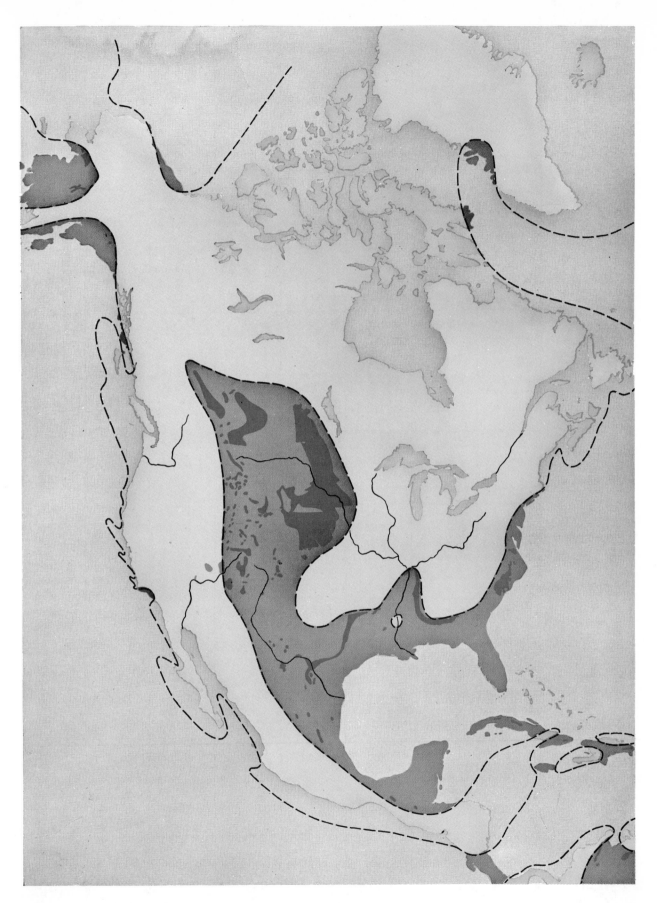

LATE CRETACEOUS SEA covered large portions of Central and North America (*dark gray*). Fossil-bearing rocks laid down at that time, and now visible at the surface of the earth, are shown in dark color. The approximate outline of North America in the Cretaceous period is represented by the broken line. The map is based on the work of the late Charles Schuchert of Yale University.

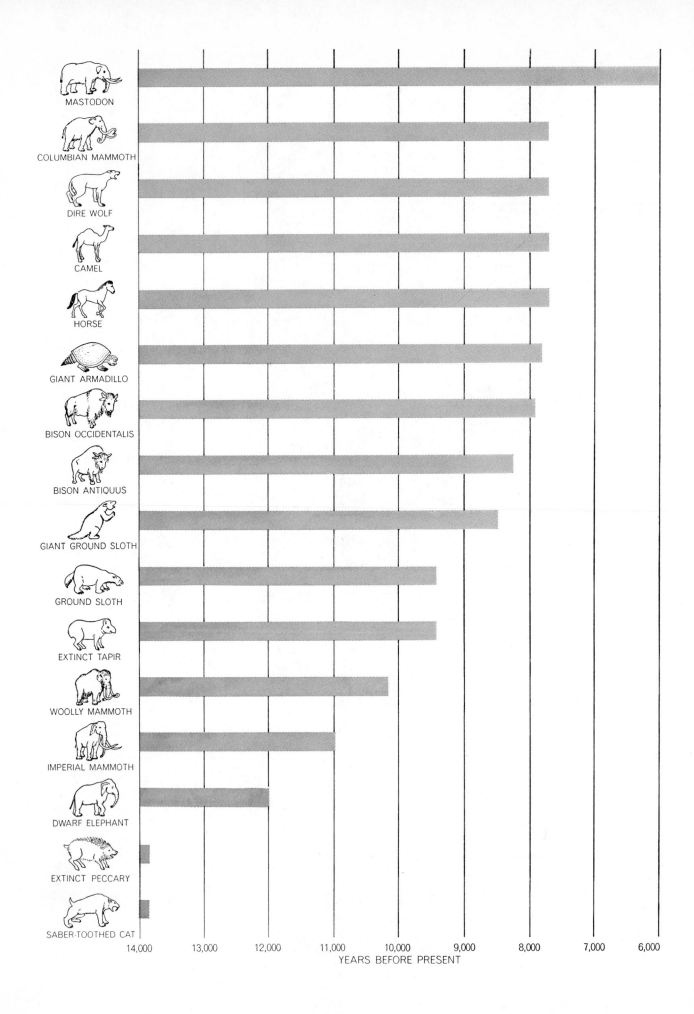

MASTODON

COLUMBIAN MAMMOTH

DIRE WOLF

CAMEL

HORSE

GIANT ARMADILLO

BISON OCCIDENTALIS

BISON ANTIQUUS

GIANT GROUND SLOTH

GROUND SLOTH

EXTINCT TAPIR

WOOLLY MAMMOTH

IMPERIAL MAMMOTH

DWARF ELEPHANT

EXTINCT PECCARY

SABER-TOOTHED CAT

14,000   13,000   12,000   11,000   10,000   9,000   8,000   7,000   6,000

YEARS BEFORE PRESENT

satisfactory because they lack testable corollaries and are designed to explain only one episode of extinction. For example, the extinction of the dinosaurs at the end of the Cretaceous period has been attributed to a great increase in atmospheric oxygen and alternatively to the explosive evolution of pathogenic fungi, both thought to be by-products of the dramatic spread of the flowering plants during late Cretaceous time.

The possibility that pathogenic fungi may have helped to destroy the dinosaurs was a recent suggestion of my own. I was aware, of course, that it would not be a very useful suggestion unless a way could be found to test it. I was also aware that disease is one of the most popular hypotheses for explaining mass extinctions. Unfortunately for such hypotheses, pathogenic organisms normally attack only one species or at most a few related species. This has been interpreted as an indication of a long antecedent history of coadaptation during which parasite and host have become mutually adjusted. According to this theory parasites that produce pathological reactions are not well adapted to the host. On first contact the pathogenic organism might destroy large numbers of the host species; it is even possible that extinction of a species might follow a pandemic, but there is no record that this has happened in historical times to any numerous and cosmopolitan group of species.

It is well to keep in mind that living populations studied by biologists generally are large, successful groups in which the normal range of variation provides tolerance for all the usual exigencies, and some unusual ones. It is for this reason that the eelgrass was not extinguished by the epidemic of the 1930's and that the human race was not eliminated by the influenza pandemic following World War I. Although a succession of closely spaced disasters of various kinds might have brought about extinction, the particular virus strains responsible for these diseases did not directly attack associated species.

Another suggestion, more ingenious

ICE-AGE MAMMALS provided North America with a fauna of large herbivores comparable to that existing in certain parts of Africa today. Most of them survived a series of glacial periods only to become extinct about 8,000 years ago, when the last glaciers were shrinking. The chart is based on a study by Jim J. Hester of the Museum of New Mexico in Santa Fe.

than most, is that mass extinctions were caused by bursts of high-energy radiation from a nearby supernova. Presumably the radiation could have had a dramatic impact on living organisms without altering the climate in a way that would show up in the geological record. This hypothesis, however, fails to account for the patterns of extinction actually observed. It would appear that radiation would affect terrestrial organisms more than aquatic organisms, yet there were times when most of the extinctions were in the sea. Land plants, which would be more exposed to the radiation and are more sensitive to it, were little affected by the changes that led to the animal extinctions at the close of the Permian and Cretaceous periods.

Another imaginative suggestion has been made recently by M. J. Salmi of the Geological Survey of Finland and Preston E. Cloud, Jr., of the University of Minnesota. They have pointed out that excessive amounts or deficiencies of certain metallic trace elements, such as copper and cobalt, are deleterious to organisms and may have caused past extinctions. This interesting hypothesis, as applied to marine organisms, depends on the questionable assumption that deficiencies of these substances have occurred in the ocean, or that a lethal concentration of metallic ions might have diffused throughout the oceans of the world more rapidly than the substances could be concentrated and removed from circulation by organisms and various common chemical sequestering agents. To account for the disappearance of land animals it is necessary to postulate further that the harmful elements were broadcast in quantity widely over the earth, perhaps as a result of a great volcanic eruption. This is not inconceivable; there have probably been significant variations of trace elements in time and place. But it seems unlikely that such variations sufficed to produce worldwide biological effects.

Perhaps the most popular of all hypotheses to explain mass extinctions is that they resulted from sharp changes in climate. There is no question that large-scale climatic changes have taken place many times in the past. During much of geologic time shallow seas covered large areas of the continents; climates were consequently milder and less differentiated than they are now. There were also several brief episodes of continental glaciation at low latitudes, but it appears that mass extinctions did not coincide with ice ages.

It is noteworthy that fossil plants,

which are good indicators of past climatic conditions, do not reveal catastrophic changes in climate at the close of the Permian, Triassic and Cretaceous periods, or at other times coincident with mass extinctions in the animal kingdom. On theoretical grounds it seems improbable that any major climatic zone of the past has disappeared from the earth. For example, climates not unlike those of the Cretaceous period probably have existed continuously at low latitudes until the present time. On the other hand, it is certain that there have been great changes in distribution of climatic zones. Severe shrinkage of a given climatic belt might adversely affect many of the contained species. Climatic changes almost certainly have contributed to animal extinctions by destruction of local habitats and by inducing wholesale migrations, but the times of greatest extinction commonly do not clearly correspond to times of great climatic stress.

Finally, we must consider the evidence that so greatly impressed Cuvier and many geologists. They were struck by the frequent association between the last occurrence of extinct animals and unconformities, or erosional breaks, in the geological record. Cuvier himself believed that the unconformities and the mass extinctions went hand in hand, that both were products of geologic revolutions, such as might be caused by paroxysms of mountain building. The idea still influences some modern thought on the subject.

It is evident that mountains do strongly influence the environment. They can alter the climate, soils, water supply and vegetation over adjacent areas, but it is doubtful that the mountains of past ages played dominant roles in the evolutionary history of marine and lowland organisms, which constitute most of the fossil record. Most damaging to the hypothesis that crustal upheavals played a major role in extinctions is the fact that the great crises in the history of life did not correspond closely in time with the origins of the great mountain systems. Actually the most dramatic episodes of mass extinction took place during times of general crustal quiet in the continental areas. Evidently other factors were involved.

Fluctuations of Sea Level

If mass extinctions were not brought about by changes in atmospheric oxygen, by disease, by cosmic radiation, by trace-element poisoning, by climatic changes or by violent upheavals of the earth's crust, where is one to look for a satisfactory—and testable—hypothesis?

The explanation I have come to favor, and which has found acceptance among many students of the paleontological record, rests on fluctuations of sea level. Evidence has been accumulating to show an intimate relation between many fossil zones and major advances and retreats of the seas across the continents. It is clear that diastrophism, or reshaping, of the ocean basins can produce universal changes in sea level. The evidence of long continued sinking of the sea floor under Pacific atolls and guyots (flat-topped submarine mountains) and the present high stand of the continents indicate that the Pacific basin has been subsiding differentially with respect to the land at least since Cretaceous time.

During much of Paleozoic and Mesozoic time, spanning some 540 million years, the land surfaces were much lower than they are today. An appreciable rise in sea level was sufficient to flood large areas; a drop of a few feet caused equally large areas to emerge, producing major environmental changes. At least 30 major and hundreds of minor oscillations of sea level have occurred in the past 600 million years of geologic time.

Repeated expansion and contraction of many habitats in response to alternate flooding and draining of vast areas of the continents unquestionably created profound ecological disturbances among offshore and lowland communities, and repercussions of these changes probably extended to communities deep inland and far out to sea. Intermittent draining of the continents, such as occurred at the close of many of the geologic epochs and periods, greatly reduced or eliminated the shallow inland seas that pro-

vided most of the fossil record of marine life. Many organisms adapted to the special estuarine conditions of these seas evidently could not survive along the more exposed ocean margins during times of emergence and they had disappeared when the seas returned to the continents. There is now considerable evidence that evolutionary diversification was greatest during times of maximum flooding of the continents, when the number of habitats was relatively large. Conversely, extinction and natural selection were most intense during major withdrawals of the sea.

It is well known that the sea-level oscillations of the Pleistocene epoch caused by waxing and waning of the continental glaciers did not produce numerous extinctions among shallow-water marine communities, but the situation was quite unlike that which prevailed during much of geological history. By Pleistocene times the continents stood high above sea level and the warm interior seas had long since disappeared. As a result the Pleistocene oscillations did not produce vast geographic and climatic changes. Furthermore, they were of short duration compared with major sea-level oscillations of earlier times.

Importance of Key Species

It might be argued that nothing less than the complete destruction of a habitat would be required to eliminate a world-wide community of organisms. This, however, may not be necessary. After thousands of years of mutual accommodation, the various organisms of a biological community acquire a high order of compatibility until a nearly steady state is achieved. Each species

plays its own role in the life of the community, supplying shelter, food, chemical conditioners or some other resource in kind and amount needed by its neighbors. Consequently any changes involving evolution or extinction of species, or the successful entrance of new elements into the community, will affect the associated organisms in varying degrees and result in a wave of adjustments.

The strength of the bonds of interdependence, of course, varies with species, but the health and welfare of a community commonly depend on a comparatively small number of key species low in the community pyramid; the extinction of any of these is sure to affect adversely many others. Reduction and fragmentation of some major habitats, accompanied by moderate changes in climate and resulting shrinkage of populations, may have resulted in extinction of key species not necessarily represented in the fossil record. Disappearance of any species low in the pyramid of community organization, as, for example, a primary food plant, could lead directly to the extinction of many ecologically dependent species higher in the scale. Because of this interdependence of organisms a wave of extinction originating in a shrinking coastal habitat might extend to more distant habitats of the continental interior and to the waters of the open sea.

This theory, in its essence long favored by geologists but still to be fully developed, provides an explanation of the common, although not invariable, parallelism between times of widespread emergence of the continents from the seas and episodes of mass extinction that closed many of the chapters of geological history.

Plate Tectonics and the History of Life in the Oceans

19

by James W. Valentine and Eldridge M. Moores
April 1974

The breakup of the ancient supercontinent of Pangaea triggered a long-term evolutionary trend that has led to the unprecedented variety of the present biosphere

During the 1960's a conceptual revolution swept the earth sciences. The new world view fundamentally altered long established notions about the permanency of the continents and the ocean basins and provided fresh perceptions of the underlying causes and significance of many major features of the earth's mantle and crust. As a consequence of this revolution it is now generally accepted that the continents have greatly altered their geographic position, their pattern of dispersal and even their size and number. These processes of continental drift, fragmentation and assembly have been going on for at least 700 million years and perhaps for more than two billion years.

Changes of such magnitude in the relative configuration of the continents and the oceans must have had far-reaching effects on the environment, repatterning the world's climate and influencing the composition and distribution of life in the biosphere. These more or less continual changes in the environment must also have had profound effects on the course of evolution and accordingly on the history of life.

Natural selection, the chief mechanism by which evolution proceeds, is a very complex process. Although it is constrained by the machinery of inheritance, natural selection is chiefly an ecological process based on the relation between organisms and their environment. For any species certain heritable variations are favored because they are particularly well suited to survive and to reproduce in their prevailing environment. To answer the question of why any given group of organisms has evolved, then, one needs to understand two main factors. First, it is necessary to

know what the ancestral organisms were that formed the "raw material" on which selection worked. And second, one must have some idea of the sequence of environmental conditions that led the ancestral stock to evolve along a particular pathway to a descendant group. Given these factors, one can then infer the organism-environment interactions that gave rise to the evolutionary events. The study of the relations between ancient organisms and their environment is called paleoecology.

The new ideas of continental drift that came into prominence in the 1960's revolve around the theory of plate tectonics. According to this theory, new sea floor and underlying mantle are currently being added to the crust of the earth at spreading centers under deep-sea ridges and in small ocean basins at rates of up to 10 centimeters per year. The sea floor spreads laterally away from these centers and eventually sinks into the earth's interior at subduction zones, which are marked by deep-sea trenches. Volcanoes are created by the consumption process and flank the trenches. The lithosphere, or rocky outer shell of the earth, therefore comprises several major plates that are generated at spreading centers and consumed at subduction zones. Most lithospheric plates bear one continent or more, which passively move with the plate on which they rest. Because the continents are too light to sink into the trenches they remain on the surface. Continents can fragment at new ridges, however, and hence oceans may appear across them. Conversely, continents can be welded together when they collide at the site of a trench. Thus continents may be assembled into supercontinents, fragmented into small continents

and generally moved about the earth's surface as passive riders on plates. In tens or hundreds of millions of years entire oceans may be created or destroyed, and the number, size and dispersal pattern of continents may be vastly altered.

The record of such continental fragmentation and reassembly is evident as deformed regions in the earth's mountain belts, particularly those mountain belts that contain the rock formations known as ophiolites. These formations are characterized by a certain sequence of rocks consisting (from bottom to top) of ultramafic rock (a magnesium-rich rock composed mostly of olivine), gabbro (a coarse-grained basaltic rock), volcanic rocks and sedimentary rocks. The major ophiolite belts of the earth are believed to represent preserved fragments of vanished ocean basins [*see illustration on pages 194 and 195*]. The existence of such a belt within a continent (for example the Uralian belt in the U.S.S.R.) is evidence for the former presence there of an ocean basin separating two continental fragments that at some time in the past collided with each other and were welded into the single larger continent. The timing of such events as the opening of ocean basins, the dispersal of continents and the closing of oceans by continental collisions can accordingly be "read" from the geology of a given mountain system.

Of course, the biological environment is constantly being altered as well. For example, the changes in continental configuration will greatly affect the ocean currents, the temperature, the nature of seasonal fluctuations, the distribution of nutrients, the patterns of productivity and many other factors of fundamental importance to living organisms. Therefore evolutionary trends in marine animals must have varied through geologic time in response to the major environmental changes, as natural selection acted to adapt organisms to the new conditions.

It should in principle be possible to detect these changes in the fossil record. Indeed, paleontologists have long recognized that vast changes in the composition, distribution and diversity of marine life are well documented by the fossil record. Now for the first time, however, it is possible to reconstruct the sequence of environmental changes based on the theory of plate tectonics, to determine their environmental consequences and to attempt to correlate them with the sequence of faunal changes that is seen in the fossil record. Such a thorough reconstruction ultimately may explain many of the enigmatic faunal changes known for many years. Even at this early stage paleontologists have succeeded in shedding much new light on a number of major extinctions and diversifications of the past.

As a first step toward understanding the relation between plate tectonics and the history of life it is helpful to investigate the relations that exist today between marine life, the present pattern of continental drift and plate-tectonic theory. The vast majority of marine species (about 90 percent) live on the continental shelves or on shallow-water portions of islands or subsurface "rises" at depths of less than about 200 meters

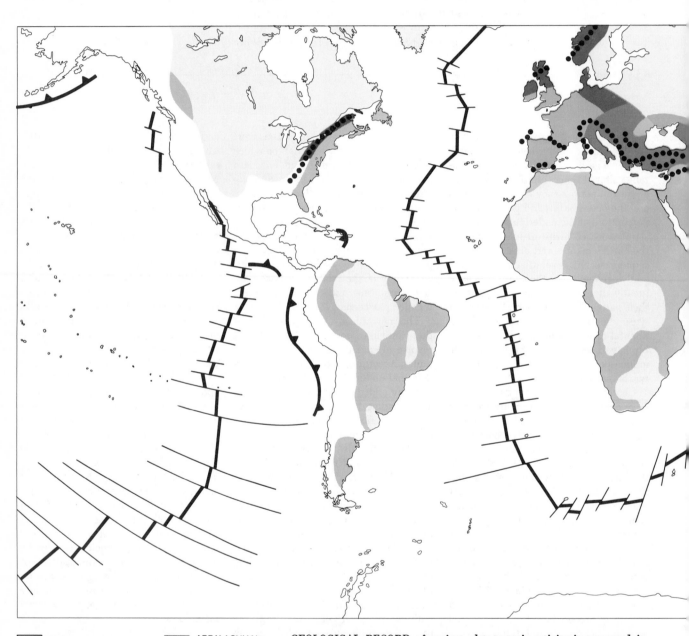

PRECAMBRIAN

PAN-AFRICAN-
BAIKALIAN

CALEDONIAN

APPALACHIAN-
HERCYNIAN

URALIAN

CORDILLERAN-
TETHYAN

GEOLOGICAL RECORD of ancient plate-tectonic activity is preserved in certain deformed mountain belts (color), particularly those that contain the characteristic rock sequences known as ophiolites (black dots). The Pan-African–Baikalian belt, for example, is made up of rocks dating from 873 to 450 million years ago and may represent the assembly of all or nearly all the landmasses near the beginning of Phanerozoic time. This supercontinent may then have fragmented into four or more smaller continents, sometime just before and during the Cambrian period. The Caledonian mountain system may represent the collision of two continents at about late Silurian or early De-

(660 feet); most of the fossil record also consists of these faunas. Therefore it is the pattern of shallow-water sea-floor animal life that is of particular interest here.

The richest shallow-water faunas are found today at low latitudes in the Tropics, where communities are packed with vast numbers of highly specialized species. Proceeding to higher latitudes, diversity gradually falls; in the Arctic or Antarctic regions less than a tenth as many animals are living as in the Trop-

ics, when comparable regions are considered [see illustration, pages 196 and 197]. The diversity gradient correlates well with a gradient in the stability of food supplies; as the seasons become more pronounced, fluctuations in primary productivity become greater. Although this strong latitudinal gradient dominates the earth's overall diversity pattern, there are important longitudinal diversity trends as well. In regions of similar latitude, for example, diversity is lower where there are sharp seasonal

changes (such as variations in the surface-current pattern or in the upwelling of cold water) that affect the nutrient supply by causing large fluctuations in productivity.

At any given latitude, therefore, diversity is highest off the shores of small islands or small continents in large oceans, where fluctuations in nutrient supplies are least affected by the seasonal effects of landmasses, whereas diversity is lowest off large continents, particularly when they face small oceans, where shallow-water seasonal variations are greatest. In short, whereas latitudinal diversity increases generally from high latitudes to low, longitudinal diversity increases generally with distance from large continental landmasses. In both of these trends the increase in diversity is correlated with increasing stability of food resources. The resource-stability pattern depends largely on the shape of the continents and should also be sensitive to the extent of inland seas and to the presence of coastal mountains. Seas lying on continental platforms are particularly important: not only do extensive shallow seas provide much habitat area for shallow-water faunas but also such seas tend to damp seasonal climatic changes and to have an ameliorating influence on the local environment.

Today shallow marine faunas are highly provincial, that is, the species living in different oceans or on opposite sides of the same ocean tend to be quite different. Even along continuous coastlines there are major changes in species composition from place to place that generally correspond to climatic changes. The deep-sea floor, generated at oceanic ridges, forms a significant barrier to the dispersal of shallow-water organisms, and latitudinal climatic changes clearly form other barriers. The present dominantly north-south series of ridges forms a pattern of longitudinally alternating oceans and continents, thereby creating a series of barriers to shallow-water marine organisms. The steep latitudinal climatic gradient, on the other hand, creates chains of provinces along north-south coastlines. As a result the marine faunas today are partitioned into more than 30 provinces, among which there is in general only a low percentage of common species [see illustration on pages 198 and 199]. It is estimated that the shallow-water marine fauna represents more than 10 times as many species today as would be present in a world with only a single province, even a highly diverse one.

The volcanic arcs that appear over subduction zones form fairly continuous

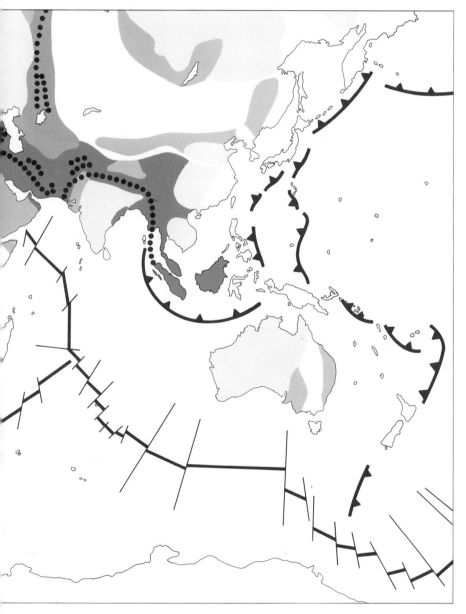

vonian time (approximately 400 million years ago). The Appalachian-Hercynian system may represent a two-continent collision during the late Carboniferous period (300 million years ago). The Uralian mountains may represent a similar collision at about Permo-Triassic time (220 million years ago). The Cordilleran-Tethyan system represents regions of Mesozoic mountain-building and includes the continental collisions that resulted in the Alpine-Himalayan mountain system. The ophiolite belts shown are the preserved remnants of ocean floor exposed in the mountain systems in question. Spreading ridges such as the Mid-Atlantic Ridge are indicated by heavy lines cut by lighter lines, which correspond to transform faults. Subduction zones are marked by heavy black curved lines with triangles.

island chains and provide excellent dispersal routes. When long island chains are arranged in an east-west pattern so as to lie within the same climatic zone, they are inhabited by wide-ranging faunas that are highly diverse for their latitude. Indeed, the widest ranging marine province, and also by far the most diverse, is the Indo-Pacific province, which is based on island arcs in its central regions. The faunal life of this province spills from these arcs onto tropical continental shelves in the west (India and East Africa) and also onto tropical intraplate volcanoes (the Polynesian and Micronesian islands) that are reasonably close to them. This vast tropical biota is cut off from the western American mainland by the East Pacific Barrier, a zoogeographic obstruction formed by a spreading ridge.

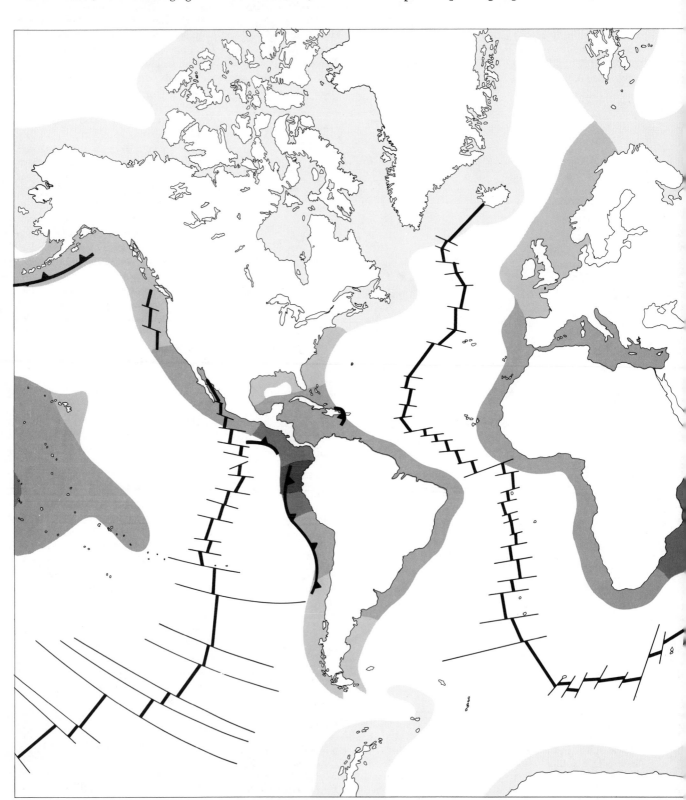

RELATIVE DIVERSITY of shallow-water, bottom-dwelling species in the present oceans is suggested by the colored patterns in this world map. The diversity classes are not based on absolute counts but are inferred from the diversity patterns of the best-

Since current patterns of marine provinciality and diversity fit closely with the present oceanic and continental geography and the resulting environmental patterns, one would expect ancient provinces and ancient diversity patterns also to fit past geographies. One of the best-established of ancient geographies is the one that existed near the beginning of the Triassic period, about 225 million years ago. The continents were then assembled into a single supercontinent named Pangaea, which must have had a continuous shallow-water margin running all the way around it, with no major physical barriers to the dispersal of shallow-water marine animals [see illustration on page 200]. Therefore provinciality must have been low compared with today, and it must have been attributable entirely to climatic effects. It is likely that the marine climate was quite mild and that even in high latitudes water temperatures were much warmer than they are today. As a result climatic provinciality must have been greatly reduced also. Furthermore, the seas at that time were largely confined to the ocean basins and did not extend significantly over the continental shelves. Thus the habitat area for shallow-water marine organisms was greatly reduced, first by the diminution of coastline that accompanies the creation of a supercontinent from smaller continents, and second by the general withdrawal of seas from continental platforms. The reduced habitat area would make for low species diversity. Finally, the extreme emergence of such a supercontinent would provide unstable nearshore conditions, with the result that food resources would have been very unstable compared with those of today. All these factors tend to reduce species diversity; hence one would expect to find that Triassic biotas were widespread and were made up of comparatively few species. That is precisely what the fossil record indicates.

Prior to the Triassic period, during the late Paleozoic, diversity appears to have been much higher [see top illustration on page 202]. It was sharply reduced again near the close of the Permian period during a vast wave of extinction that on balance is the most severe known to have been suffered by the marine fauna. The late Paleozoic species that were the more elaborately adapted specialists became virtually extinct, whereas the surviving descendants tended to have simple skeletons. A high proportion of these survivors appear to have been detritus feeders or suspension feeders that harvested the water layers just above the sea floor. These successful types seem to be ecologically similar to the populations found today in unstable environments, for instance in high latitudes; the unsuccessful specialists, on the other hand, seem ecologically similar to the populations found in stable environments, for instance in the Tropics. Thus the extinctions appear to have been caused by the reduced potential for diversity of the shallow seas, a trend associated with less provinciality, less habitat area and less stable environmental conditions.

In the period following the great extinction, as Pangaea broke up and the

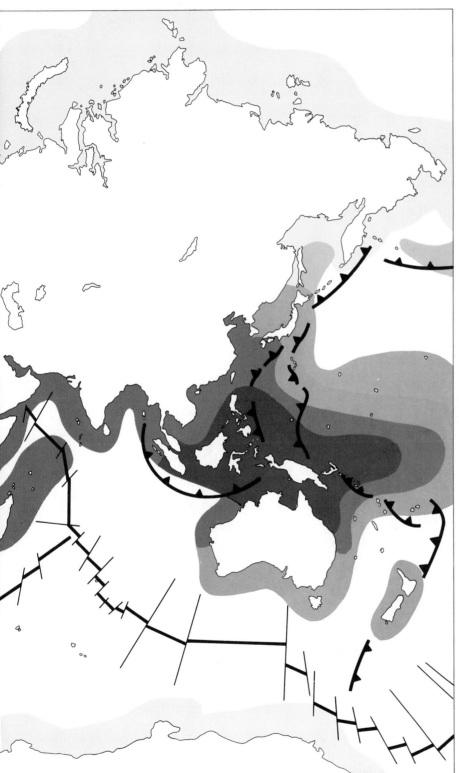

known skeletonized groups, chiefly the bivalves, gastropods, echinoids and corals. The highest class (*darkest color*) is about 20 times as diverse as the lowest (*lightest color*).

resulting continents themselves gradually fragmented and migrated to their present positions, provinciality increased, communities in stabilized regions became filled with numerous specialized animals and the overall diversity of species in the world ocean rose to un-

precedented heights, even though occasional waves of extinctions interrupted this long-range trend.

There is another time in the past besides the early Triassic period when low provinciality and low diversity were

coupled with the presence of a high proportion of detritus feeders and near-bottom suspension feeders. That is in the late Precambrian and Cambrian periods, when a widespread, soft-bodied fauna of low diversity gave way to a slightly provincialized, skeletonized fauna of some-

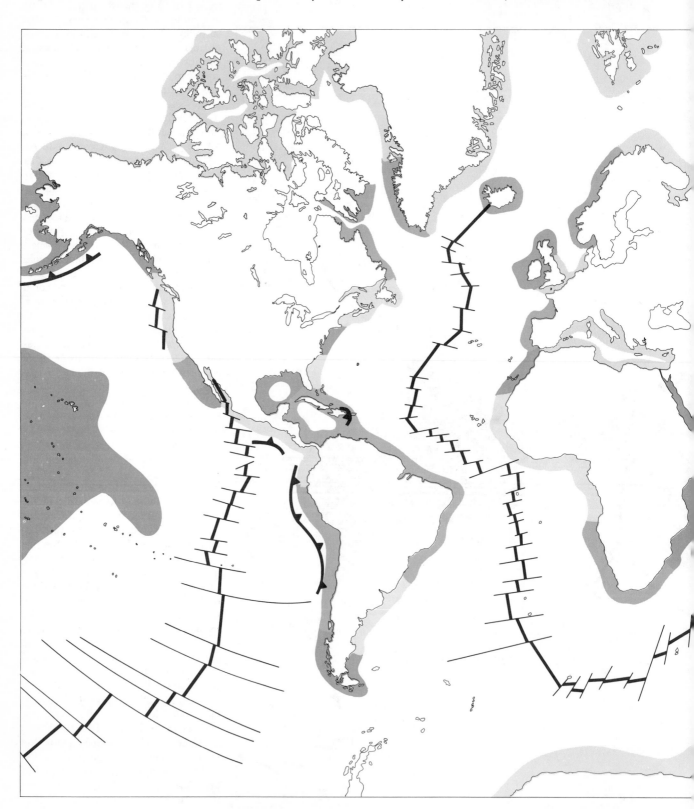

PRINCIPAL SHALLOW-WATER MARINE PROVINCES at present are indicated by the colored areas. The dominant north-

south chains of provinces along the continental coastlines are created by the present high latitudinal gradient in ocean temperature

what higher diversity. It seems likely that the late Precambrian environment was quite unstable and that there may well have been a supercontinent in existence, or at least that the continents then were collected into a more compact assemblage than at present. In the late

Precambrian period one finds the first unequivocal records of invertebrate life, including burrowing forms that were probably coelomic, or hollow-bodied, worms. In the Cambrian four continents may have existed although they were not arranged in the present pattern. During

the Cambrian a skeletonized fauna appears that is at first almost entirely surface-dwelling and that includes chiefly detritus-feeding and suspension-feeding forms, probably with some browsers.

It seems possible, therefore, that the late Precambrian species were adapted to highly unstable conditions and became diversified chiefly as a bottom-living, detritus-feeding assemblage. The coelomic body cavity, evidently a primitive adaptation for burrowing, was developed and diversified into a variety of forms, perhaps as many as five basic ones: highly segmented worms that lived under the ocean floor and were detritus feeders; slightly segmented worms that lived attached to the ocean floor and were suspension feeders; slightly segmented worms that lived attached to the ocean floor and were detritus feeders; "pseudosegmented" worms that lived on the ocean floor and were detritus feeders or browsers, and nonsegmented worms that lived under the ocean floor and fed by means of an "introvert." In addition to these coelomates there were a number of coelenterate stocks (such as corals, sea anemones and jellyfishes) and probably also flatworms and other noncoelomate worms.

From the chiefly wormlike coelomate stocks higher forms of animal life have originated; many of them appear in the Cambrian period, when they evidently first became organized into the groups that characterize them today. Animals with skeletons appeared in the fossil record at that time. Presumably the invasion of the sea-floor surface by coelomates and the origin of numerous skeletonized species accompanied a general amelioration of environmental conditions as the continents became dispersed; the skeletons themselves can be viewed as adaptations required for worms to lead various modes of life on the surface of the sea floor rather than under it. The sudden appearance of skeletons in the fossil record therefore is associated with a generalized elaboration of the bottom-dwelling members of the marine ecosystem. Later, free-swimming and underground lineages developed from the skeletonized ocean-floor dwellers, with the result that skeletons became general in all marine environments.

The correlation of major events in the history of life with major environmental changes inferred from plate-tectonic processes is certainly striking. Even though details of the interpretation are still provisional, it seems certain that further work on this relation will prove fruitful. Indeed, the ability of geologists

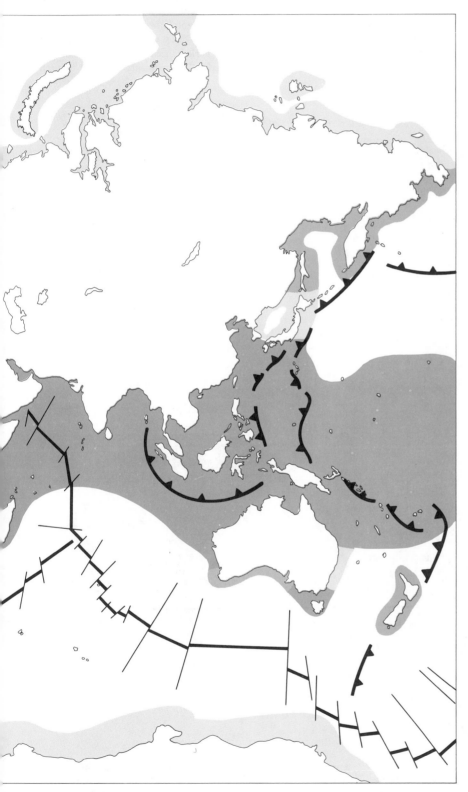

and by the undersea barriers formed by spreading ridges. The vast Indo-Pacific province (*darkest color*) spills out onto scattered islands as indicated. There are 31 provinces shown.

to determine past continental geographies should provide the basis for reconstructing the historical sequence of global environmental conditions for the first time. That sequence can then be compared with the sequence of organisms revealed in the fossil record. The following tentative account of such a comparison, on the broadest scale and without detail, will indicate the kind of history that is emerging; it is based on the examples reviewed above and on similar considerations.

Before about 700 million years ago bottom-dwelling, multicellular animals had developed that somewhat resembled flatworms. As yet no fossil evidence for their evolutionary pathways exists, but evidence from embryology and comparative anatomy suggests that they arose from swimming forms, possibly larval jellyfish, which in turn evolved from primitive single-celled animals.

Approximately 700 million years ago, perhaps in response to the onset of fluctuating environmental conditions

brought about by continental clustering, a true coelomic body cavity was evolved to act as a hydrostatic skeleton in roundworms; this adaptation allowed burrowing in soft sea floors and led to the diversification of a host of worm architectures as that mode of life was explored. Burrows of this type are still preserved in some late Precambrian rocks. As the environment later became more stable, several of the worm lineages evolved more varied modes of life. The changes in body plan necessary to adapt to such

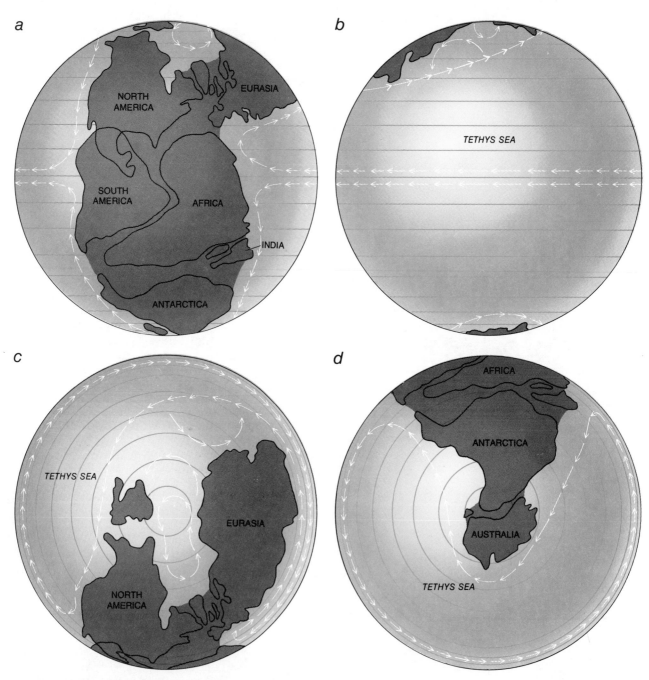

ANCIENT OCEAN CURRENTS in the vicinity of Pangaea, the single "supercontinent" that is believed to have existed near the beginning of the Triassic period some 225 million years ago, are indicated here in two equatorial views (a, b) and two polar views (c, d). Owing to a combination of geographic and environmental factors, including the predominantly warm-water currents shown, one would expect the continuous shallow-water margin that surrounded Pangaea to have been populated by comparatively few but widespread species. Such low species diversity combined with low provinciality is precisely what the fossil record indicates.

a life commonly involved the development of a skeleton. There were evidently three or four main types of worms that are represented by skeletonized descendants today. One type was highly segmented like earthworms, and presumably burrowed incessantly for detrital food; these were represented in the Cambrian period by the trilobites and related species. A second type was segmented into two or three coelomic compartments and burrowed weakly for domicile, afterward filtering suspended food from the seawater just above the ocean floor; these evolved into such forms as brachiopods and bryozoans. A third type consisted of long-bodied creepers with a series of internal organs but without true segmentation; from these the classes of mollusks (such as snails, clams and cephalopods) have descended. Probably a fourth type consisted of unsegmented burrowers that fed on surface detritus and gave rise to the modern sipunculid worms. These may also have given rise to the echinoderms (which include the sea cucumber and the spiny sea urchin), and eventually to the chordates and to man. Although the lines of descent are still uncertain among these primitive and poorly known groups, the adaptive steps are becoming clearer.

The major Cambrian radiation of the underground species into sea-floor surface habitats established the basic evolutionary lineages and occupied the major marine environments. Further evolutionary episodes tended to modify these basic animals into more elaborate structures. After the Cambrian period shallow-water marine animals became more highly specialized and richer in species, suggesting a continued trend toward resource stabilization. Suspension feeders proliferated and exploited higher parts of the water column, and predators also became more diversified. This trend seems to have reached a peak (or perhaps a plateau) in the Devonian period, some 375 million years ago. The characteristic Paleozoic fauna was finally swept away during the reduction in diversity that accompanied the great Permian-Triassic extinctions. Thus the rise of the Paleozoic fauna accompanied an amelioration in environmental conditions and increased provinciality, whereas the decline of the fauna accompanied a reestablishment of severe, unstable conditions and decreased provinciality. The subsequent breakup and dispersal of the continents has led to the present biosphere.

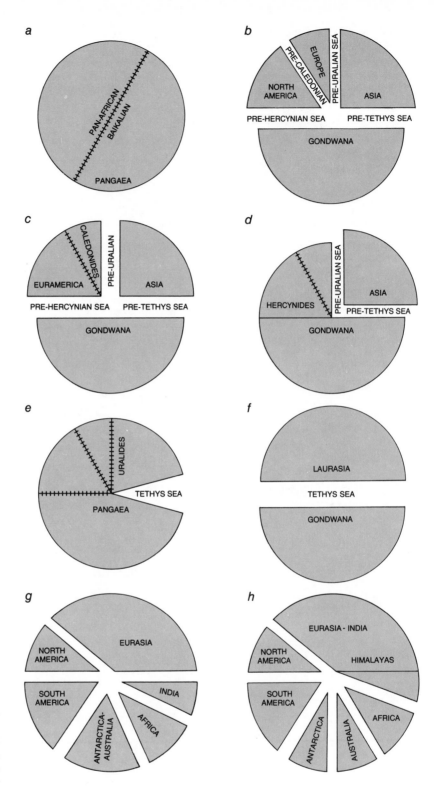

SIMPLIFIED DIAGRAMS are employed to suggest the relative configuration of the continents and the oceans during the past 700 million years. The late Precambrian supercontinent (a), which probably existed some 700 million years ago, may have been formed from previously separate continents. The Cambrian world (b) of about 570 million years ago consisted of four continents. The Devonian period (c) of about 390 million years ago was distinguished by three continents following the collapse of the pre-Caledonian Ocean and the collision of ancient Europe and North America. In the late Carboniferous period (d), about 300 million years ago, Euramerica became welded to Gondwana along the Hercynian belt. In the late Permian period (e), about 225 million years ago, Asia was welded to the remaining continents along the Uralian belt to form Pangaea. In early Mesozoic time (f), about 190 million years ago, Laurasia and Gondwana were more or less separate. In the late Cretaceous period (g), about 70 million years ago, Gondwana was highly fragmented and Laurasia partially so. The present continental pattern (h) shows India welded to Eurasia.

Today we live in a highly diverse world, probably harboring as many species as have ever lived at any time, associated in a rich variety of communities and a large number of provinces, probably the richest and largest ever to have existed at one time. We have been furnished with an enviably diverse and interesting biosphere; it would be a tragedy if we were to so perturb the environment as to return the biosphere to a low-diversity state, with the concomitant extinction of vast arrays of species. Of course, natural processes might eventually recoup the lost diversity, if we waited patiently for perhaps a few tens of millions of years. Alternatively we can work to preserve the environment in its present state and therefore to preserve the richness and variety of nature.

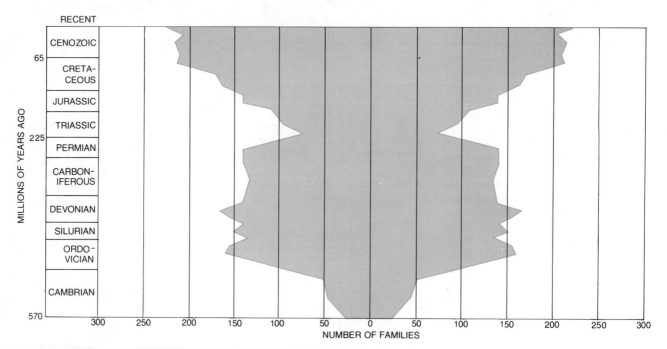

FLUCTUATIONS in the number of families, and hence in the level of diversity, of well-skeletonized invertebrates living on the world's continental shelves during the past 570 million years are plotted by geologic epoch in this graph. Time proceeds upward.

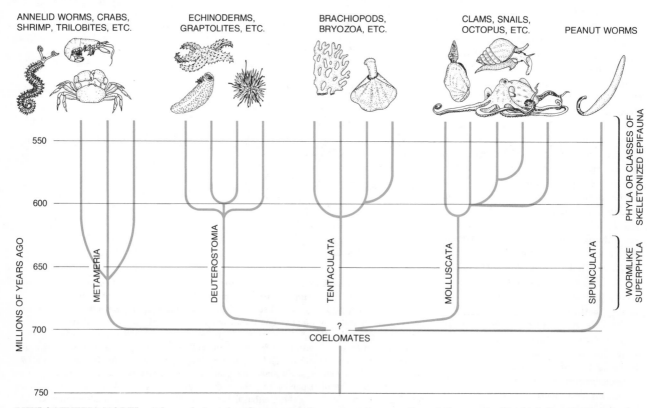

PHYLOGENETIC MODEL of the evolution of coelomate, or hollow-bodied, marine organisms is based on inferred adaptive pathways. The late Precambrian lineages were chiefly worms, which gave rise to epifaunal (bottom-dwelling), skeletonized phyla during the Cambrian period. The organisms depicted in the drawings at top are modern descendants of the major Cambrian lineages.

Continental Drift and Evolution

by Björn Kurtén
March 1969

*The breakup of ancient supercontinents would have
had major effects on the evolution of living organisms.
Does it explain the difference in the diversification of
reptiles and mammals?*

The history of life on the earth, as it is revealed in the fossil record, is characterized by intervals in which organisms of one type multiplied and diversified with extraordinary exuberance. One such interval is the age of reptiles, which lasted 200 million years and gave rise to some 20 reptilian orders, or major groups of reptiles. The age of reptiles was followed by our own age of mammals, which has lasted for 65 million years and has given rise to some 30 mammalian orders.

The difference between the number of reptilian orders and the number of mammalian ones is intriguing. How is it that the mammals diversified into half again as many orders as the reptiles in a third of the time? The answer may lie in the concept of continental drift, which has recently attracted so much attention from geologists and geophysicists [see "The Confirmation of Continental Drift," by Patrick M. Hurley; SCIENTIFIC AMERICAN Offprint 874]. It now seems that for most of the age of reptiles the continents were collected in two supercontinents, one in the Northern Hemisphere and one in the Southern. Early in the age of mammals the two supercontinents had apparently broken up into the continents of today but the present connections between some of the continents had not yet formed. Clearly such events would have had a profound effect on the evolution of living organisms.

The world of living organisms is a world of specialists. Each animal or plant has its special ecological role. Among the mammals of North America, for instance, there are grass-eating prairie animals such as the pronghorn antelope, browsing woodland animals such as the deer, flesh-eating animals specializing in large game, such as the mountain lion, or in small game, such as the fox, and so on. Each order of mammals

comprises a number of species related to one another by common descent, sharing the same broad kind of specialization and having a certain physical resemblance to one another. The order Carnivora, for example, consists of a number of related forms (weasels, bears, dogs, cats, hyenas and so on), most of which are flesh-eaters. There are a few exceptions (the aardwolf is an insect-eating hyena and the giant panda lives on bamboo shoots), but these are recognized as late specializations.

Radiation and Convergence

In spite of being highly diverse, all the orders of mammals have a common origin. They arose from a single ancestral species that lived at some unknown time in the Mesozoic era, which is roughly synonymous with the age of reptiles. The American paleontologist Henry Fairfield Osborn named the evolution of such a diversified host from a single ancestral type "adaptive radiation." By adapting to different ways of life—walking, climbing, swimming, flying, plant-eating, flesh-eating and so on—the descendant forms come to diverge more and more from one another. Adaptive radiation is not restricted to mammals; in fact we can trace the process within every major division of the plant and animal kingdoms.

The opposite phenomenon, in which stocks that were originally very different gradually come to resemble one another through adaptation to the same kind of life, is termed convergence. This too seems to be quite common among mammals. There is a tendency to duplication—indeed multiplication—of orders performing the same function. Perhaps the most remarkable instance is found among the mammals that have specialized in large-scale predation on termites

and ants in the Tropics. This ecological niche is filled in South America by the ant bear *Myrmecophaga* and its related forms, all belonging to the order Edentata. In Asia and Africa the same role is played by mammals of the order Pholidota: the pangolins, or scaly anteaters. In Africa a third order has established itself in this business: the Tubulidentata, or aardvarks. Finally, in Australia there is the spiny anteater, which is in the order Monotremata. Thus we have members of four different orders living the same kind of life.

One can cite many other examples. There are, for instance, several living and extinct orders of hoofed herbivores. There are two living orders (the Rodentia, or rodents, and the Lagomorpha, or rabbits and hares) whose chisel-like incisor teeth are specialized for gnawing. Some extinct orders specialized in the same way, and an early primate, an ice-age ungulate and a living marsupial have also intruded into the "rodent niche" [see top illustration on page 207]. This kind of duplication, or near-duplication, is an essential ingredient in the richness of the mammalian life that unfolded during the Cenozoic era, or the age of mammals. Of the 30 or so orders of land-dwelling mammals that appeared during this period almost two-thirds are still extant.

The Reptiles of the Cretaceous

The 65 million years of the Cenozoic are divided into two periods: the long Tertiary and the brief Quaternary, which includes the present [see illustration on page 204]. The 200-million-year age of reptiles embraces the three periods of the Mesozoic era (the Triassic, the Jurassic and the Cretaceous) and the final period (the Permian) of the preceding era. It is instructive to compare the number

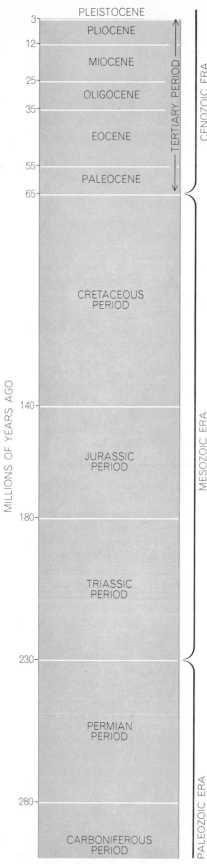

MILLIONS OF YEARS AGO

PLEISTOCENE
3
PLIOCENE
12
MIOCENE
25
OLIGOCENE
35
EOCENE
55
PALEOCENE
65

CRETACEOUS
PERIOD

140

JURASSIC
PERIOD

180

TRIASSIC
PERIOD

230

PERMIAN
PERIOD

280

CARBONIFEROUS
PERIOD

TERTIARY PERIOD — CENOZOIC ERA

MESOZOIC ERA

PALEOZOIC ERA

SIX PERIODS of earth history were occupied by the age of reptiles and the age of mammals. The reptiles' rise began 280 million years ago, in the final period of the Paleozoic era. Mammals replaced reptiles as dominant land animals 65 million years ago.

of reptilian orders that flourished during some Mesozoic interval about as long as the Cenozoic era with the number of mammalian orders in the Cenozoic. The Cretaceous period is a good candidate. Some 75 million years in duration, it is only slightly longer than the age of mammals. Moreover, the Cretaceous was the culmination of reptilian life and its fossil record on most continents is good. In the Cretaceous the following orders of land reptiles were extant:

Order Crocodilia: crocodiles, alligators and the like. Their ecological role was amphibious predation; their size, medium to large.

Order Saurischia: saurischian dinosaurs. These were of two basic types: bipedal upland predators (Theropoda) and very large amphibious herbivores (Sauropoda).

Order Ornithischia: ornithischian dinosaurs. Here there were three basic types: bipedal herbivores (Ornithopoda), heavily armored quadrupedal herbivores (Stegosauria and Ankylosauria) and horned herbivores (Ceratopsia).

Order Pterosauria: flying reptiles.

Order Chelonia: turtles and tortoises.

Order Squamata: The two basic types were lizards (Lacertilia) and snakes (Serpentes). Both had the same principal ecological role: small to medium-sized predator.

Order Choristodera (or suborder in the order Eosuchia): champsosaurs. These were amphibious predators.

One or two other reptilian orders may be represented by rare forms. Even if we include them, we get only eight or nine orders of land reptiles in Cretaceous times. One could maintain that an order of reptiles ranks somewhat higher than an order of mammals; some reptilian orders include two or even three basic adaptive types. Even if these types are kept separate, however, the total rises only to 12 or 13. Furthermore, there seems to be only one clear-cut case of ecological duplication: both the crocodilians and the champsosaurs are sizable amphibious predators. (The turtles cannot be considered duplicates of the armored dinosaurs. For one thing, they were very much smaller.) A total of somewhere between seven and 13 orders over a period of 75 million years seems a sluggish record compared with the mammalian achievement of perhaps 30 orders in 65 million years. What light can paleogeography shed on this matter?

The Mesozoic Continents

The two supercontinents of the age of reptiles have been named Laurasia (after

Laurentian and Eurasia) and Gondwanaland (after a characteristic geological formation, the Gondwana). Between them lay the Tethys Sea (named for the wife of Oceanus in Greek myth, who was mother of the seas). Laurasia, the northern supercontinent, consisted of what would later be North America, Greenland and Eurasia north of the Alps and the Himalayas. Gondwanaland, the southern one, consisted of the future South America, Africa, India, Australia and Antarctica. The supercontinents may have begun to split up as early as the Triassic period, but the rifts between them did not become effective barriers to the movement of land animals until well into the Cretaceous, when the age of reptiles was nearing its end.

When the mammals began to diversify in the late Cretaceous and early Tertiary, the separation of the continents appears to have been at an extreme. The old ties were sundered and no new ones had formed. The land areas were further fragmented by a high sea level; the waters flooded the continental margins and formed great inland seas, some of which completely partitioned the continents. For example, South America was cut in two by water in the region that later became the Amazon basin, and Eurasia was split by the joining of the Tethys Sea and the Arctic Ocean. In these circumstances each chip of former supercontinent became the nucleus for an adaptive radiation of its own, each fostering a local version of a balanced fauna. There were at least eight such nuclei at the beginning of the age of mammals. Obviously such a situation is quite different from the one in the age of reptiles, when there were only two separate land masses.

Where the Reptiles Originated

The fossil record contains certain clues to some of the reptilian orders' probable areas of origin. The immense distance in time and the utterly different geography, however, make definite inferences hazardous. Let us see what can be said about the orders of Cretaceous reptiles (most of which, of course, arose long before the Cretaceous):

Crocodilia. The earliest fossil crocodilians appear in Middle Triassic formations in a Gondwanaland continent (South America). The first crocodilians in Laurasia are found in Upper Triassic formations. Thus a Gondwanaland origin is suggested.

Saurischia. The first of these dinosaurs appear on both supercontinents in the Middle Triassic, but they are more

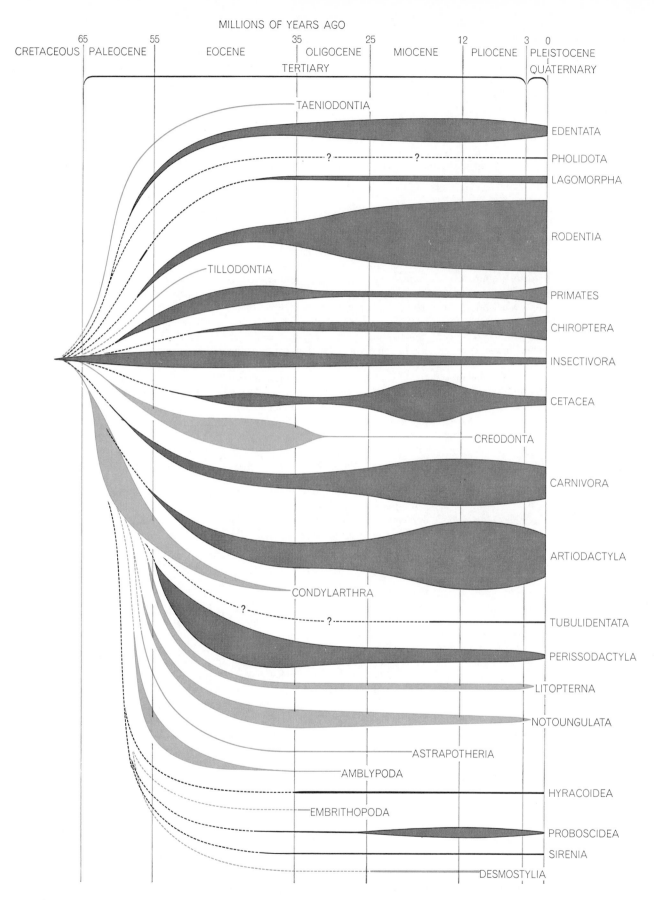

MILLIONS OF YEARS AGO

65   55   35   25   12   3   0
CRETACEOUS | PALEOCENE   EOCENE   OLIGOCENE   MIOCENE   PLIOCENE | PLEISTOCENE
                                                                      QUATERNARY
                              TERTIARY

TAENIODONTIA

EDENTATA

PHOLIDOTA

LAGOMORPHA

RODENTIA

TILLODONTIA

PRIMATES

CHIROPTERA

INSECTIVORA

CETACEA

CREODONTA

CARNIVORA

ARTIODACTYLA

CONDYLARTHRA

TUBULIDENTATA

PERISSODACTYLA

LITOPTERNA

NOTOUNGULATA

ASTRAPOTHERIA

AMBLYPODA

HYRACOIDEA

EMBRITHOPODA

PROBOSCIDEA

SIRENIA

DESMOSTYLIA

ADAPTIVE RADIATION of the mammals has been traced from its starting point late in the Mesozoic era by Alfred S. Romer of Harvard University. Records for 25 extinct and extant orders of placental mammals are shown here. The lines increase and decrease in width in proportion to the abundance of each order. Extinct orders are shown in color; broken lines mean that no fossil record exists during the indicated interval and question marks imply doubt about the suggested ancestral relation between some orders.

varied in the south. A Gondwanaland origin is very tentatively suggested.

Ornithischia. These dinosaurs appear in the Upper Triassic of South Africa (Gondwanaland) and invade Laurasia somewhat later. A Gondwanaland origin is indicated.

Pterosauria. The oldest fossils of flying reptiles come from the early Jurassic of Europe. They represent highly specialized forms, however, and their antecedents are unknown. No conclusion seems possible.

Chelonia. Turtles are found in Triassic formations in Laurasia. None are found in Gondwanaland before Cretaceous times. This suggests a Laurasian origin. On the other hand, a possible forerunner of turtles appears in the Permian of South Africa. If the Permian form was in fact ancestral, a Gondwanaland origin would be indicated. In any case, the order's main center of evolution certainly lay in the northern supercontinent.

Squamata. Early lizards are found in the late Triassic of the north, which may suggest a Laurasian origin. Unfortunately the lizards in question are aberrant gliding animals. They must have had a long history, of which we know nothing at present.

Choristodera. The crocodile-like champsosaurs are found only in North America and Europe, and so presumably originated in Laurasia.

The indications are, then, that three orders of reptiles—the crocodilians and the two orders of dinosaurs—may have originated in Gondwanaland. Three others—the turtles, the lizards and snakes and the champsosaurs—may have originated in Laurasia. The total number of basic adaptive types in the Gondwanaland group is six; the Laurasia group has four. The Gondwanaland radiation may well have been slightly richer than the Laurasian because it seems that the southern supercontinent was somewhat larger and had a slightly more varied climate. Laurasian climates seem to have been tropical to temperate. Southern parts of Gondwanaland were heavily glaciated late in the era preceding the Mesozoic, and its northern shores (facing the Tethys Sea) had a fully tropical climate.

Although some groups of reptiles, such as the champsosaurs, were confined to one or another of the supercontinents, most of the reptilian orders sooner or later spread into both of them. This means that there must have been ways for land animals to cross the Tethys Sea. The Tethys was narrow in the west and wide to the east. Presumably whatever land connection there was—a true land bridge or island stepping-stones—was located in the western part of the sea. In any case, migration along such routes meant that there was little local differentiation among the reptiles of the Mesozoic era. It was over an essentially uniform reptilian world that the sun finally set at the end of the age of reptiles.

Early Mammals of Laurasia

The conditions of mammalian evolution were radically different. In early and middle Cretaceous times the connections between continents were evidently close enough for primitive mammals to spread into all corners of the habitable world. As the continents drifted farther apart, however, populations of these primitive forms were gradually isolated from one another. This was particularly the case, as we shall see, with the mammals that inhabited the daughter continents of Gondwanaland. Among the Laurasian continents North America was drifting away from Europe, but at the beginning of the age of mammals the distance was not great and there is good evidence that some land connection remained well into the early Tertiary. North American and European mammals were practically identical as late as early Eocene times. Furthermore, throughout the Cenozoic era there was a connection between Alaska and Siberia, at least intermittently, across the Bering Strait. On the other hand, the inland sea extending from the Tethys to the Arctic Ocean formed a complete barrier to direct migration between Europe and Asia in the early Tertiary. Migrations could take place only by way of North America.

In this way the three daughter continents of ancient Laurasia formed three semi-isolated nuclear areas. Many orders of mammals arose in these Laurasian nuclei, among them seven orders that are now extinct but that covered a wide spectrum of specialized types, including primitive hoofed herbivores, carnivores, insectivores and gnawers. The orders of mammals that seem to have arisen in the northern daughter continents and that are extant today are:

Insectivora: moles, hedgehogs, shrews and the like. The earliest fossil insectivores are found in the late Cretaceous of North America and Asia.

Chiroptera: bats. The earliest-known bat comes from the early Eocene of North America. At a slightly later date bats were also common in Europe.

Primates: prosimians (for example, tarsiers and lemurs), monkeys, apes, man. Early primates have recently been found in the late Cretaceous of North America. In the early Tertiary they are common in Europe as well.

Carnivora: cats, dogs, bears, weasels and the like. The first true carnivores appear in the Paleocene of North America.

Perissodactyla: horses, tapirs and other odd-toed ungulates. The earliest forms appear at the beginning of the Eocene in the Northern Hemisphere.

Artiodactyla:· cattle, deer, pigs and other even-toed ungulates. Like the odd-toed ungulates, they appear in the early Eocene of the Northern Hemisphere.

Rodentia: rats, mice, squirrels, beavers and the like. The first rodents appear in the Paleocene of North America.

Lagomorpha: hares and rabbits. This order makes its first appearance in the Eocene of the Northern Hemisphere.

Pholidota: pangolins. The earliest come from Europe in the middle Tertiary.

The fact that a given order of mammals is found in older fossil deposits in North America than in Europe or Asia does not necessarily mean that the order arose in the New World. It may simply reflect the fact that we know much more about the early mammals of North America than we do about those of Eurasia. All we can really say is that a total of 16 extant or extinct orders of mammals probably arose in the Northern Hemisphere.

Early Mammals of South America

The fragmentation of Gondwanaland seems to have started earlier than that of Laurasia. The rifting certainly had a much more radical effect. Looking at South America first, we note that at the beginning of the Tertiary this continent was tenuously connected to North America but that for the rest of the period it was completely isolated. The evidence for the tenuous early linkage is the presence in the early Tertiary beds of North America of mammalian fossils representing two predominantly South American orders: the Edentata (the order that includes today's ant bears, sloths and armadillos) and the Notoungulata (an order of extinct hoofed herbivores).

Four other orders of mammals are exclusively South American: the Paucituberculata (opossum rats and other small South American marsupials), the Pyrotheria (extinct elephant-like animals), the Litopterna (extinct hoofed herbivores, including some forms resembling

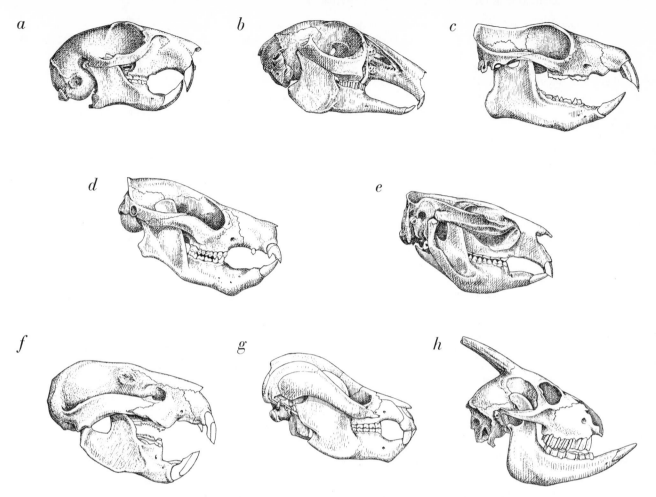

CHISEL-LIKE INCISORS, specialized for gnawing, appear in animals belonging to several extinct and extant orders in addition to the rodents, represented by a squirrel (*a*), and the lagomorphs, represented by a hare (*b*), which are today's main specialists in this ecological role. Representatives of other orders with chisel-like incisor teeth are an early tillodont, *Trogosus* (*c*), an early primate, *Plesiadapis* (*d*), a living marsupial, the wombat (*e*), one of the extinct multituberculate mammals, *Taeniolabis* (*f*), a mammal-like reptile of the Triassic, *Bienotherium* (*g*), and a Pleistocene cave goat, *Myotragus* (*h*), whose incisor teeth are in the lower jaw only.

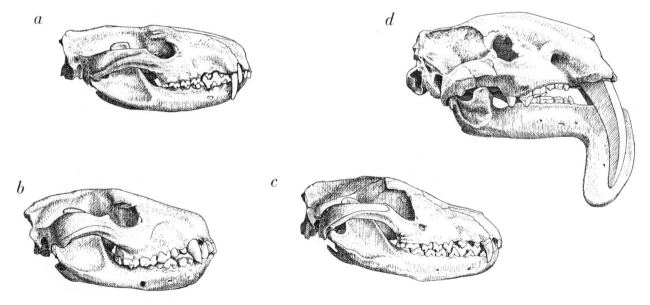

CARNIVOROUS MARSUPIALS, living and extinct, fill an ecological niche more commonly occupied by the placental carnivores today. Illustrated are the skulls of two living forms, the Australian "cat," *Dasyurus* (*a*), and the Tasmanian devil, *Sarcophilus* (*b*). The Tasmanian "wolf," *Thylacinus* (*c*), has not been seen for many years and may be extinct. A tiger-sized predator of South America, *Thylacosmilus* (*d*) became extinct in Pliocene times, long before the placental sabertooth of the Pleistocene, *Smilodon*, appeared.

horses and camels) and the Astrapotheria (extinct large hoofed herbivores of very peculiar appearance). Thus a total of six orders, extinct or extant, probably originated in South America. Still another order, perhaps of even more ancient origin, is the Marsupicarnivora. The order is so widely distributed, with species found in South America, North America, Europe and Australia, that its place of origin is quite uncertain. It includes, in addition to the extinct marsupial carnivores of South America, the opossums of the New World and the native "cats" and "wolves" of the Australian area.

The most important barrier isolating South America from North America in the Tertiary period was the Bolívar Trench. This arm of the sea cut across the extreme northwest corner of the continent. In the late Tertiary the bottom of the Bolívar Trench was lifted above sea level and became a mountainous land area. A similar arm of sea, to which I have already referred, extended across the continent in the region that is now the Amazon basin. This further enhanced the isolation of the southern part of South America.

Africa's role as a center of adaptive radiation is problematical because practically nothing is known of its native mammals before the end of the Eocene. We do know, however, that much of the continent was flooded by marginal seas, and that in the early Tertiary, Africa was cut up into two or three large islands. Still, there must have been a land route to Eurasia even in the Eocene; some of the African mammals of the following epoch (the Oligocene) are clearly immigrants from the north or northeast. Nonetheless, the majority of African mammals are of local origin. They include the following orders:

Proboscidea: the mastodons and elephants.

Hyracoidea: the conies and their extinct relatives.

Embrithopoda: an extinct order of very large mammals.

Tubulidentata: the aardvarks.

In addition the order Sirenia, consisting of the aquatic dugongs and manatees, is evidently related to the Proboscidea and hence presumably also originated in Africa. The same may be true of another order of aquatic mammals,

the extinct Desmostylia, which also seems to be related to the elephants. The one snag in this interpretation is that desmostylian fossils are found only in the North Pacific, which seems rather a long way from Africa. Nonetheless, once they were waterborne, early desmostylians might have crossed the Atlantic, which was then only a narrow sea, navigated the Bolívar Trench and, rather like Cortes (but stouter), found themselves in the Pacific.

Early Mammals of Africa

Thus there are certainly four, and possibly six, mammalian orders for which an African origin can be postulated. Here it should be noted that Africa had an impressive array of primates in the Oligocene. This suggests that the order Primates had a comparatively long history in Africa before that time. Even though the order as such does not have its roots in Africa, it is possible that the higher primates—the Old World monkeys, the apes and the ancestors of man—may have originated there. Most of the fossil primates found in the Oligocene

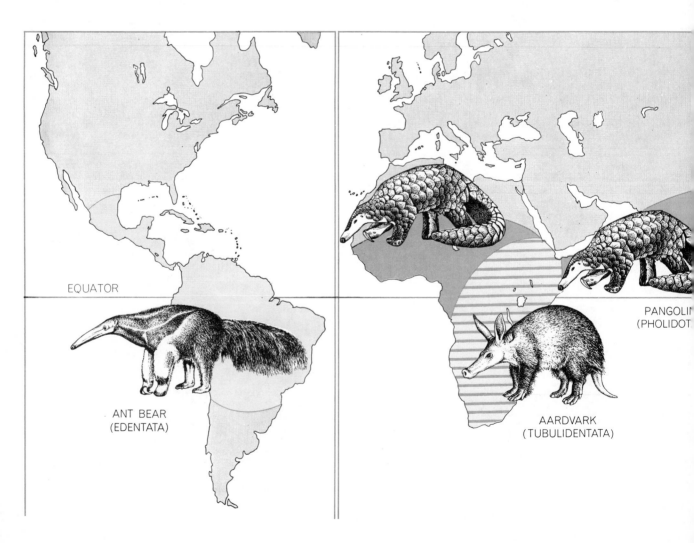

EQUATOR

ANT BEAR
(EDENTATA)

AARDVARK
(TUBULIDENTATA)

PANGOLIN
(PHOLIDOTA)

formations of Africa are primitive apes or monkeys, but there is at least one form (*Propliopithecus*) whose dentition looks like a miniature blueprint of a set of human teeth.

The Rest of Gondwanaland

We know little or nothing of the zoogeographic roles played by India and Antarctica in the early Tertiary. Mammalian fossils from the early Tertiary are also absent in Australia. It may be assumed, however, that the orders of mammals now limited to Australia probably originated there. These include two orders of marsupials: the Peramelina, comprised of several bandicoot genera, and the Diprotodonta, in which are found the kangaroos, wombats, phalangers and a number of extinct forms. In addition the order Monotremata, a very primitive group of mammals that includes the spiny anteater and the platypus, is likely to be of Australian origin. This gives us a total of three orders probably founded in Australia.

Summing up, we find that the three Laurasian continents produced a total

of 16 orders of mammals, an average of five or six orders per continent. As for Gondwanaland, South America produced six orders, Africa four to six and Australia three. The fact that Australia is a small continent probably accounts for the lower number of orders founded there. Otherwise the distribution—the average of five or six orders per subdivision—is remarkably uniform for both the Laurasian and Gondwanaland supercontinents. The mammalian record should be compared with the data on Cretaceous reptiles, which show that the two supercontinents produced a total of 12 or 13 orders (or adaptively distinct suborders). A regularity is suggested, as if a single nucleus of radiation would tend in a given time to produce and support a given amount of basic zoological variation.

As the Tertiary period continued new land connections were gradually formed, replacing those sundered when the old supercontinents broke up. Africa made its landfall with Eurasia in the Oligocene and Miocene epochs. Laurasian orders of mammals spread into Africa and crowded out some of the local forms, but at the same time some African mammals (notably the mastodons and elephants) went forth to conquer almost the entire world. In the Western Hemisphere the draining and uplifting of the Bolívar Trench was followed by intense intermigration and competition among the mammals of the two Americas. In the process much of the typical South American mammal population was exterminated, but a few forms pressed successfully into North America to become part of the continent's spectacular ice-age wildlife.

India, a fragment of Gondwanaland that finally became part of Asia, must have made a contribution to the land fauna of that continent but just what it was cannot be said at present. Of all the drifting Noah's arks of mammalian evolution only two—Antarctica and Australia—persist in isolation to this day. The unknown mammals of Antarctica have long been extinct, killed by the ice that engulfed their world. Australia is therefore the only island continent that still retains much of its pristine mammalian

fauna. [*see illustration on pages 210–211*].

If the fragmentation of the continents at the beginning of the age of mammals promoted variety, the amalgamation in the latter half of the age of mammals has promoted efficiency by means of a large-scale test of the survival of the fittest. There is a concomitant loss of variety; 13 orders of land mammals have become extinct in the course of the Cenozoic. Most of the extinct orders are island-continent productions, which suggests that a system of semi-isolated provinces, such as the daughter continents of Laurasia, tends to produce a more efficient brood than the completely isolated nuclei of the Southern Hemisphere. Not all the Gondwanaland orders were inferior, however; the edentates were moderately successful and the proboscidians spectacularly so.

As far as land mammals are concerned, the world's major zoogeographic provinces are at present four in number: the Holarctic-Indian, which consists of North America and Eurasia and also northern Africa; the Neotropical, made up of Central America and South America; the Ethiopian, consisting of Africa south of the Sahara, and the Australian. This represents a reduction from seven provinces with about 30 orders of mammals to four provinces with about 18 orders. The reduction in variety is proportional to the reduction in the number of provinces.

In conclusion it is interesting to note that we ourselves, as a subgroup within the order Primates, probably owe our origin to a radiation within one of Gondwanaland's island continents. I have noted that an Oligocene primate of Africa may have been close to the line of human evolution. By Miocene times there were definite hominids in Africa, identified by various authorities as members of the genus *Ramapithecus* or the genus *Kenyapithecus*. Apparently these early hominids spread into Asia and Europe toward the end of the Miocene. The cycle of continental fragmentation and amalgamation thus seems to have played an important part in the origin of man as well as of the other land mammals.

SPINY ANTEATER
(MONOTREMATA)

FOUR ANT-EATING MAMMALS have become adapted to the same kind of life although each is a member of a different mammalian order. Their similar appearance provides an example of an evolutionary process known as convergence. The ant bears of the New World Tropics are in the order Edentata. The aardvark of Africa is the only species in the order Tubulidentata. Pangolins, found both in Asia and in Africa, are members of the order Pholidota. The spiny anteater of Australia, a very primitive mammal, is in the order Monotremata.

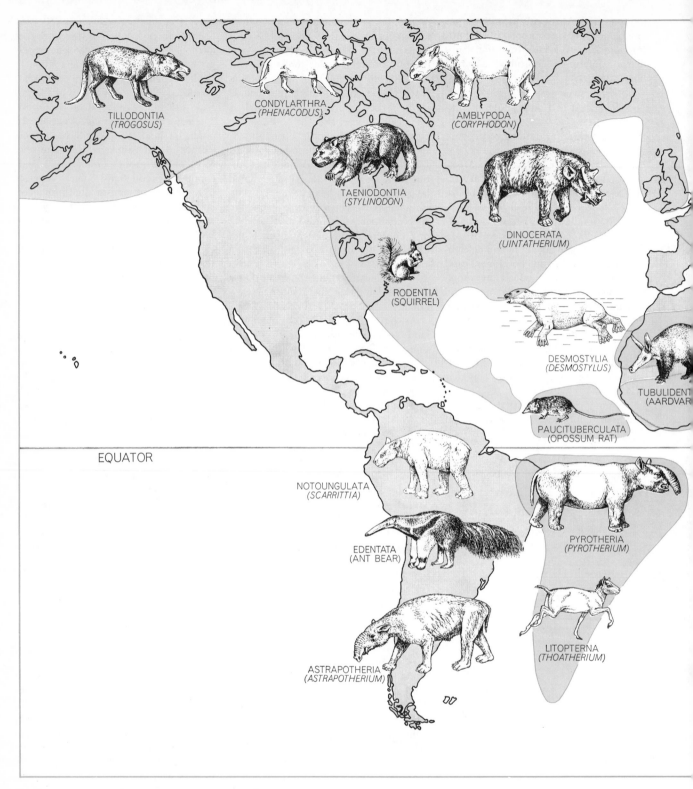

TILLODONTIA
(TROGOSUS)

CONDYLARTHRA
(PHENACODUS)

AMBLYPODA
(CORYPHODON)

TAENIODONTIA
(STYLINODON)

DINOCERATA
(UINTATHERIUM)

RODENTIA
(SQUIRREL)

DESMOSTYLIA
(DESMOSTYLUS)

TUBULIDENT
(AARDVAR

PAUCITUBERCULATA
(OPOSSUM RAT)

EQUATOR

NOTOUNGULATA
(SCARRITTIA)

PYROTHERIA
(PYROTHERIUM)

EDENTATA
(ANT BEAR)

LITOPTERNA
(THOATHERIUM)

ASTRAPOTHERIA
(ASTRAPOTHERIUM)

CONTINENTAL DRIFT affected the evolution of the mammals by fragmenting the two supercontinents early in the Cenozoic era. In the north, Europe and Asia, although separated by a sea, remained connected with North America during part of the era. The

free migration that resulted prevents certainty regarding the place of origin of many orders of mammals that evolved in the north. The far wider rifting of Gondwanaland allowed the evolution of unique groups of mammals in South America, Africa and Australia.

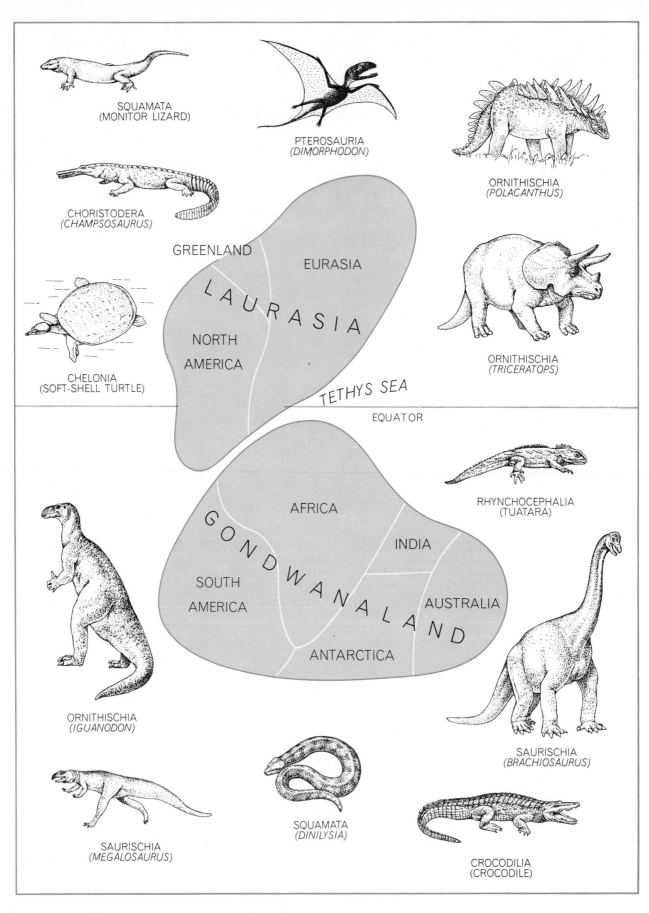

SQUAMATA
(MONITOR LIZARD)

PTEROSAURIA
(*DIMORPHODON*)

ORNITHISCHIA
(*POLACANTHUS*)

CHORISTODERA
(*CHAMPSOSAURUS*)

CHELONIA
(SOFT-SHELL TURTLE)

GREENLAND

EURASIA

LAURASIA

NORTH
AMERICA

ORNITHISCHIA
(*TRICERATOPS*)

TETHYS SEA

EQUATOR

AFRICA

RHYNCHOCEPHALIA
(TUATARA)

GONDWANALAND

INDIA

SOUTH
AMERICA

AUSTRALIA

ANTARCTICA

ORNITHISCHIA
(*IGUANODON*)

SAURISCHIA
(*BRACHIOSAURUS*)

SAURISCHIA
(*MEGALOSAURUS*)

SQUAMATA
(*DINILYSIA*)

CROCODILIA
(CROCODILE)

TWO SUPERCONTINENTS of the Mesozoic era were Laurasia in the north and Gondwanaland in the south. The 12 major types of reptiles, represented by typical species, are those whose fossil remains are found in Cretaceous formations. Most of the orders inhabited both supercontinents; migrations were probably by way of a land bridge in the west, where the Tethys Sea was narrowest.

BIBLIOGRAPHIES

I THE PHENOMENON OF EVOLUTION

1. Charles Darwin

THE FOUNDATIONS OF THE ORIGIN OF SPECIES. Charles Darwin. Cambridge University Press, 1909.

THE LIFE AND LETTERS OF CHARLES DARWIN. Edited by Francis Darwin. D. Appleton and Company, 1888.

2. The Genetic Basis of Evolution

GENETICS AND THE ORIGIN OF SPECIES. Theodosius Dobzhansky. Columbia University Press, 1937.

3. Darwin's Missing Evidence

THE CONTRIBUTION OF INDUSTRIAL MELANISM IN THE LEPIDOPTERA TO OUR KNOWLEDGE OF EVOLUTION. H. B. D. Kettlewell in *The Advancement of Science*, Vol. 13, No. 52, pages 245–252; March, 1957.

FURTHER SELECTION EXPERIMENTS ON INDUSTRIAL MELANISM IN THE LEPIDOPTERA. H. B. D. Kettlewell in *Heredity*, Vol. 10, Part 3, pages 287–301; December, 1956.

A RÉSUMÉ OF INVESTIGATIONS ON THE EVOLUTION OF MELANISM IN THE LEPIDOPTERA. H. B. D. Kettlewell in *Proceedings of the Royal Society of London*, Series B, Vol. 145, No. 920, pages 297–303; July 24, 1956.

SELECTION EXPERIMENTS ON INDUSTRIAL MELANISM IN THE LEPIDOPTERA. H. B. D. Kettlewell in *Heredity*, Vol. 9, Part 3, pages 323–342; December, 1955.

A SURVEY OF THE FREQUENCIES OF *Biston Betularia* (L.) (LEP.) AND ITS MELANIC FORMS IN GREAT BRITAIN. H. B. D. Kettlewell in *Heredity*, Vol. 12, Part 1, pages 51–72; February, 1958.

4. Darwin's Finches

DARWIN'S FINCHES. David Lack. Cambridge University Press, 1947.

SYSTEMATICS AND THE ORIGIN OF SPECIES. Ernst Mayr. Columbia University Press, 1942.

II EARLIEST TRACES OF LIFE

5. The Oldest Fossils

MICROORGANISMS FROM THE GUNFLINT CHERT. Elso S. Barghoorn and Stanley A. Tyler in *Science*, Vol. 147, No. 3658, pages 563–577; February 5, 1965.

MICROORGANISMS THREE BILLION YEARS OLD FROM THE PRECAMBRIAN OF SOUTH AFRICA. Elso S. Barghoorn and J. William Schopf in *Science*, Vol. 152, No. 3723, pages 758–763; May 6, 1966.

PRECAMBRIAN MICRO-ORGANISMS AND EVOLUTIONARY EVENTS PRIOR TO THE ORIGIN OF VASCULAR PLANTS.

J. William Schopf in *Biological Reviews*, Vol. 45, No. 3, pages 319–352; August, 1970.

CHEMICAL EVOLUTION AND THE ORIGIN OF LIFE: A COMPREHENSIVE BIBLIOGRAPHY. Compiled by Martha W. West and Cyril Ponnamperuma in *Space Life Sciences*, Vol. 2, No. 2, pages 225–295; September, 1970.

6. Pre-Cambrian Animals

The Geology and Late Precambrian Fauna of the Ediacara Fossil Reserve. M. F. Glaessner and B. Daily in *Records of the South Australian Museum,* Vol. XIII, No. 3, pages 369–401; July 2, 1959.

The Oldest Fossil Faunas of South Australia. M. F. Glaessner in *Sonderdruck aus der Geologischen Rundschau,* Vol. 47, No. 2, pages 522–531; 1958.

Search for the Past. J. R. Beerbower. Prentice-Hall, 1960.

Time, Life and Man: The Fossil Record. R. A. Stirton. John Wiley & Sons, 1959.

III   HOW FOSSILS OCCUR AND WHAT THEY TELL US

7. Corals as Paleontological Clocks

Changes in the Earth's Moment of Inertia. S. K. Runcorn in *Nature,* Vol. 204, No. 4961, pages 823–825; November 28, 1964.

Coral Growth and Geochronometry. John W. Wells in *Nature,* Vol. 197, No. 4871, pages 948–950; March 9, 1963.

Coral Growth-Rate, an Environmental Indicator. E. A. Shinn in *Journal of Paleontology,* Vol. 40, No. 2, pages 233–240; March, 1966.

Periodicity in Devonian Coral Growth. Colin T. Scrutton in *Palaeontology,* Vol. 7, Part 4, pages 552–558; January, 1965.

8. The Petrified Forests of Yellowstone Park

Cenozoic Stratigraphy and Structural Geology, Northeast Yellowstone National Park, Wyoming and Montana. Charles W. Brown in *The Geological Society of America Bulletin,* Vol. 72, No. 8, pages 1173–1193; August, 1961.

Fossil Flora of the Yellowstone National Park. Frank Hall Knowlton in *United States Geological Survey Monographs,* Vol. 32, Part 2, pages 651–791. Government Printing Office, 1899.

Tertiary Fossil Forests of Yellowstone National Park, Wyoming. Erling Dorf in *Billings Geological Society Guidebook of the Eleventh Annual Field Conference,* 1960.

9. Insects in Amber

Ancient Insects; Fossils in Amber and Other Deposits. Charles T. Brues in *The Scientific Monthly,* Vol. 17, No. 4, pages 289–304; October, 1923.

Progressive Change in the Insect Population of Forests Since the Early Tertiary. Charles T. Brues in *The American Naturalist,* Vol. 67, No. 712, pages 385–406; September–October, 1933.

10. Fossil Behavior

Biogenic Sedimentary Structures. Adolf Seilacher in *Approaches to Paleoecology,* edited by John Imbrie and Norman Newell. John Wiley and Sons, 1964.

Paleontological Studies on Turbidite Sedimentation and Erosion. Adolf Seilacher in *The Journal of Geology,* Vol. 70, No. 2, pages 227–234; March, 1962.

Trace Fossils and Problematica. Walter Häntzschel in *Treatise on Invertebrate Paleontology: Part W,* edited by Raymond C. Moore. Geological Society of America and University of Kansas Press, 1962.

Vorzeitliche Lebensspuren. Othenio Abel. Verlag von Gustav Fischer, 1935.

11. Micropaleontology

Atlantic Deep-Sea Sediment Cores. David B. Ericson, Maurice Ewing, Goesta Wollin and Bruce C. Heezen in *Bulletin of the Geological Society of America,* Vol. 72, No. 2, pages 193–286; February, 1961.

Catalogue of Foraminifera. Edited by Brooks Fleming Ellis and Angelina R. Messina. American Museum of Natural History, 1940––.

Ecology and Distribution of Recent Foraminifera. Fred B. Phleger. Johns Hopkins Press, 1960.

Introduction to Microfossils. Daniel J. Jones. Harper & Row, 1956.

Principles of Micropalaeontology. Martin F. Glaessner. John Wiley & Sons, 1947.

IV REEFS, DINOSAURS, MAMMALS, AND HUMANS

12. The Evolution of Reefs

REVOLUTIONS IN THE HISTORY OF LIFE. Norman D. Newell in *Uniformity and Simplicity: A Symposium on the Principle of the Uniformity of Nature.* The Geological Society of America, Special Paper 89, edited by Claude C. Albritton, Jr., 1967.

AN OUTLINE HISTORY OF TROPICAL ORGANIC REEFS. Norman D. Newell in *American Museum Novitates,* No. 2465; September 21, 1971.

REEF ORGANISMS THROUGH TIME. Proceedings of the North American Paleontological Convention, edited by Ellis Yochelson. Allen Press, 1971.

13. Dinosaur Renaissance

ECOLOGY OF THE BRONTOSAURS. Robert T. Bakker in *Nature,* Vol. 229, No. 5281, pages 172–174; January 15, 1971.

DINOSAUR MONOPHYLY AND A NEW CLASS OF VERTEBRATES. Robert T. Bakker and Peter M. Galton in *Nature,* Vol. 248, No. 5444, pages 168–172; March 8, 1974.

EXPERIMENTAL AND FOSSIL EVIDENCE OF THE EVOLUTION OF TETRAPOD BIOENERGETICS. Robert T. Bakker in *Perspectives in Biophysical Ecology,* edited by David Gates and Rudolf Schmerl. Springer-Verlag, 1975.

14. The Ancestors of Mammals

THE MAMMAL-LIKE REPTILE *Lycaenops.* Edwin H. Colbert in the *Bulletin of the American Museum of Natural History,* Vol. 89, pages 353–404; 1948.

THE MAMMAL-LIKE REPTILES OF SOUTH AFRICA AND THE ORIGIN OF MAMMALS. Robert Broom. H. F. and G. Witherby, 1932.

15. The Cave Bear

DIE DRACHENHÖHLE BEI MIXNITZ. Edited by O. Abel and G. Kyrle. Speläologische Monographien, Vienna, 1931.

A REVIEW OF FOSSIL AND RECENT BEARS OF THE OLD WORLD. D. P. Erdbrink. Deventer, 1953.

PLEISTOCENE MAMMALS OF EUROPE. Björn Kurtén. Aldine Publishing Company, 1968.

CAVE BEARS. Björn Kurtén in *Studies in Speleology,* Vol. 2, Part 1, pages 13–24; July, 1969.

16. *Homo erectus*

MANKIND IN THE MAKING, rev. ed. William W. Howells. Doubleday & Company, 1967.

THE NOMENCLATURE OF THE HOMINIDAE. Bernard G. Campbell. Occasional Paper No. 22, Royal Anthropological Institute of Great Britain and Ireland, 1965.

THE TAXONOMIC EVOLUTION OF FOSSIL HOMINIDS. Ernst Mayr in *Classification and Human Evolution,* edited by Sherwood L. Washburn. Viking Fund Publications in Anthropology, No. 37, 1963.

17. The Distribution of Man

HUMAN ANCESTRY FROM A GENETICAL POINT OF VIEW. Reginald Ruggles Gates. Harvard University Press, 1948.

MANKIND IN THE MAKING. William White Howells. Doubleday & Company, 1959.

RACES: A STUDY OF THE PROBLEMS OF RACE FORMATION IN MAN. Carleton S. Coon, Stanley M. Garn and Joseph B. Birdsell. Charles C. Thomas, 1950.

THE STORY OF MAN. Carleton Stevens Coon. Alfred A. Knopf, 1954.

V SOME MAJOR PATTERNS IN THE HISTORY OF LIFE

18. Crises in the History of Life

BIOTIC ASSOCIATIONS AND EXTINCTION. David Nicol in *Systematic Zoology,* Vol. 10, No. 1, pages 35–41; March, 1961.

EVOLUTION OF LATE PALEOZOIC INVERTEBRATES IN RESPONSE TO MAJOR OSCILLATIONS OF SHALLOW SEAS. Raymond C. Moore in *Bulletin of the Museum of Comparative Zoology at Harvard College,* Vol. 112, N 3, pages 259–286; October, 1954.

PALEONTOLOGICAL GAPS AND GEOCHRONOLOGY. Norman D. Newell in *Journal of Paleontology,* Vol. 36, No. 3, pages 592–610; May, 1962.

TETRAPOD EXTINCTIONS AT THE END OF THE TRIASSIC PERIOD. Edwin H. Colbert in *Proceedings of the National Academy of Sciences of the U.S.A.,* Vol. 44, No. 9, pages 973–977; September, 1958.

19. Plate Tectonics and the History of Life in the Oceans

DYNAMICS IN METAZOAN EVOLUTION. R. B. Clark. Oxford University Press, 1964.

GLOBAL TECTONICS AND THE FOSSIL RECORD. James W. Valentine and Eldridge M. Moores in *The Journal of Geology*, Vol. 80, No. 2, pages 167–184; March, 1972.

EVOLUTIONARY PALEOECOLOGY OF THE MARINE BIOSPHERE. J. W. Valentine. Prentice-Hall, 1973.

A REVOLUTION IN THE EARTH SCIENCES: FROM CONTINENTAL DRIFT TO PLATE TECTONICS. A. A. Hallam. Oxford University Press, 1973.

20. Continental Drift and Evolution

VERTEBRATE PALEONTOLOGY. Alfred Sherwood Romer. University of Chicago Press, 1966.

THE AGE OF THE DINOSAURS. Björn Kurtén. World University Library, 1968.

INDEX